Politics of Citizenship and Migration

Series Editor
Leila Simona Talani, Department of European and International Studies,
King's College London, London, UK

The *Politics of Citizenship and Migration* series publishes exciting new research in all areas of migration and citizenship studies. Open to multiple approaches, the series considers interdisciplinary as well political, economic, legal, comparative, empirical, historical, methodological, and theoretical works. Broad in its coverage, the series promotes research on the politics and economics of migration, globalization and migration, citizenship and migration laws and policies, voluntary and forced migration, rights and obligations, demographic change, diasporas, political membership or behavior, public policy, minorities, border and security studies, statelessness, naturalization, integration and citizen-making, and subnational, supranational, global, corporate, or multilevel citizenship. Versatile, the series publishes single and multi-authored monographs, short-form Pivot books, and edited volumes.

For an informal discussion for a book in the series, please contact the series editor Leila Simona Talani (leila.talani@kcl.ac.uk), or Palgrave editor Anne Birchley-Brun (anne-kathrin.birchley-brun@palgrave.com).

This series is indexed in Scopus.

More information about this series at
https://link.springer.com/bookseries/15403

Matilde Rosina

The Criminalisation of Irregular Migration in Europe

Globalisation, Deterrence, and Vicious Cycles

Matilde Rosina
European Institute
London School of Economics
and Political Science
London, UK

The research at the basis of this work was supported by the Fondazione Luigi Einaudi through the Roberto Einaudi Scholarship, as well as by the Fondation Jean Monnet pour l'Europe through the Henri Rieben Scholarship. The views expressed in this book are the author's, and not those of the funders.

ISSN 2520-8896 ISSN 2520-890X (electronic)
Politics of Citizenship and Migration
ISBN 978-3-030-90346-6 ISBN 978-3-030-90347-3 (eBook)
https://doi.org/10.1007/978-3-030-90347-3

© The Editor(s) (if applicable) and The Author(s), under exclusive license to Springer Nature Switzerland AG 2022

This work is subject to copyright. All rights are solely and exclusively licensed by the Publisher, whether the whole or part of the material is concerned, specifically the rights of translation, reprinting, reuse of illustrations, recitation, broadcasting, reproduction on microfilms or in any other physical way, and transmission or information storage and retrieval, electronic adaptation, computer software, or by similar or dissimilar methodology now known or hereafter developed.

The use of general descriptive names, registered names, trademarks, service marks, etc. in this publication does not imply, even in the absence of a specific statement, that such names are exempt from the relevant protective laws and regulations and therefore free for general use.

The publisher, the authors and the editors are safe to assume that the advice and information in this book are believed to be true and accurate at the date of publication. Neither the publisher nor the authors or the editors give a warranty, expressed or implied, with respect to the material contained herein or for any errors or omissions that may have been made. The publisher remains neutral with regard to jurisdictional claims in published maps and institutional affiliations.

Cover credit: Ludovica Musso Piantelli (photograph) and dikobraziy/IStock (map)

This Palgrave Macmillan imprint is published by the registered company Springer Nature Switzerland AG
The registered company address is: Gewerbestrasse 11, 6330 Cham, Switzerland

To my grandparents.

Acknowledgements

Many people have contributed to making this book possible. My deepest gratitude goes to all my interviewees, for taking the time to meet me and sharing their experiences. I also want to thank all those who filled out my surveys, for agreeing to tell their story.

Several of the interviews and questionnaires could not have been carried out without the help of many people, who bent over backwards to aid my project. Claudia Chiappori, Manuela Dogliotti, Graziella Rizza, Maria Laura Innocenti, Ahmed Osman, Emanuela Siniscalchi, Vincent Picard: I want to thank them all for their invaluable support. I am also grateful to Prof. Romano Prodi, as well as to the staff at Maavar Association, Coronata Campus, Antoniano Onlus and the Community of Sant'Egidio, for helping with the research and questionnaires.

I am indebted to many academics who have provided feedback and advice throughout the years, many of whom I am lucky to call friends. I am particularly thankful to Simona Talani, for believing in my research since the very beginning, for her continuous encouragement, and insightful advice. I am also especially grateful to Anna Sergi and Henk Overbeek, who provided precious feedback on my work. From my years at King's College London, I want to thank Roberto Roccu, Cristina Juverdeanu, Marianna Griffini, Margherita de Candia, Malte Laub, Alfio Puglisi, and Stella Gianfreda, for their comments on earlier drafts and the many thought-provoking discussions. From my time at LSE, I am grateful to Eiko Thielemann, Oula Kadhum, and the colleagues at the

European Institute and the Migration Studies Unit, for their support and the stimulating exchanges.

I would like to thank the Fondazione Luigi Einaudi/San Giacomo Charitable Foundation, for putting their trust in my project, and for their generous support through the Roberto Einaudi Scholarship. I am also grateful to the Fondation Jean Monnet pour l'Europe for making the archives in Lausanne available to me, through the Henri Rieben Scholarship. The Centre for Doctoral Studies, the Faculty of Social Sciences and Public Policy, and the European and International Studies Department at King's College London, as well as the European Institute at LSE, have all supported my participation in several conferences, whose discussions greatly enriched the contents of this book. Finally, the editorial team at Palgrave Macmillan has provided patient and constructive support. I wish to thank Anne-Kathrin Birchley-Brun, Ashwini Elango, Nandakini Lahiri and Liam McLean, who have been a great team to work with.

My acknowledgements would not be complete without thanking my family. I am truly grateful to my parents, to my brother, and to Cristina, for always being by my side. I am most indebted to my grandparents, whose strength and kindness are my inspiration. Finally, I am deeply grateful to my partner, Nicolò, for his daily encouragement and positivity, without which, this research would have been a much harder endeavour.

About This Book

The book discusses the criminalisation of irregular migration. In particular, it investigates the meaning, purpose, and consequences of criminalising unauthorised entry and stay, paying specific attention to the European context. From a theoretical perspective, the book adds to the debate on the persistence of irregular migration, despite governments' attempts at deterring it, by taking an interdisciplinary approach that draws from international political economy and criminology. Using Italy and France as case studies, and relying on previously unreleased data and interviews, it argues that criminalisation has no deterrent effect on migratory flows, and that this is due to factors including the latter's structural determinants, and the creation of substitution effects. Furthermore, criminalisation is found to lead to adverse consequences, including by contributing to vicious cycles of irregularity and insecurity.

Contents

1 **Introduction** — 1
   Introduction — 1
   Key Questions and Implications of the Research — 3
   Methodological Notes — 9
   Plan of the Book — 13
   References — 16

2 **Globalisation and Irregular Migration** — 21
   Introduction — 21
   The IPE of Irregular Migration: Theoretical Interpretations — 22
     The Realist Approach — 22
     The Liberal Institutionalist Approach — 29
     The Transnationalist Approach — 34
   Migration Policies and Their 'Gaps' — 41
   Controlling Migration Through Deterrence? — 44
   Conclusion — 50
   References — 51

3 **Deterrence and Criminalisation: Between Migration, IPE, and Criminology** — 57
   Introduction — 57
   The Functioning of Deterrence — 58
     Definition — 58
     Legal Costs: Certainty and Severity — 61

Bounded Rationality and Perceptions	65
Social Costs: Stigmatisation and Social Control	67
The Pitfalls of Deterrence	71
Do Costs Outweigh Benefits?	71
Availability of Alternatives	73
Multiple Audiences: The Political Dimension	75
Information and Communication	78
Criminalising Migration in Europe	81
Conclusion	84
References	86

4 Italy: Tough Rhetoric, Underground Economy, and Counterproductive Consequences — 93

Introduction	93
From Rhetoric to Paper: Discursive Gaps	95
Legal Aspects and Judicial Controversies	98
Demanded Repeal and Political Implications	100
From Paper to Enforcement: Implementation Gaps	104
The Unfolding of Criminal Trials in the Italian Reception System	104
The Numbers of Criminalisation	109
From Enforcement to Overall Outcomes: Efficacy Gaps	121
Efficacy (I): Criminal Law and Expulsions	121
Efficacy (II): Criminal Law and Landings	123
Efficiency: The Cost of Criminalisation	149
Coherence: Irrational Sanctions and Legal Concerns	155
Sustainability: The Long-Term Implications	156
Utility: Who Benefits from Criminalisation?	157
Conclusion	160
References	161

5 France: Instrumentalisation, Courts, and Marginalisation — 173

Introduction	173
The Evolution of Criminalisation	174
1938 to 2004: Introducing and Strengthening Criminal Sanctions for Irregular Migration	174
Post-2004: The Influence of the ECJ	178
A Univocal Trend Towards Decriminalisation?	180
Rhetoric and Instrumentalisation	182

Implementation	184
Criminalisation in Practice	184
Criminalisation in Numbers	187
Effectiveness: Beyond Goals and Outcomes	198
Irregular Migration in France: 2008–2017	198
Efficacy (I): Criminalisation and Expulsions	202
Efficacy (II): Criminalisation and Irregular Migration	203
Efficiency: Prison, Expulsions, and Fines	216
Coherence: The Contrast with the Return Directive	220
Sustainability: Long-Term Effects of Criminalisation	221
Utility: Law-Enforcement and Political Concerns	222
Conclusion	227
References	229
6 The Effects and Counter-Effects of Criminalisation: On Skinny Balloons and Vicious Cycles	239
Introduction	239
Primary and Secondary Migration Flows	241
Discursive Gaps	244
Implementation Gaps	247
Evaluation and Efficacy Gaps	254
(In)Efficacy: Stalling Returns, Undiminishing Migration	254
(In)efficiency: The Duplication of Processes and Costs	259
(In)Coherence: The Contradictory Nature of Sanctions, and Other Problems	261
(Un)Sustainability: The Trade-Off Between Certainty and Efficiency	263
(Political) Utility	264
Deterrence Pitfalls and Criminalisation	267
Deterrence and the Drivers of Migration	267
Does Criminalisation Deter, or Simply Divert? on Alternatives and Skinny Balloons	270
The Politicisation of Migration	277
The Role of Information	285
Conclusion	286
References	290
7 Conclusion	303
Introduction	303

The Criminalisation of Irregular Migration: Between Policy Gaps and Vicious Cycles   304
Criminalisation, Migration, and Globalisation: Theoretical Observations   311
What Alternatives to Criminalisation and Deterrence?   318
References   321

Appendices   325

Index   331

About the Author

Dr. Matilde Rosina is LSE Fellow in International Migration, at the European Institute at the London School of Economics and Political Science. Her research focuses on irregular migration, with specific reference to the European context. Matilde received her Ph.D. in International Political Economy from King's College London, having been awarded the King's Outstanding Thesis Prize for 2021, and scholarships by the Luigi Einaudi Foundation and the Jean Monnet Foundation for Europe. Before joining LSE, Matilde lectured at City, University of London, and at Fordham University's London Centre.

ABBREVIATIONS

CESEDA	Code de l'Entrée et du Séjour des étrangers et du Droit d'Asile—Code of Foreigners' Entry and Stay and of the Right to Asylum
ECJ	European Court of Justice
EMN	European Migration Network
EU	European Union
EU MS	EU Member State
FN	Front National
FR	France (in tables)
GAV	Garde à Vue—Criminal Custody in the French system
IOM	International Organisation for Migration
IPE	International Political Economy
ISTAT	Italian National Institute of Statistics
IT	Italy (in tables)
ITF	Interdiction du territoire français—Ban from French Territory
JP	Justice of the Peace
LN	Lega Nord
M5S	Movimento Cinque Stelle
MoI	Ministry of the Interior
MoJ	Ministry of Justice
MP	Member of Parliament
NGO	Non-Governmental Organisation
OECD	Organisation for Economic Co-operation and Development
PAF	Police aux Frontières—French Border Police
PD	Partito Democratico
PM	Prime Minister
PP	Public Prosecutor

SAR	Search and Rescue
TCN	Third Country Nationals
TUI	Testo Unico sull'Immigrazione—Consolidated Text of Provisions Governing Immigration and the Status of Aliens
UK	United Kingdom
UN	United Nations
UNHCR	United Nations High Commissioner for Refugees

List of Figures

Fig. 4.1	Criminal and administrative procedures for irregular entry or stay in Italy (since 2009) (*Source* Author's elaboration. Criminal procedures in light grey; administrative ones in dark grey)	108
Fig. 4.2	Total known cases in relation to art. 10-bis TUI, as divided by initiated trials and dismissed cases, 2009–2015 (*Source* Author's elaboration on ISTAT [2018])	110
Fig. 4.3	Condemnations of foreigners for article 10-bis TUI, 2009–2016 (*Source* Author's elaboration on ISTAT [2018])	112
Fig. 4.4	Condemnations of foreigners for article 10-bis by type of sanction, 2009–2016 (*Source* Author's elaboration on ISTAT [2018])	113
Fig. 4.5	Third country nationals subjectable to article 10-bis, 2009–2016 (*Source* Author's elaboration based on Ismu [2014], MoI [2017, 2018], EUROSTAT [migr_eipre and migr_asylapp], ISTAT [2018])	117
Fig. 4.6	Proportion of condemnations for article 10-bis, over total number of TNCs subjectable to the norm, 2009–2016 (*Source* Author's elaboration based on Ismu [2014], MoI [2017, 2018], EUROSTAT [migr_eipre and migr_asylapp], ISTAT [2018])	118

Fig. 4.7	Comparison of landings, TCNs found irregularly staying, and asylum applicants, 2008–2018 (*Source* Author's elaboration on Eurostat [migr_eipre, migr_asyappctza], ISMU [2014] and MoI [2017, 2018])	126
Fig. 4.8	Detection of irregular border crossings on the Central Mediterranean route by nationality, 2009–2018 (top four nationalities in overall period) (*Source* Author's elaboration on Frontex [2019])	127
Fig. 4.9	Reasons for choosing Italy rather than other countries (*Source* Author's elaboration)	130
Fig. 4.10	Reasons for choosing Italy rather than other countries: Definition of 'Other' as added by respondents (*Source* Author's elaboration)	130
Fig. 4.11	Knowledge of the sanctions for unauthorised migration in Italy (*Source* Author's elaboration)	134
Fig. 4.12	Location where sanctions for irregular migration were learnt (*Source* Author's elaboration)	134
Fig. 4.13	How migrants learnt about sanctions for irregular migration (*Source* Author's elaboration)	136
Fig. 4.14	Did knowing about the consequences of irregular migration affect your decision on how to migrate? (*Source* Author's elaboration)	138
Fig. 5.1	Irregular entry and stay in France—criminal and administrative pathways (*Source* Author's elaboration. Criminal procedures in light grey; administrative ones in dark grey)	186
Fig. 5.2	Infractions to general conditions of entry and stay registered by French police and gendarmerie, 2000–2018 (*Source* Author's elaboration based on Departmental figures registered by police and gendarmerie [2018])	188
Fig. 5.3	Persons indicted by the Police aux Frontières for infractions of entry and stay, in metropolitan France and Calais, 2005–2010 (*Source* Author's elaboration based on data from Crimes to foreigners' police [2011 and 2013] and Scherr [2011])	190
Fig. 5.4	Condemnations for irregular stay in France, 1993–2016 (*Source* Author's elaboration based on data from Condemnations by infraction [2016])	193
Fig. 5.5	Condemnations for irregular entry or stay in France, by type of sanction, 2012–2016 (*Source* Author's elaboration based on Tables of condemnations [2012–2016: Code 26101])	193

Fig. 5.6	Average fine and length of imprisonment for irregular entry or stay, 2012–2016 (*Source* Author's elaboration based on Tables of condemnations [2012–2016: Code 26101])	194
Fig. 5.7	Overview of data concerning irregular entry and stay in France, 1993–2018 (*Source* Author's elaboration based on Departmental figures registered by police and gendarmerie [2018], Crimes to foreigners' police [2011 and 2013: Index 69 of état 4001], and Condemnations by infraction [2016])	196
Fig. 5.8	Third country nationals found irregularly present in France, and refused entry at French external borders, 2008–2018 (*Source* Author's elaboration based on Eurostat migr_eipre and migr_eirfs)	199
Fig. 5.9	Third country nationals found to be irregularly present in France by nationality (rounded, selected nationalities only), 2008–2018 (*Source* Author's elaboration based on migr_eipre)	201
Fig. 5.10	Third country nationals (TCNs) ordered to leave and returned to third countries by France, 2008–2017 (*Source* Author's elaboration based on Eurostat data)	203
Fig. 6.1	Irregular migration in Italy and France, by apprehension pathway (irregular entry or stay) (stacked graph), 2008–2018 (*Source* Author's elaboration based on Eurostat [migr_eipre, migr_eirfs], ISMU [2014], MoI [2017, 2018])	242
Fig. 6.2	Registered crimes and condemnations for irregular entry and/or stay in Italy and France, 2000–2018 (Author's elaboration on Departmental figures registered by police and gendarmerie [2018], Condemnations by infraction [2016], and ISTAT [2018])	248
Fig. 6.3	Punishment for irregular entry/stay in Italy and France: Foreseen vs sanctioned, 2009–2018 (*Source* Author's elaboration on Tables of condemnations [2012–2016] and MoJ [2018])	250
Fig. 6.4	Proportion of women in irregular migratory flows and in condemnations for irregular entry/stay, 2008–2016 (*Source* Author's elaboration based on Annuaires statistiques [2012], Tables of condemnations [2012–2016], ISTAT [2018], Eurostat migr_eipre [2018])	251

| Fig. 6.5 | Effective return rates to third countries, 2008–2017 (*Source* Author's elaboration based on Eurostat migr_eirtn [2018]) | 256 |
| Fig. 6.6 | Criminalisation and trade-offs (*Source* Author's elaboration) | 264 |

List of Tables

Table 2.1	Freeman's model of migration policy	27
Table 4.1	Factors contributing to the lack of effectiveness of the criminalisation of migration in Italy	148
Table 6.1	The criminalisation of migration and its discursive gaps	246
Table 6.2	The criminalisation of migration and its implementation gaps	253
Table 6.3	The criminalisation of migration and its efficacy gaps	259
Table 6.4	Overview of the criminalisation of migration in Italy and France, by evaluation parameter	266

CHAPTER 1

Introduction

INTRODUCTION

International migration is neither a new nor an unprecedented phenomenon. In contemporary political debates, however, people's movement across borders (especially when unauthorised) is often presented as an exceptional event or challenge, in need of a likewise extra-ordinary solution. The latter, in turn, is frequently framed in terms of security, contributing to what has been termed the 'securitisation' of migration.[1]

Aiming to address unauthorised entries, governments throughout Europe have in recent years placed significant emphasis on deterring migration, by threatening harsh sanctions. They have often done so by *criminalising* unauthorised mobility, to the extent that, today, all but two EU member states sanction irregular migration through criminal penalties.

As a matter of fact, criminalisation and, more broadly, deterrence, have a great appeal for policy-makers, due to their ease of applicability in both explaining irregular migration, and providing a solution to it[2]: Because

---

[1] Huysmans (2000). Following Huysmans (2000, p. 752), the securitisation of migration refers to the 'social construction of migration as a security question'.

[2] Cf. Pratt et al. (2008, p. 367).

individuals are expected to migrate irregularly when the anticipated benefits are greater than the foreseen costs, increasing the latter should make the conduct less profitable and appealing, and therefore reduce its occurrence. In this way, irregular migration is presented as a phenomenon that can not only be controlled, but also prevented.

But is criminalisation effective in deterring migratory flows?

In investigating the effectiveness of measures aimed at deterring migration, vast part of the literature adopts a quantitative perspective, researching evidence of a correlation between stricter measures and migratory flows. Although several studies obtain results in favour of limited effectiveness,[3] some contradictory results also emerge.[4] On the contrary, qualitative studies, which may better capture the nuances of deterrence strategies, as well as of migrants' perceptions of them, are less common, but seem more open to the possibility that deterrence might be partly effective.[5] Often, however, they focus more on the (lack of) legitimacy of the strategy, rather than its outcomes.[6]

Ultimately, the debate on the effectiveness of criminalisation is rooted in the broader theoretical discussion on states' ability to control migration, in the context of globalisation. If for realist scholars, states can control migration, if it is in their interest to do so,[7] liberal institutionalist and transnationalist academics question such argument. For institutionalists, states' capacity to control migration is in fact constrained by legal and supranational institutions.[8] For transnationalists, on the contrary, globalisation processes imply systemic changes that make migration a structural and largely uncontrollable phenomenon.[9]

In view of the above, the objective of this study is to shed light on the meaning, purpose, and consequences of the criminalisation of irregular migration in Europe. With the term 'criminalisation', I specifically refer to the use of the criminal law to sanction migration-related offences and,

---

[3] See, for example, Thielemann (2003, 2011), Cornelius (2005), and Cornelius and Salehyan (2007).

[4] See, for example, Espenshade (1994) and Dávila et al. (2002).

[5] See, for example, Godenau and López-Sala (2016, p. 15).

[6] See Stumpf (2006), Aliverti (2012), Hassan (2000), and Briskman (2008).

[7] See, for instance, Weiner (1995) and Freeman (1998).

[8] See, for instance, Hollifield (1998) and Joppke (1998).

[9] See, for instance, Sassen (1996) and Overbeek (1996).

in particular, foreigners' irregular entry or stay. Throughout the book, I study criminalisation as an example of deterrence, namely a strategy aiming to discourage irregular migration to a country, through potential migrants' fear of negative consequences.[10]

In this introductory chapter, I present the key goals and contents of the book. I start by outlining the research questions, defining the objectives and contribution of the study, and justifying the case selection. Subsequently, I discuss the methodological approach, and then conclude by providing an outline of the book's contents.

KEY QUESTIONS AND IMPLICATIONS OF THE RESEARCH

This book investigates the effects and counter-effects of the criminalisation of migration. Specifically, the questions lying at the core of this work are the following: Does criminalisation succeed at deterring irregular migration? From a theoretical perspective, can it ever do so? In the cases characterised by lack of success, do the reasons for such an outcome lie in the inadequate application of deterrence principles, or in other, more structural, problems? Finally, what are the consequences of criminalising unauthorised entry and stay, and who benefits from it?

Using Italy and France as case studies, due to the exceptional severity of the measures foreseen in the two countries, the book also poses the following empirical questions: Did criminalisation curb irregular flows to the two countries? How did the size and composition of the flows change, in relation to developments in the criminal law? To what extent is it possible to argue that criminalisation was effective, and did it lead to any counterproductive outcomes?

To investigate the above, the study draws from the 'policy gap' hypothesis.[11] In particular, it looks for potential inconsistencies between politicians' rhetoric and the actual design of the measures ('discursive gaps'), the measures' design and eventual enforcement ('implementation gaps'), and, finally, the latter's enforcement and overall results ('efficacy gaps').[12]

---

[10] See Chapter 3 for a discussion of criminalisation and deterrence.
[11] See Cornelius et al. (1994).
[12] Based on Czaika and de Haas (2013).

To anticipate the core argument of the book, this is that criminalisation has had no deterrent effect on migratory flows, and that this can be explained by factors including the latter's structural determinants, and the likely creation of substitution effects. Furthermore, criminalisation is found to lead to adverse consequences, including by contributing to vicious cycles of irregularity and insecurity.

Delving deeper into the contribution and relevance of the research, the main theoretical goal of the book is to add to the international political economy (IPE) debate on the persistence of irregular migration, despite states' attempts at regulating it. Specifically, I aim to do so by examining the criminalisation of unauthorised entry and stay through an interdisciplinary analysis. The study thus approaches the above-mentioned research questions by drawing from both IPE and criminology: While international political economy represents the basis of the analysis to discuss the degree of states' ability to control migration in times of globalisation, criminology becomes crucial to understand the specific functioning of deterrence (broadly) and criminalisation (specifically). Indeed, the latter discipline has vastly researched such strategies, from both theoretical and empirical perspectives, asking whether and when they can aid to reduce offences. By investigating the debates in the literature and applying the emerging principles to irregular migration, it is possible to gain better understanding of what elements make deterrent threats likely to succeed or fail, revealing, in particular, whether their inefficiencies (if any) are caused by the inappropriate application of deterrence principles, or by deeper problems. It is worth stressing that the use of criminological studies is not aimed at supporting the classification of irregular migration as a crime, but simply takes advantage of the well-established basis of research that the field possesses.

The analysis of the effectiveness of criminalisation and deterrence, in turn, serves a very pragmatic goal, that is the exposure of the actual interests behind the adoption of such measures. While politicians tend to present them as a way to avoid the problem of irregular migration in the first place (as potential newcomers would not, supposedly, cross the state's border in fear of the consequences), such choices bear important repercussions on migrants' experiences: Once migration does take place, in spite of sanctions or threats, the measures that were originally intended as deterrents are indeed enforced. The number of people willing to risk their lives in the hope of a better existence,[13] the hardships they endure during their journey, and the increasing interconnectedness between smuggling and trafficking in human beings,[14] are all crucial factors that increase the importance of reflecting on the consequences of measures meant to control access. If these are to be continued, we should at least know whether they can be effective, and, if not, why that is the case.

Although attention to migration control through deterrence is gradually increasing, few scholars research the determinants and mechanisms that are specific to it, and, those who do, mostly focus on either enhanced border controls or detention.[15] In particular, limited consideration is given to the deterrent effect stemming from the use of the criminal law to govern immigration offences, despite this being an increasingly common measure in European countries.[16]

In recent decades, the number of EU member states criminalising irregular migration has risen: As of 2014, 26 of them considered either irregular *entry* or *stay* (or both) as a crime.[17] While it could be argued that this is now an established trend, it appears that the implications of such choices have only rarely been fully assessed. As revealed by a 2013 survey conducted by the European Migration Network (EMN), only two EU member states declared having carried out an evaluation of the effectiveness of their legislation concerning criminal penalties for irregular migrants.[18] Conversely, thirteen reported never having assessed the effects of the norm, despite using it. Among the latter were both Italy and France.

Empirically, therefore, this research represents (to the best of the author's knowledge) the first evaluation of the effectiveness of the criminalisation of irregular migration in Italy and France, from an IPE perspective: No government analysis has ever been conducted, and existing academic studies tend to either lack data, or focus predominantly on

---

[13] As an example, in 2016, over 5000 people lost their lives while crossing the Mediterranean Sea (IOM 2017).

[14] See, for example, Amnesty International (2017) and Bartolo (2018).

[15] With some notable exceptions, including Godenau and López-Sala (2016), López-Sala (2015), and Thielemann (2003, 2011).

[16] See FRA (2014).

[17] FRA (2014).

[18] EMN (2013). Only 19 countries made their answers publicly available.

legal aspects.[19] Through the above-mentioned interdisciplinary approach, previously unreleased data, and interviews with both stakeholders and migrants, this study intends to overcome such issues, providing an in-depth assessment of the measure, and aspiring to inform policy-making processes.

It should be mentioned that the criminalisation of irregular migration has often been discussed in the context of what has become known as the 'crimmigration law'. Crimmigration studies denounce the ever-closer merger of immigration and criminal law, especially visible in terms of overlap in the substance of the law, similarities in enforcement, and procedural parallels.[20] However, most of them are not so much concerned with the deterrent effect of criminalisation, but rather with questioning its legitimacy and respect of migrants' rights. While the present book is more concerned with the outcomes of the norm, its findings may prove of relevance for crimmigration scholars too, by providing evidence on the extent to which criminalisation proved a successful measure. Should the effectiveness of the norm in discouraging further unauthorised migration emerge as absent, then its moral justification would be further questioned too.[21]

In this context, the study of the criminalisation of migration is also significant in view of the fact that the criminal law should only be used as a last resort.[22] Due to its potential to 'subject persons to state punishment',[23] the criminal law should, as argued by Blanchard and Saas, 'only intervene to protect a sufficient legitimate interest, when no other coercive method allows to achieve the set goals'.[24] This is especially the case since irregular migration has often been depicted as a *mala prohibita*, that

---

[19] See, for example, Ferrero (2010), Henriot (2013), GISTI (2012), and Lochak (2016). Partial exceptions are di Martino et al. (2013) and Scherr (2011), although the former predominantly takes a legal approach to the (Italian) criminalisation of migration, and the latter mainly focuses on the implementation of the norm (in France).

[20] See Stumpf (2006), Kanstroom (2000), and Weber (2002).

[21] Similarly: Banks and Pagliarella (2004, pp. 106–107).

[22] Blanchard and Saas (2012, p. 5).

[23] Husak (2004, p. 211).

[24] Blanchard and Saas (2012, p. 5). Author's translation.

is to say an infraction that neither harms society nor is morally wrong, but that is simply declared 'illegal' by statute.[25]

Finally, some authors limit their investigations to measures specifically aimed at deterring unfounded asylum applications. Yet, it is often problematic to draw a neat line between asylum seekers and 'economic migrants': Not only do most people leave their countries in view of a mix of economic and political reasons, but refugees use smugglers and irregular means to get to Europe, and people who escape poverty apply for asylum,[26] in lack of alternatives to regularise their status. In this study, I thus focus predominantly on third country nationals who have entered or stayed irregularly in Italy or France.

With the above in mind, the analysis and comparison of the cases of Italy and France are especially instructive to approach the criminalisation of irregular entry and stay as a tool of deterrence, for several reasons. To start with, although both countries employ the criminal law to address irregular migration, they do so in very different ways. On one hand, Italy is among the EU member states sanctioning unauthorised entry and stay with the highest fines (ranging from €5000 to 10,000).[27] On the other hand, France imposes more moderate financial penalties (€3750), but also commands the imprisonment of the offender, becoming among the very few countries in the Union that inflict a mandatorily joint punishment.[28] At the same time, both countries have witnessed amendments to their legislation criminalising migration in recent years, with Italy introducing the crime of irregular entry and stay in 2009, and France partially repealing it in 2012 and 2018. These represent key turning points, to shed light on how changing deterrent measures may influence migratory flows.

Additionally, while the two countries are comparable in terms of their supranational legislative framework (being bound by common EU laws including the Return Directive, Dublin Regulation, and Schengen Agreement), they have substantially different traditions and patterns of

---

[25] Mark Thornton 'Social Harm', in McLaughlin and Muncie (2001, p. 277) and Ryo (2015, p. 641).

[26] Triandafyllidou (2016, p. 34).

[27] Together with Spain and Latvia. See FRA (2014, annex).

[28] Together with Belgium, Bulgaria, and Greece. See L- 621-1 of Ceseda, initial version; FRA (2014, annex).

immigration. Indeed, France is an example of the north-western European countries that saw sustained levels of immigration throughout the twentieth century.[29] On the contrary, Italy is among the southern European ones that only started experiencing significant migratory inflows in the 1980s.[30] Interestingly, even though, today, irregular migration is generally more 'visible' in Italy, due to the easily observable sea landings, it should be noted that, between 2008 and 2015, France consistently reported a higher number of third country nationals residing irregularly on its territory.[31] As a matter of fact, the latter is among the five EU member states with the highest number of apprehensions of foreigners' irregular presence.[32]

Finally, the criminalisation of irregular migration has been extremely controversial in Italy, but generally less so in France,[33] where most of the public discussion has focused instead on the criminalisation of migrant helpers. Even in the latter country, however, the symbolism of employing the criminal law against immigrants is very powerful, as was revealed by a politician's statement in 2006, arguing that repealing the crime of irregular stay would have been interpreted as a reduced willingness of the Government to fight against irregular migration.[34] The varying degree of politicisation of the norm may thus bring insight into the political rationales for its adoption or amendment, as well as into the significance of discursive gaps.

In sum, the study and comparison of the criminalisation of irregular migration in France and Italy are helpful to deepen the understanding of the norm in the European context, due to the different degree of severity of the sanctions foreseen, varying historical and contemporary trends in migratory flows, and divergent extent in the politicisation of the measure.

---

[29] Ambrosini (2018, p. 62) and Schain (2012, p. 39).
[30] Ambrosini (2018, p. 62) and Boswell and Geddes (2011, p. 91).
[31] Eurostat migr_eipre (2018).
[32] Ibid.
[33] Except for specific legal debates. See, for example, Gisti (2012).
[34] Pascal Clément, then Minister of Justice, cited in Buffet (2006).

METHODOLOGICAL NOTES

In assessing the success or failure of criminalisation, several methodological challenges emerge. Indeed, figures on irregular migration are, by definition, never completely accurate, to the extent of being defined as migration studies' Achilles heel.[35] In Italy, for instance, the EMN estimates internal controls to roughly concern only 10% of irregularly staying foreigners.[36] Furthermore, establishing causality is not easy: Even when there appears to be a correlation between an intervention and the number of irregular arrivals, it is possible that the change in the latter was caused by external factors (such as alterations in third countries' economic or political situation).

To limit the above-mentioned problematics, the book adopts a policy-evaluation approach based on a set of parameters derived from EU guidelines. It also relies on over 50 interviews with key stakeholders, more than 100 questionnaires with third country nationals, and two previously unreleased official datasets, triangulating data to increase the validity of findings. In this section, I expand on the methodological choices, before moving on to outline the structure of the book in the last part of the chapter.

Starting by delving deeper into the book's policy-evaluation approach, this relies on five criteria developed by the European Union to draw ex-post evaluations.[37] Understanding effectiveness as made up of different aspects (rather than just its ability to achieve the desired goals) allows not only to isolate problematic aspects, but also to point to potential reasons of weaknesses as due to, for instance, inconsistencies with supranational legislation, or stakeholders' intervention.

As may be expected, the first aspect to be examined, when evaluating the effectiveness of a measure, is its *efficacy*,[38] that is to say the extent to which its goals have been achieved, and the degree to which these can

---

[35] Skeldon (2012, p. 229).

[36] EMN (2012, p. 84).

[37] Cf BR Toolbox.

[38] The actual term used is 'effectiveness', but I refer here to 'efficacy', to distinguish this specific parameter from the evaluation of the policies' 'effectiveness', a more comprehensive term that I use to define the overall degree of success, summarising the five criteria presented in this section.

be attributed to the intervention itself.[39] This is however only the first parameter to be analysed, as further criteria are equally important. Specifically, by assessing the *efficiency* of the norm, the latter's administrative, compliance, and implementation costs are balanced against the benefits generated by the intervention.[40] Moreover, by analysing the *coherence* of the measure, both the *internal* consistency among the various articles, inputs, or actions included in the intervention, as well as the initiative's overall *external* congruence with national and supranational pieces of legislation, are evaluated.[41] While efficacy, efficiency, and coherence are indicated by the European Commission as core parameters to consider in policy evaluations,[42] it is useful to add two more criteria, which are proposed by the institution as additional ones, to be used depending on the intervention being studied: Sustainability and utility. On one hand, through the study of *sustainability*, it becomes possible to appreciate whether the changes generated can endure after the initial implementation period, and in the aftermath of the intervention.[43] On the other hand, through the analysis of *utility*, the varying degrees of satisfaction of different stakeholders with the results of the norm are assessed.[44] This last parameter offers valuable insight, by addressing the (possibly diverging) interests of pressure groups and beneficiaries, as well as having the potential to explain contrasting or unclear developments in terms of political pressure, and to evaluate the effectiveness of the intervention from the perspective of multiple actors.

In short, to overcome the methodological issues mentioned above, this book analyses the effectiveness of the deterrent threat stemming from the

---

[39] BR Toolbox, p. 272.

[40] Ibid., pp. 272, 278.

[41] Ibid., p. 274.

[42] Two more parameters are highlighted by the European Commission as core, but omitted from this analysis (BR Toolbox, pp. 273–276). The first is the 'EU added value' (investigating the changes that can be attributed to the intervention of the EU), which is omitted here, as the criminalisation of migration is based on national (rather than EU) laws. The second parameter considers the 'relevance' of interventions, to understand whether the objectives are appropriate to solve the need or problem. This too has been omitted, since it corresponds to the very question at the basis of this research, whose answer depends on the overall assessment of the norm.

[43] BR Toolbox, p. 276.

[44] Ibid., p. 275.

criminalisation of migration, by relying on five policy-evaluation parameters: Efficacy, efficiency, coherence, sustainability, and utility. Together, these criteria are particularly helpful to shed light on whether a gap between the goals of the criminalisation of migration, and its eventual outcomes, exists, and on the reasons underlying it.

To strengthen and enrich the policy evaluation, the study relies on primary sources including interviews with key stakeholders, questionnaires with migrants, and previously unreleased official datasets.

To begin with, 54 interviews with stakeholders were carried out across Italy, France, and Belgium, between June 2017 and October 2018. The interviews were meant to gain first-hand understanding on the development, implementation, and effectiveness of the criminalisation of migration, and targeted three groups of respondents. First, I contacted politicians and bureaucrats involved with the design and amendment of the norm, to gather information on the political rationales underpinning criminalisation. I also had meetings with officials working for the European Commission and national Permanent Representations in Brussels, to assess the EU perspective. Second, I interviewed police officers, public prosecutors, judges, lawyers, and border guards who had direct experience with cases of irregular migration, to obtain details on how the norm is implemented in practice, as well as on its possible effects and counter-effects. Third, I approached humanitarian associations, cultural mediators, and journalists engaged with migrants, to understand their interpretation of the application of criminalisation, as well as of migrants' attitudes towards it. All conversations were semi-structured, and revolved around two main themes: (1) The status of the practical implementation of the criminalisation of migration, and (2) respondents' perceptions of the effectiveness and potential consequences of the norm. Both purposive and snowball techniques were used as sampling tools, by targeting key stakeholders who could provide helpful insight, and then relying on a chain-effect to reach further individuals.[45]

In addition to the interviews, and aiming to include third country nationals' perspectives too, I conducted 104 written questionnaires with migrants in Italy and France, between November 2017 and October 2018.[46] Eligible participants were third country nationals who had moved

---

[45] See Tansey (2007, pp. 770–771).

[46] Out of the 104 questionnaires, 99 were conducted in Italy, 5 in France. Initially, a higher number of respondents was foreseen for the latter country too, and numerous

to Italy and/or France, and who were 16 or older at the time of the survey. In particular, the surveys comprised four parts: The first section focused on respondents' general information, such as their demographic characteristics and education level. Following this introductory phase, the second and third parts centred on respondents' entry and stay in the receiving country (either France or Italy). Here, questions concerned the way in which migrants arrived, the support they found in the receiving country, and the reasons for choosing that specific state. Lastly, the fourth section investigated respondents' knowledge of migration-related documents and measures, as well as the potential effect that such awareness had on their willingness to migrate. The questionnaires were anonymous, and based on non-probability sampling methods, including purposive and snowball sampling. Specifically, several associations involved with migrants were contacted, and asked to act as gatekeepers and administer the questionnaires.[47] A chain-effect was also relied upon, to obtain contacts of further organisations or potential respondents.

Finally, to provide an accurate outline of the use of the criminalisation of migration in Italy and France, several datasets were sought and relied upon. These include EU-level databases (released by Eurostat and Frontex)[48] and national ones. While the French police and Government release information regarding the number of registered cases and types of condemnations, including as specifically related to the offence of irregular entry or stay,[49] however, the Italian counterparts do not. To overcome such obstacle, two previously unreleased databases were

---

attempts were made at contacting multiple NGOs. However, limited engagement constrained the distribution of the surveys. To reduce the impact of this divergence and increase the reliability of findings, questionnaire results are triangulated with official datasets, interviews, and secondary sources throughout the book.

[47] The questionnaires were conducted at the following locations: Maavar Association (Paris: https://www.maavar.com); Coronata Campus, managed by Un'Altra Storia and Migrantes-Caritas (Genoa, Liguria: http://www.coopunaltrastoria.it/about-us/); Sant'Egidio School of Italian Language (Genoa, Liguria: https://www.santegidio.org/pageID/30104/langID/it/SCUOLE-DI-LINGUA-E-CULTURA.html); Coompany & (Alessandria, Piedmont: http://www.coompany.it); Antoniano Onlus centres for asylum seekers (Bologna, Emilia-Romagna: https://www.antoniano.it/progetti/accoglienza/).

[48] See Eurostat (2018) and Frontex (2017, 2019).

[49] See Annuaire Statistique (2006, 2007, 2012), Condemnations by infraction (2016), and Departmental figures registered by police and gendarmerie (2018).

obtained from ISTAT and the Ministry of Justice in 2018.[50] As will be discussed, although the two datasets do not always coincide, together, they enable an in-depth analysis of how the criminalisation of migration is implemented in the country. The databases are also at times integrated by figures on local dynamics, provided by the offices of the justices of the peace and public prosecutors interviewed, which help validate interviewees' statements and identify further relevant trends.

In sum, this study takes a policy-evaluation approach, and relies on over 50 interviews with key stakeholders across France, Italy, and Belgium, more than 100 surveys with migrants, and two previously unreleased databases. While the interviews aim to better understand respondents' perceptions of the deterrent effect of apprehending, processing, and sanctioning irregular entry and stay through the criminal law, the questionnaires are meant to bring in migrants' own perspectives, and in particular their knowledge and understanding of specific migration-control measures. Figures are then intended to provide evidence on the use of the norm in Italy and France.

To increase the validity of findings, information is triangulated. Specifically, not only are interviews, questionnaires and figures discussed in relation to each other, but they are also analysed in conjunction with further documents and data, including legislative and judicial texts, policy documents, academic material, and news articles. Triangulation is aimed 'to provide a parallax view upon events',[51] and indeed has the potential to both add details to the understanding of criminalisation, and increase the reliability of the research conclusions.[52]

Plan of the Book

To shed light on the effects and consequences of the criminalisation of irregular migration, the book is organised in two parts. The first one (Chapters 2–3) provides an overview of the theoretical context in which the research is embedded. Specifically, it presents the main IPE approaches regarding states' ability to control migration, and provides an interdisciplinary analysis of deterrence and criminalisation, combining IPE studies

---

[50] ISTAT (2018) and MoJ (2018).
[51] Davies (2001, p. 75).
[52] Similarly: Castles (2012, p. 15) and Natow (2020, p. 168).

with criminological ones. The second part (Chapters 4–6) analyses empirical evidence on the criminalisation of migration, focusing specifically on the cases of Italy and France. It then compares findings, and studies how the theoretical arguments advanced in the first part of the book applied in practice. Finally, the conclusion offers some theoretical reflections and policy recommendations.

More specifically, Chapter 2 examines the IPE debate on states' ability to control migration and the persistence of irregular flows, in the context of globalisation. Starting by addressing the matter at the broader theoretical level, it then looks at the 'policy gap' hypothesis, according to which a gap has emerged between the stated goals of migration policies and their outcomes. Specifically, the author proposes to follow Czaika and de Haas's threefold conceptualisation of the hypothesis, thus distinguishing between discursive, implementation, and efficacy gaps.[53] The final section of the chapter examines studies of deterrence in the migration literature, setting the basis for a more elaborate discussion of the strategy in the following chapter.

Deterrence is, together with criminalisation, the main subject of investigation of Chapter 3, which deepens the analysis of the functioning and weaknesses of the strategy, drawing from IPE and criminological studies. The latter discipline enables in-depth understanding of the way in which deterrence works, and highlights in particular the relevance of the certainty and (to a lesser extent) severity of sanctions, of how such factors are perceived, and of the social costs involved in penalties. Building on such and the previous chapter's discussion, the author hypothesizes deterrence strategies to be affected by four 'pitfalls', namely: The inability of costs to outweigh the benefits of migration, the possible generation of 'substitution effects', the prioritisation of domestic concerns (rather than migrants'), and potential newcomers' lack of information. The last part of the chapter analyses theoretical aspects linked to the criminalisation of migration. Specifically, it discusses the meaning and purpose of the decision to use the criminal law to sanction infractions, and argues that criminalisation emerges as a key example of deterrence, thanks to its double aim to simultaneously increase the legal and social costs involved.

Transitioning to the empirical analysis, the second part of the book deals with the criminalisation of migration in practice. Specifically,

---

[53] Czaika and de Haas (2013).

Chapter 4 sheds light on the evolution, implementation, and effects of the norm in the Italian context. The structure follows the threefold understanding of policy inconsistencies mentioned above, by investigating the presence of potential discursive, implementation, and efficacy gaps. First, the chapter builds upon legislative and judicial texts to reconstruct the adoption of the norm and eventual legal changes, finding that a gap between politicians' rhetoric and the actual norm on paper appears evident in Italy. Second, it analyses data relating to the implementation of criminalisation (such as the number of trials and condemnations), arguing that the actual enforcement of the norm has been limited. Finally, it relies on primary and secondary sources to investigate the actual effectiveness of the measure (and whether any efficacy gaps occurred), organising the discussion around the five evaluation parameters derived from EU ex-post evaluations (efficacy, efficiency, coherence, sustainability, and utility). The chapter argues that criminalisation not only scored poorly on all such parameters, but also led to multiple counterproductive effects, including by lengthening investigations against smugglers, and duplicating legal processes. Reflecting on the utility of the measure for different groups of actors, the chapter suggests that criminalisation may be understood as a case of deliberate malintegration, in which electoral concerns were balanced with economic interests.

Chapter 5 deepens the analysis on the criminalisation of migration in France, following a structure that is parallel to that of the previous chapter. Here, the author finds that, while the discursive gap emerges as likely more limited, both implementation and efficacy ones were significant. First, the chapter provides data supporting the view that criminalisation was used instrumentally in the country: While criminal custody was employed extensively, full criminal trial was only rarely pursued in practice. Second, it argues that criminalisation in France (like in Italy) was both ineffective and counterproductive, including by promoting migrants' marginalisation, and a significant duplication of expenses. Concerning the utility of the measure, the chapter suggests that this emerged primarily from the instrumental use of the norm, meant to facilitate the meeting of law-enforcement actors' repatriation targets.

In Chapter 6, the author brings together the theoretical and empirical analyses, to compare criminalisation in Italy and France, and examine how the pitfalls hypothesised in Chapter 3 applied in practice. Specifically, the analysis pays particular attention to the counterproductive effects of criminalisation, suggesting that the latter may promote vicious cycles, by contributing to both more irregularity in migratory flows, and greater insecurity among the domestic public. The chapter concludes by discussing the effectiveness of criminalisation in light of the theoretical IPE discussion on states' ability to control migration.

The concluding chapter (Chapter 7) reflects on the findings of the study, stressing the inability of criminalisation to decrease migration, and its parallel contradictory results. It also expands on the theoretical implications of the study, arguing that the three IPE perspectives discussed in Chapter 2 seem, in fact, to each address a different type of gaps: While realists tend to focus on discursive gaps, liberal institutionalists more often address implementation ones, and transnationalist scholars are inclined to focus their attention on efficacy gaps. In closing, the chapter proposes alternative solutions to criminalisation and deterrence, including the opening of legal pathways.

References

Aliverti, Ana. 2012. Making People Criminal: The Role of the Criminal Law in Immigration Enforcement. *Theoretical Criminology* 16 (4): 417–434.

Ambrosini, Maurizio. 2018. *Irregular Immigration in Southern Europe: Actors, Dynamics and Governance*. Cham: Palgrave Macmillan.

Amnesty International. 2017. *Libya's Dark Web of Collusion: Abuses Against Europe-Bound Refugees and Migrants*. London: Amnesty International.

Banks, Cyndi, and Irene Pagliarella. 2004. *Criminal Justice Ethics: Theory and Practice*. Thousand Oaks: Sage.

Bartolo, Pietro. 2018. *Le stelle di Lampedusa: La storia di Anila e di altri bambini che cercano il loro futuro fra noi*. Milano: Mondadori.

Blanchard, Emmanuel, and Claire Saas. 2012. Introduction. In *Immigration: un régime pénal d'exception*, ed. Gisti, 5–13. Paris: Gisti. https://www.gisti.org/publication_som.php?id_article=2781#1lutte. Accessed 23 Feb 2017.

Boswell, Christine, and Andrew Geddes. 2011. *Migration and Mobility in the European Union*. New York: Palgrave Macmillan.

Briskman, Linda. 2008. Detention as Deterrence. In *Asylum Seekers: International Perspectives on Interdiction and Deterrence*, ed. Alperhan Babacan and Linda Briskman, 128–142. Newcastle: Cambridge Scholars Publishing.

Buffet, François-Noël. 2006. Immigration Clandestine: une réalité inacceptable, une réponse ferme, juste et humaine. Report of the Inquiry Commission on Clandestine Migration, N. 300, April 6.

Castles, Stephen. 2012. Understanding the Relationship Between Methodology and Methods. In *Handbook of Research Methods in Migration*, ed. Carlos Vargas-Silva, 7–25. Cheltenham: Edward Elgar.

Cornelius, Wayne A. 2005. Controlling 'Unwanted' Immigration: Lessons from the United States, 1993–2004. *Journal of Ethnic and Migration Studies* 31 (4): 775–794.

Cornelius, Wayne A., and Idean Salehyan. 2007. Does Border Enforcement Deter Unauthorised Immigration? The Case of Mexican Migration to the United States of America. *Regulation & Governance* 1: 139–153.

Cornelius, Wayne, Philip Martin, and James Hollifield. 1994. Introduction: The Ambivalent Quest for Immigration Control. In *Controlling Immigration: A Global Perspective*, ed. Wayne Cornelius, Philip Martin, and James Hollifield, 3–41. Stanford: Stanford University Press.

Czaika, Mathias, and Hein de Haas. 2013. The Effectiveness of Immigration Policies. *Population and Development Review* 39 (3): 487–508.

Davies, Philip. 2001. Spies as Informants: Triangulation and the Interpretation of Elite Interview Data in the Study of the Intelligence and Security Services. *Politics* 21 (1): 73–80.

Dávila, Alberto, José A. Pagán, and Gökçe Soydemir. 2002. The Short-Term and Long-Term Deterrence Effects of INS Border and Interior Enforcement on Undocumented Immigration. *Journal of Economic Behavior and Organization* 49: 459–472.

di Martino, Alberto, Francesca Biondi Dal Monte, Ilaria Boiano, and Rosa Raffaelli. 2013. *La criminalizzazione dell'immigrazione irregolare: legislazione e prassi in Italia*. Pisa: Pisa University Press.

EMN, ed. 2012. *Practical Responses to Irregular Migration: The Italian Case*. Rome: EMN Italy: Callia, Raffaele; Giuliani, Marta; Pasztor, Zsuzsanna; Pittau, Franco; Ricci, Antonio. https://ec.europa.eu/home-affairs/sites/homeaffairs/files/what-we-do/networks/european_migration_network/reports/docs/emn-studies/irregular-migration/it_20120105_practicalmeasurestoirregularmigration_en_version_final_en.pdf. Accessed 5 Jan 2020.

EMN. 2013. Ad-Hoc Query on Criminal Penalties for Illegally Entering/Staying TCNs. Requested by NL EMN NCP on 4 April 2013, Compilation produced on 6 August 2013, Responses from Austria, Belgium, Cyprus, Estonia, Finland, France, Germany, Greece, Hungary, Italy, Latvia, Lithuania, Luxembourg, Netherlands, Poland, Portugal, Slovak Republic, Spain, Sweden, United Kingdom plus, Norway (21 in Total). https://ec.europa.eu/home-affairs/sites/homeaffairs/files/what-we-do/networks/european_migration_network/reports/docs/ad-hoc-queries/illegal-immigration/467._emn_ad-hoc_query_on_criminal_penalties_wider_dissemination_en.pdf. Accessed 5 Jan 2020.

Espenshade, Thomas J. 1994. Does the Threat of Border Apprehension Deter Undocumented US Immigration? *Population and Development Review* 20 (4): 871–892.

European Union: European Commission. 2015. Better Regulation "Toolbox" (BR Toolbox). http://ec.europa.eu/info/law/law-making-process/better-regulation-why-and-how_en. Accessed 10 Jan 2017.

Eurostat. 2018. Statistics on enforcement of immigration legislation (including migr_eipre, migr_eirfs, migr_eirtn, migr_eiord, migr_dubro, migr_asyapp, migr_asyapp, migr_eirt_ass). http://appsso.eurostat.ec.europa.eu/nui/show.do?dataset=migr_eipre&lang=en (downloaded December 2018).

Ferrero, Giancarlo. 2010. *Contro il Reato di Immigrazione Clandestina: Un'Inutile, Immorale, Impraticabile Minaccia*, 2nd ed. Roma: Ediesse.

FRA. 2014. *Criminalisation of Migrants in an Irregular Situation and of Persons Engaging with Them*. Vienna: FRA.

Freeman, Gary P. 1998. The Decline of Sovereignty? Politics and Immigration Restriction in Liberal States. In *Challenge to the Nation-State: Immigration in Western Europe and the United States*, ed. Christian Joppke, 86–108. Oxford: Oxford University Press.

French Parliament, Code de l'Entrée et du Séjour des Étrangers et du Droit d'Asile, as of 1 January 2017 (Ceseda). https://www.legifrance.gouv.fr/affichCode.do?cidTexte=LEGITEXT000006070158. Accessed 22 Jan 2016.

Frontex. 2017, 2019. *Detections of illegal border-crossings statistics*. https://frontex.europa.eu/along-eu-borders/migratory-map/ (downloaded on 7 October 2017 and 6 August 2019).

Gisti, ed. 2012. *Immigration: un régime pénal d'exception*. Paris: Gisti. https://www.gisti.org/publication_som.php?id_article=2781#llutte. Accessed 23 Feb 2017.

Godenau, Dirk, and Ana López-Sala. 2016. Multi-Layered Migration Deterrence and Technology in Spanish Maritime Border Management. *Journal of Borderlands Studies* 31: 1–19.

Hassan, Lisa. 2000. Deterrence Measures and the Preservation of Asylum in the United Kingdom and United States. *Journal of Refugee Studies* 13 (2): 184–204.

Henriot, Patrick. 2013. Dépénalisation du séjour irrégulier des étrangers: l'opiniâtre résistance des autorités françaises. *La Revue des droits de l'homme* [online] 3: 1–14.

Hollifield, James F. 1998. Migration, Trade, and the Nation-State: The Myth of Globalization. *The UCLA Journal of International Law & Foreign Affairs* 3: 595–636.

Husak, Douglas. 2004. The Criminal Law as Last Resort. *Oxford Journal of Legal Studies* 24 (2): 207–235.

Huysmans, Jef. 2000. The European Union and the Securitization of Migration. *Journal of Cutaneous Medicine and Surgery: Incorporating Medical and Surgical Dermatology* 38 (5): 751–777.

IOM Italy. 2017. La Tratta di Esseri Umani attraverso la Rotta del Mediterraneo Centrale: Dati, Storie e Informazioni Raccolte dall'Organizzazione Internazionale per le Migrazioni. http://www.italy.iom.int/sites/default/files/news-documents/RAPPORTO_OIM_Vittime_di_tratta_0.pdf. Accessed 9 Oct 2017.

ISTAT. 2018. Data related to article 10-bis TUI, 2009–2016 (obtained in May 2018).

Italian Ministry of Justice (MoJ). 2018. Data related to article 10-bis TUI, 2009–2018 (obtained in May 2018).

Joppke, Christian. 1998. Why Liberal States Accept Unwanted Immigration. *World Politics* 50 (2): 266–293.

Kanstroom, Daniel. 2000. Deportation, Social Control, and Punishment: Some Thoughts About Why Hard Laws Make Bad Cases. *Harvard Law Review* 11: 1889–1935.

Lochak, Danièle. 2016. 'Pénalisation', in 'L'étranger et le Droit Pénal'. *AJ Pénal*, Janvier, pp. 10–12. https://hal-univ-paris10.archives-ouvertes.fr/hal-01674295/document. Accessed 5 Jan 2019.

López-Sala, Ana. 2015. Exploring Dissuasion as a (Geo)Political Instrument in Irregular Migration Control at the Southern Spanish Maritime Border. *Geopolitics* 20 (3): 513–534.

McLaughlin, Eugene, and John Muncie, eds. 2001. *The SAGE Dictionary of Criminology*. London: Sage.

Natow, Rebecca. 2020. The Use of Triangulation in Qualitative Studies Employing Elite Interviews. *Qualitative Research* 20 (2): 160–173.

Overbeek, Henk. 1996. L'Europe en quête d'une politique de migration: les contraintes de la mondialisation et de la restructuration des marchés du travail. *Études internationales* 27 (1): 53–80.

Pratt, Travis C., Francis T. Cullen, Kristie R. Blevins, Leah E. Daigle, and Tamara D. Madensen. 2008. The Empirical Status of Deterrence Theory: A Meta-Analysis. In *Taking Stock: The Status of Criminological Theory*, ed. Francis Cullen, John Wright, and Kristie Blevins, 367–395. New Brunswick, NJ: Transaction Publishers.

Ryo, Emily. 2015. Less Enforcement, More Compliance: Rethinking Unauthorized Migration. *UCLA Law Review* 62: 622–670.

Sassen, Saskia. 1996. *Losing Control? Sovereignty in an Age of Globalization*. New York: Columbia University Press.

Schain, Martin A. 2012. *The Politics of Immigration in France, Britain, and the United States*, 2nd ed. New York: Palgrave Macmillan.

Scherr, Mickaël. 2011. L'évolution des phénomènes d'immigration illégale à travers les statistiques sur les personnes mises en cause par la police et la gendarmerie nationale. Dossier INHESJ, 144–158. https://inhesj.fr/sites/

default/files/ondrp_files/contributions-exterieures/m_scherr.pdf. Accessed 12 Dec 2018.

Skeldon, Ronald. 2012. Migration and Its Measurement: Towards a More Robust Map of Bilateral Flows. In *Handbook of Research Methods in Migration*, ed. Carlos Vargas-Silva, 229–248. Cheltenham: Edward Elgar.

Stumpf, Juliet. 2006. The Crimmigration Crisis: Immigrants Crime and Sovereign Power. *American University Law Review* 56 (2): 367–419.

Tansey, Oisin. 2007. Process Tracing and Elite Interviewing: A Case for Non-Probability Sampling. *PS Online: Political Science & Politics* 40 (4): 765–772.

Thielemann, Eiko. 2003. 'Does Policy Matter? On Governments' Attempts to Control Unwanted Migration. IIIS Discussion Paper No. 09. https://www.tcd.ie/triss/assets/PDFs/iiis/iiisdp09.pdf. Accessed 13 Dec 2021.

Thielemann, Eiko. 2011. How Effective Are Migration and Non Migration Policies That Affect Forced Migration? LSE Migration Study Unit Working Paper No. 2011/14.

Triandafyllidou, Anna. 2016. Governing Irregular Migration: Transnational Networks and National Borders. In *Europe: No Migrant's Land?* ed. Maurizio Ambrosini. Novi Ligure: ISPI.

Weber, Leanne. 2002. The Detention of Asylum Seekers: 20 Reasons Why Criminologists Should Care. *Current Issues in Criminal Justice* 14 (9): 9–30.

Weiner, Myron. 1995. *The Global Migration Crisis: Challenge to States and to Human Rights*. New York: Harper Collins College Publishers.

CHAPTER 2

Globalisation and Irregular Migration

INTRODUCTION

The question of whether states can control migration, in the context of contemporary globalisation processes, has long been a major source of concern for Western governments, and a subject of intense discussion for academics. Whereas some scholars support the idea that governments are still the key actors (realist approach),[1] others argue that a relevant role is played by legal and supranational institutions (liberal institutionalist approach),[2] and others yet maintain that migration is to a large extent an uncontrollable phenomenon (transnationalist approach).[3]

This chapter presents and discusses the different theoretical positions on such debate. Starting with an analysis of the IPE literature on migration control, it then moves on to the study of the 'policy gap' hypothesis, according to which a gap has emerged between the stated goals of measures meant to reduce migration, and their actual outcomes. Finally,

---

[1] Weiner (1985, 1995, 1996), Borjas (1989), Teitelbaum (2002), Freeman (1995, 1998), and Freeman and Kessler (2008).

[2] Hollifield (1998, 2004), Geddes (2003), and Geddes and Korneev (2015).

[3] Sassen (1996), Mittelman (2000), Dicken (2011), León and Overbeek (2015), Castles (2004a, b), and Talani (2010, 2015).

it examines migration studies addressing the specific topic of control through deterrence, thus setting the basis for a more elaborate discussion of the strategy in the subsequent chapter.

THE IPE OF IRREGULAR MIGRATION: THEORETICAL INTERPRETATIONS

The Realist Approach

Realism has a long-standing tradition in international relations. In the context of migration, authors subscribing to such theoretical viewpoint support the idea that the rules of entry and exit, as designed and implemented by both countries of origin and destination, play a key role in forcing, promoting, or impeding migration.[4] Globalisation, they argue, does not nullify states' capacity to control migration.[5] In Patrick Weil's words:

> [T]he transformation of migration into a world-wide phenomenon has not automatically involved an increase in immigration flows: When this has happened, it was due primarily to legal windows still open to entry. The closure of these windows has demonstrated that regulation is possible for a democratic state.[6]

Among the most prominent exponents of the realist school is Myron Weiner, who maintains that, despite globalisation processes, both receiving and sending states are able to control the flows of people traversing their borders, albeit with varying degree of effectiveness.[7]

Specifically, receiving countries can affect immigration in three main ways: By adopting controlling measures, defining the degree of inclusiveness of their political system, and intervening in sending countries. Concerning the first option, the academic argues that ease of access is more relevant than market forces (such as labour demand) in shaping migratory flows, highlighting in particular the utility of identity cards and

---

[4] Borjas (1989), Freeman (1995, 1998), Weiner (1995, p. 134), and Teitelbaum (2002).

[5] Freeman (1998) and Weiner (1996).

[6] Weil (1998, p. 18), cited in Teitelbaum (2002, p. 166).

[7] Weiner (1996, pp. 184–185).

checks on irregular employment.[8] Moving on to the second argument, Weiner maintains that the political empowerment of migrant communities and, consequently, the extent to which they can exert pressures against restrictive policies, is directly related to the inclusiveness of the host state. In other words, a state is able to affect the degree of influence of its foreign population, by determining the amount of inclusion of its political system.[9] Lastly, for the scholar, intervention in another country's (political or economic) situation can also be considered as a strategy to reduce migration, although it should be subject to what is 'morally justifiable', and results should not be expected to be achievable in a short time.[10]

If receiving countries thus have several options available to control migration, in Weiner's view, the role of sending countries should not be underestimated: By either encouraging (sometimes forcing) certain categories of people to leave, or preventing others from doing so, they can exert notable influence on the flows.[11]

In similar fashion, although based on quantitative reasoning, economist George Borjas argues that migration policies in both countries of origin and destination 'can have a major impact' on migrants' calculations of the utility of mobility.[12] As shown by his model, when deciding whether (and where) to migrate, individuals compare alternative options and choose the one that provides them with the highest benefits, given financial and legal constraints.[13] It is indeed the latter type of limitations that includes immigration regulations as imposed by countries of origin and destination, and that determines therefore the feasibility of migration (together with personal financial availability).[14] As a result, the academic argues, immigration policies on both entry and exit effectively influence the size and composition of migratory flows.[15]

---

[8] Weiner (1995, pp. 205–207; 1996, p. 184).

[9] Weiner (1995, p. 125).

[10] Weiner (1995, pp. 213–216; 1996, p. 189).

[11] Weiner (1985, pp. 444–445; 1995, pp. 34, 205–207; 1996, p. 184).

[12] Borjas (1989, pp. 460–461).

[13] Ibid.

[14] Ibid.

[15] Ibid.

A key argument proposed by realist authors is that the persistence of some irregular migration does not necessarily imply states' inability to regulate the phenomenon: In the absence of controls, unauthorised border crossings would be higher. In this context, Michael Teitelbaum depicts the effectiveness of controlling measures as a continuum between authoritarian regimes, able to exercise 'substantial' control, and democratic ones, capable of 'variably effective' control.[16] Yet, both he and Weiner argue, even in the latter case it would be wrong to say that state action is ineffective, as, without any controls and constraints, more people would attempt to migrate.[17] In other words, from this perspective, the control of irregular migration is not a matter of stopping the flows altogether, but rather of containing them to an acceptable level.[18]

Like in traditional international relations theory, the realist approach has been highly influential in the debate on states' ability to control migration. At the same time, however, it has also often been criticised for employing a monolithic conception of the state and failing to appreciate the pluralism of interests that are involved in migration issues.[19] This has two major implications: First, realism does not clarify how to determine national interest, in pursuit of which states are expected to act.[20] Second, the theory cannot explain how those migration policies that are criticised for being irrational, or inadequate, came to be adopted.[21] On the contrary, a focus on domestic politics, such as the one suggested by Freeman, may be better suited to appreciate such matters, through the study of the decision-making process and of the influence of domestic actors.[22]

Gary Freeman approaches the processes that lead to the adoption of immigration policies from an economic-interests point of view. While he

---

[16] Teitelbaum (2002, pp. 165–166).

[17] Teitelbaum (2002, pp. 160, 166) and Weiner (1996, p. 185). Similarly, Espenshade (1994).

[18] Espenshade (1994, p. 889).

[19] Samers (2015, p. 377).

[20] Freeman (2011, pp. 1549–1550).

[21] Meyers (2000, p. 1265).

[22] Ibid.

can be considered part of the realist school due to his focus on the state as the main unit of analysis,[23] he slightly departs from it, stressing the importance of the multiple actors involved in the decision-making and lobbying processes.

The core argument of the academic is that the management of migration still belongs to the realm of state sovereignty, and it would be wrong to assume a general loss of control over inflows, since it is domestic politics itself that decides the degree of openness it wants to allow.[24] In other words, states' ability to control migration is not decreasing, but actually increasing, and the main obstacles to achieve such a goal lie in political factors.

More specifically, Freeman's model argues that it is possible to produce 'theoretically driven expectations' of the kind of political model that will be in place in liberal democracies, by examining whether political mobilisation is concentrated (cM) or diffuse (dM), and whether effects of migration on the host society are large (lE) or small (sE). Combining these elements, four types of politics emerge: client (cM, lE), entrepreneurial (cM, sE), interest-group (dM, lE), and majoritarian (dM, sE) (Table 2.1).[25]

In the West, for example, immigration involves large costs primarily shouldered by the individuals who feel in competition with newcomers for the allocation of scarce resources, be they of economic, social, or financial nature.[26] Still, the public at large is too scattered to mobilise, and the actors who eventually do so are, instead, a restricted niche of people, such as economic interests, previous migrants, and intellectuals. Consequently, the client mode of politics prevails.[27]

As a matter of fact, socio-economic interests may play an important role in asking for openness. In Western countries, the fast ageing and low fertility rate of the population, the little rural workforce left available to recruit,[28] and the increasing unwillingness of citizens to undertake dirty

---

[23] Talani (2015, p. 21).

[24] Freeman (1998, p. 103). For a similar argument, see Messina (2007).

[25] Freeman and Kessler (2008, pp. 671–672).

[26] Freeman (1995, p. 885) and Freeman and Kessler (2008).

[27] Freeman (1995, p. 885; 1998, p. 103) and Freeman and Kessler (2008).

[28] Weiner (1995, p. 199).

and dangerous occupations, often even when unemployed,[29] have created a 'structurally embedded' demand for alien labour.[30] The need for foreign labour is particularly visible in the sectors of agriculture, constructions, and domestic help,[31] and takes advantage of the latter's low cost, little trade union participation and easy dismissal.[32] According to Overbeek, this is to the extent that '[t]he employment of undocumented foreign labour has [...] in many cases become a condition for the continued existence of small and medium size firms, creating a substantial economic interest in continued (illegal) immigration'.[33] As a result, employers in the aforementioned labour-intensive industries, as well as those relying on unskilled workers, together with the businessmen involved in the 'migration industry', may contribute to sustain mobility, either by exerting influence on governments to allow for expansive policies,[34] or by directly taking action to organise migration.[35] In this respect, although, according to Salt and Clarke (writing in 2000), the facilitation of migration is to a large extent a legal business,[36] the illegal side of the migration industry should not be neglected, as human smuggling activities are thought to generate profits that are only second to drug traffics.[37]

Delving deeper into Freeman's theoretical model, the effects of migration on host societies can be distinguished between those on wage and income, on one hand, and on the fiscal system, on the other hand. Economic models are useful to make predictions regarding the former, and highlight that class cleavages between skilled/unskilled, and organised/unorganised labour, are at the basis of political action.[38] Different

---

[29] Ibid.

[30] Cornelius and Tsuda (2004, p. 9). Similarly, Overbeek (1996, p. 78).

[31] Boswell (2003, p. 36), Migration Policy Centre team, and Peter Bosch (2015, p. 3), and Hollifield et al. (2014, p. 4).

[32] Weiner (1995, p. 199).

[33] Overbeek (2002, p. 3).

[34] Freeman (1995), Castles (2004a, p. 208), and Wagstyl (2015).

[35] Castles (2004a, p. 209).

[36] Salt and Clark (2000), cited in Castles (2004a, p. 209).

[37] Di Nicola and Musumeci (2014, p. 40).

[38] Freeman and Kessler (2008, p. 670).

Table 2.1 Freeman's model of migration policy

		Political mobilisation	
		Concentrated	Diffuse
Wage/income effects	Large	Client politics	Interest-group politics
	Small	Entrepreneurial politics	Majoritarian politics

Source Freeman and Kessler (2008, p. 672)

economic hypotheses, however, forecast different outcomes.[39] According to the standard production function, migration creates class-based political cleavages, as while local workers face growing competition, businesses profit from the increased supply of labour. Trade models such as the Heckscher-Ohlin one, on the other hand, forecast newcomers to be easily integrated in the labour market according to their skills, without any significant long-term impact on the receiving society, and with consequent little reason for political mobilisation. Specific-factors models, yet, predict migration to exert downward pressures on the wages of people in direct competition with aliens, but also to increase the income of other socio-economic actors, therefore creating an array of possible coalitions.[40]

As for the fiscal effects of migration, they transcend traditional class cleavages and sectoral lines created by wage and income effects.[41] Among the interests involved, crucial are those of taxpayers, and of the local areas that are affected the most by migration. Indeed, geographical concentration may prompt locals to resent the investment of public resources for the benefit of newcomers, and therefore incentivise the mobilisation of voters who would otherwise be in favour of migration, or even a cross-class coalition requesting relief from fiscal burden.[42]

Overall, if people react to the economic and fiscal stimuli generated by migration, the above models should forecast the political cleavages and coalitions formed on issues of migration. Yet, two observations are key, for Freeman. First, perceptions of the effects of migration are more

---

[39] Ibid., pp. 660–663.
[40] Ibid.
[41] Ibid., pp. 661, 672.
[42] Ibid.

important than actual facts, and can be shaped by an array of actors, including lobbies, the media, politicians, and think tanks.[43] Second, because migration policy is not decided through direct-democracy tools, the interests of organised groups are stronger than the stances of individual citizens, in terms of the influence exercised on the legislative, administrative, and electoral processes.[44] As a consequence, the role of trade unions and employer federations is crucial. While, in theory, trade unions would be expected to oppose migration on the basis of its depressing effect on wages, in practice, the cleavage between skilled and unskilled workers complicates the matter.[45] In particular, their eventual stance may depend on their composition in terms of skills, the density of workforce organisation, and the cohesion of trade federations.[46]

Freeman's model is key in emphasising the role of the political processes that lead to the adoption of immigration policies, as well as of the groups that influence such processes. Yet, the client mode of politics, which the academic ascribes to Western countries, has been criticised for not being applicable to Europe, where there has never been an established tradition of influential lobbies in favour of immigration (either for humanitarian or business reasons).[47] Further, his study neglects the role of ethnicity and identity issues,[48] which are often leveraged by populist radical right parties through 'nativist' sentiments[49] and can therefore greatly alter political dynamics. Indeed, the politicisation of migration has enabled the popular discontent against the (perceived or real) negative externalities of migratory flows to be channelled into political means that allow even the unorganised public to express its fears and concerns, a transformation that should make it possible, in theory, for the latter to contrast the influence of groups in favour of openness.

Altogether, the realist position considers states as the fundamental unit of analysis and contends that they are still able to control migration, if and when it is in their interest to do so. Yet, the number of migrants entering

---

[43] Ibid., pp. 670–673.
[44] Ibid.
[45] Ibid.
[46] Ibid., p. 671.
[47] Geddes (2003, p. 21) and Cornelius and Tsuda (2004, p. 12).
[48] Talani (2015, p. 23).
[49] Samers (2015, p. 375).

Europe irregularly suggests a less than perfect capability to restrain the phenomenon, even in times of high political pressures to do so, making numerous academics question the validity of the realist hypotheses.

The Liberal Institutionalist Approach

An alternative school of thought has sought to go beyond realism, by developing arguments aimed at combining political and economic considerations. This is liberal institutionalism, according to which, states' capacity to control migration is constrained by legal and supranational institutions.[50]

One of the main defenders of this thesis is James Hollifield, who argues that liberal democracies are limited, in their attempts to restrict migratory flows, by the legal rights that they themselves grant to individuals on their territory.[51] While he also stresses the relevance of domestic coalitions, of ideational and cultural factors in supporting liberal migration regimes,[52] his main contribution to the understanding of the debate on states' control over migration stems from his emphasis on the institutionalisation of rights.

His most fundamental argument in this context is that states are trapped in a 'liberal paradox', when dealing with migration, caused by a clash between the pressures for greater openness by international economic forces, and the push for greater closure by domestic political forces.[53] How such paradox is resolved, and therefore what the eventual degree of openness of a country to migration is, depends on the extent to which rights are institutionalised and constitutionally protected.[54] In other words, the strength of the legal system (through the protection of rights, and courts' intervention), is able to shift the balance from the political preference for closure, to the economic one for openness.[55]

---

[50] Hollifield (1998).
[51] Ibid.
[52] Ibid.
[53] Hollifield (2004, p. 886).
[54] Ibid., pp. 886, 904.
[55] Ibid., pp. 895, 904.

This is not only because citizenship is not necessary for individuals to be admitted to what he calls 'rights of membership' (which may incorporate more or less rudimentary civil and social rights, sometimes even political, depending on the country of reference), but also due to the fact that, once granted, rights are difficult to withdraw.[56] Consequently, Hollifield argues, the institutionalisation of rights represents the most significant constraint on states' capacity to control migration.[57] Because such institutionalisation of rights, in turn, depends not only on political or economic interests, but also on ideas and institutions, the academic maintains that states cannot detain control over migration in a unilateral way, but need multilateral cooperation.[58] Following from the above, Hollifield's 'liberal paradox' may in fact also be interpreted as referring to the contrast between liberal states' attempts to control migration, and their attribution of rights to foreigners (which stem from the former's liberal democratic nature itself): While granting rights to migrants would act as a pull factor for further migration, reducing them would in fact make the concerned states less 'liberal'.[59]

From an operational perspective, Hollifield investigates the above arguments in two ways: First, by calculating the impact that changing policies have on migratory flows.[60] Second, by investigating the ways in which rights 'act', chiefly through the action of courts, to weaken migration control efforts.[61]

The liberal institutionalist hypothesis is well exemplified by a judicial case taking place in Italy in early 2019. Ruling on the possibility of granting international protection to a young man from Mali, the Court of Venice acknowledged the absence of the applicant's direct persecution in his home country, but decided not to repatriate him (and instead granted him a residence permit).[62] This was done on the basis of the person's integration in the Italian society, as demonstrated by good language

---

[56] Hollifield (1998, pp. 619–620; 2004).

[57] Ibid.

[58] Hollifield (1998, p. 598; 2004, p. 903).

[59] For an example of this second interpretation, see Talani (2021, Chapter 6).

[60] Hollifield and Wong (2014, p. 243).

[61] Ibid.

[62] See Rame (2019).

skills, attendance of secondary school, active employment record, and involvement in volunteering activities.[63]

Although Hollifield's hypothesis looks promising in including both political and economic analysis, the academic regards rights as deriving from a state's legal system,[64] rather than from supranational regimes, which has led some to question whether he can be fully classified as a liberal institutionalist.[65] Indeed, it is questionable whether the fundamental element on which such theories rests, a strong international regime of migration, actually exists, or is being developed.[66] Hollifield himself acknowledges that the International Organisation for Migration (IOM) and the International Labour Organisation (ILO) are not even close to the existing international regimes that regulate trade and international finance, such as the General Agreement on Tariffs and Trade, the World Trade Organisation, the International Monetary Fund (IMF), and the World Bank.[67] An exception is of course the European Union but, even there, freedom of movement is only applicable to citizens of member states, not to third country nationals.[68]

In addition to the above, not all agree that shifting the politics of immigration to supranational venues necessarily entails a loss of sovereignty by the state. As an example, Virginie Guiraudon argues that the upward shift of the policy-making process to the EU level is actually the consequence of a strategy of 'venue-shopping', consciously implemented by domestic agencies in the attempt to overcome the constraints posed by national courts protecting liberal norms.[69] From her perspective, judicial review at the national (rather than supra- or sub-national) level is best placed to limit states' attempts to control migration.[70] As a consequence, governments, with their law and order officials, look for new venues to navigate institutional constraints and exert pressure (they

---

[63] Ibid.
[64] Hollifield (1998, p. 619).
[65] Talani (2015, p. 24).
[66] Ibid., p. 23.
[67] Hollifield (1998, p. 615).
[68] Ibid., p. 626.
[69] Guiraudon (2001).
[70] Guiraudon (2000, pp. 261–262; 2001).

'venue-shop').[71] It is indeed her emphasis on the role of institutional and legal constraints that may make it possible to categorise her approach as liberal institutionalist,[72] although she includes constructivist aspects too, for instance by arguing that, when venue-shopping, security actors 'resort to framing processes or policy images',[73] including by linking migration to security issues. Overall, for the scholar, international agreements, far from constraining states, actually 'sanction national or protectionist initiatives'.[74]

Guiraudon's argument finds support in Andrew Geddes' analysis, which, focusing on the EU too, similarly argues that cooperating and harmonising policies on immigration issues have allowed member states to escape the jurisdictional and political constraints they faced domestically, consequently increasing their capacity to control migration.[75]

In a further variant of the liberal institutionalist argument, which explicitly leaves aside the supranational element, Christian Joppke contends that, in the case of European states, it is possible to understand the acceptance of 'unwanted immigration' as a result of 'self-limited sovereignty'.[76] Starting from Freeman's model of client politics, the academic maintains that, while this may be applicable in the US context, it holds less explanatory power for Europe. In the latter region, it was legal constraints and states' moral obligations (such as those deriving from Germany's authoritarian past), rather than the political process, that were responsible for limiting migration restrictionism.[77] This is because courts, being isolated from anti-immigration demands or other socio-economic pressures, have ample room for manoeuvre to protect the residence and family rights of newcomers, thus contributing to the self-limitation of state sovereignty.[78]

Although Joppke's approach is strictly linked to Freeman's realist account, his emphasis on legal and institutional constraints, especially in

---

[71] Guiraudon (2000, p. 258).
[72] Similarly, Boswell (2007) and Talani (2015).
[73] Ibid.
[74] Lahav and Guiraudon (2006, p. 207).
[75] Geddes (2003, pp. 20, 196).
[76] Joppke (1998, p. 284).
[77] Ibid., p. 292.
[78] Ibid.

the analysis of Europe (the focus of the present study) make it possible to associate him to a liberal institutionalist perspective.[79] Still, a key difference with Hollifield's understanding of the thesis should be stressed as, in Joppke's view, it is internal, rather than external, factors that hinder liberal states' control of migration.[80] In other words, the supranational element loses importance, as stated most clearly in his depiction of the international human rights regime as 'perhaps the single most inflated construction in recent social science discourse'.[81]

Overall, as seen thus far, the core idea at the basis of the liberal institutionalist hypothesis is that legal and (according to some scholars) supranational institutions play a key role in constraining states' attempts at restricting migration. Indeed, while social structures can sometimes prevent migrants from benefiting from their entitlements despite court judgements, the importance of courts is generally accepted as playing a key role in mediating between legal inclusion and exclusion.[82]

Before introducing a third, rather contrasting, perspective on whether states can control migration, it is interesting to discuss Christina Boswell's analysis, insofar as it builds upon both realist and liberal institutionalist perspectives.

Boswell's argument concentrates on states' 'functional imperatives' (i.e. their interests) in the field of migration, and how these affect the response to societal actors' preferences and institutional constraints.[83] In particular, in her model, state legitimacy depends on four elements: (1) The provision of security for its citizens, (2) the provision of the necessary conditions to enable wealth accumulation; (3) the promotion of a fair pattern of distribution; and (4) the guarantee of institutional legitimacy (i.e. compliance with liberal democratic values).[84] Because these four goals are hard to realise simultaneously, when states face different pressures on migration, they tend to adopt one of two strategies: Either *deliberate malintegration*, that is to say intentionally pursuing contradictory goals, such as restricting entry but allowing regularisations; or

---

[79] Similarly, see Boswell (2007, p. 83).
[80] Joppke (1998, p. 271).
[81] Ibid., p. 293.
[82] Geddes (2003, pp. 21–22).
[83] Boswell (2007).
[84] Ibid., pp. 89–91.

populist mobilisation, namely channelling support towards one of the four criteria, outshining the remaining ones.[85]

Boswell's analysis may be interpreted as partly sharing some realist assumptions, including viewing the state as the key actor (as opposed to legal or supranational institutions), focusing on the role of contrasting interests, and placing more emphasis on political, rather than legal, processes. At the same time, however, her understanding of states' interests is significantly different from Freeman's. Indeed, she views them not as stemming from pressure groups' influence, but as based on 'states conceptions of the preconditions for legitimacy',[86] thus adding a constructivist element to the analysis too. In her words: 'We might paraphrase Alexander Wendt [...]: the liberal constraint is what states make of it'.[87] Although such interpretation may be commended for including ideational aspects, it involves drawbacks in the operationalisation of the concept, as the way in which such interests come to be appreciated by politicians and/or bureaucracies is not straightforward. Still, by pointing out that policy incoherence may be the result of deliberate choices to address contrasting demands, Boswell's contribution is of crucial importance, for the study of migration governance.

The Transnationalist Approach

Beyond realism and liberal institutionalism, the debate on states' ability to control migration has been approached from a third relevant perspective: The transnationalist one. According to scholars subscribing to this tradition, globalisation has gradually eroded state sovereignty, to the extent that migration is largely an uncontrollable phenomenon.[88] As globalisation flourishes, border controls become increasingly less relevant, merely causing the transformation of legal flows into irregular ones.[89] Borrowing Bhagwati's words: 'The ability to control migration has shrunk as the

---

[85] Ibid., pp. 92–93.

[86] Ibid., p. 96.

[87] Ibid. For a similar argument, see Bonjour (2011).

[88] Dicken (2011), Mittelman (2000), Sassen (1996), Léon and Overbeek (2015), Talani (2010, 2015), and Castles (2004a, b).

[89] Ghosh (2000, p. 18).

desire to do so has increased. The reality is that borders are beyond control and little can be done to really cut down on immigration'.[90]

Following Peter Dicken, it is possible to describe globalisation as an indeterminate set of processes composed by qualitative transformations.[91] In this sense, globalisation differs from internationalisation, as, while the latter only concerns the quantitative increase in cross-border traffic, the former also addresses where and how such changes occur.[92] In defining globalisation, Dicken cautions against what he calls 'sceptical-internationalist' approaches, according to which the world was more interconnected in the late nineteenth and early twentieth centuries than it is now (as exemplified by the fact that the proportion of the migrant population at the global level has never been as high as in that period), on the basis that a quantitative analysis alone cannot fully appreciate the complexity of the transformations occurring.[93]

Building on the above, how does migration come to be an uncontrollable phenomenon, for transnationalist theorists? Saskia Sassen argues that, in developing countries, the answer lies in their integration in the global political economy. By promoting the increase in foreign direct investments as well as the feminisation of the industrial workforce, and by intensifying economic relations with developed economies (following, for example, practices of off-shored production, or recruitment of foreign labour), globalisation disrupts traditional labour structures, thus incentivising out-migration.[94]

Extending the discussion to countries that are not developing, and following Dicken's emphasis on the importance of adopting a qualitative approach, the transnationalist perspective proposed by authors such as James Mittelman, Henk Overbeek, and Leila Talani views globalisation as a process that leads to the 'emergence of a new global division of labour, [and] produces mass migration from marginalised regions of the globe'.[95]

---

[90] Bhagwati (2003, p. 99).
[91] Dicken (2011, pp. 6–8).
[92] Ibid., also Léon and Overbeek (2015, p. 39).
[93] Dicken (2011, p. 6).
[94] Sassen (1996), Sassen, in Talani (2015, pp. 27–28).
[95] Talani (2015, p. 30). Similarly, Overbeek (1996, 2002) and Mittelman (2000).

Delving deeper into such thesis, Talani stresses that technological transformations are at the basis of the alterations that affect global production and its geographical reallocation, by contributing, for example, cheaper transportation costs and remote labour control systems.[96] As a consequence of such changes, the academic argues, production has moved to specialised areas, characterised by lower costs, generating dynamics of 'regionalisation within globalisation', that is to say creating economically integrated regions.[97] This process however has, as a side effect, the increase in the marginalisation of the areas not involved in the process of geographical reallocation of production, and the deterioration of such zones' living conditions.[98] As a result, both brain drain and mass migration, by skilled and unskilled workers respectively, are incentivised.[99]

Indeed, as argued by Mittelman, the restructuring of the global political economy simultaneously affects, and is affected by, labour migration.[100] If, on one hand, migration depends not only on individuals' preferences, but also, and to a significant extent, on structural factors beyond their control, on the other hand, emigration further aggravates the marginalisation of developing countries, by taking away their healthy workforce (in the case of mass migration) or their investments in human capital (in the case of brain drain).[101] Through such dynamics, a vicious cycle emerges between migration and the restructuring of the global political economy.[102]

In a related account of the transnationalist thesis, Overbeek argues that, starting in the late twentieth century, changing globalisation processes have affected both industrialised and third-world countries, therefore playing a key role in shaping migratory dynamics.[103] To start with, in OECD countries, globalisation has increased reliance on low-skilled workers.[104] In particular, the crisis of Keynesianism in the 1970s, and the

---

[96] Talani (2010, p. 39).

[97] Talani (2015, p. 31). Similarly, see Overbeek (1996, p. 64).

[98] Talani (2015, p. 31).

[99] Talani (2015, p. 31). Similarly, see Mittelman (2000, pp. 63, 66).

[100] Mittelman (2000, p. 58).

[101] Ibid., pp. 58, 62–63.

[102] Ibid., p. 58.

[103] Overbeek (1996; 2002, p. 2).

[104] Overbeek (1996, pp. 67–72; 2002, p. 3).

subsequent rise of the neoliberal emphasis on deregulation and flexibilization, have spurred the demand for unskilled and flexible labour.[105] This, in turn, has frequently taken the form of foreign (and often irregular) workers.[106] As an example, as reported by the author, several European countries rely on irregular migration to support their economies, to the extent that certain sectors (such as construction or domestic services) would collapse, without them.[107]

Simultaneously, globalisation processes have incentivised emigration from peripheral areas, such as Africa and Latin America.[108] In such areas, not only did the marginalisation from the restructuring of production increase incentives for migration, but also the end of financial support by the United States or the Soviet Union (paralleling the end of the Cold War), and the 'conditionalities' set by the IMF or World Bank in the 1980s (as prerequisites for loans), led to a contraction in the 'external sources of finance available for redistribution by the state', with two main consequences.[109] First, in some states, like Sudan, social crises and domestic conflicts erupted, leading in several instances to an increase in forced migration.[110] Second, in other states, governments sought alternative sources of income, often finding them in the form of remittances, and therefore ended up promoting emigration of their citizens.[111] In sum, Overbeek argues that the neoliberal globalisation has led to intrinsic changes both in core and peripheral areas of the world, and, in so doing, also contributed to incentivising both the reliance on cross-regional migration, and the phenomenon itself.

As a final note, by increasing the links between more and less developed countries, globalisation may also facilitate the creation of transnational

---

[105] Overbeek (2002, p. 3) and Léon and Overbeek (2015, p. 39). Similarly, see Mittelman (2000, pp. 60–61).

[106] Overbeek (2002, p. 3).

[107] Overbeek (1996, pp. 69–70).

[108] Overbeek (2002, pp. 3–4).

[109] Ibid., Overbeek (1996, pp. 65, 67).

[110] Ibid.

[111] Overbeek (2002, p. 4). A similar argument linking both refugee and migrant flows to changes in globalisation patters is made by Mittelman (2000, p. 64), according to whom 'the distinction between political and economic refugees, used by receiving countries as a screening mechanism, obscures the fact that both categories of migrants have as their origin the same *globalizing of production relations*' (emphasis in original).

communities,[112] the role of which has been found as key in facilitating the movement of people across borders, so much so that, according to Douglas Massey, once a flow starts, it is likely to become self-sustained, and difficult to interrupt through restrictive policies.[113] This is due not only to the intensity of the social relations that connect prospective migrants to previous ones,[114] but also to the revolution of communications, diffusion of easily accessible media, and decline in transportation costs, which have all contributed to reducing the social and financial costs of migration.[115] In the words of Stephen Castles, globalisation has increased the 'cultural capital and technical means' that are necessary for people to move.[116]

Notably, while the above dynamics can, in theory, facilitate both regular and irregular migration, in practice, the transnationalist view argues that the structure of the current liberal order is particularly favourable to the latter, which allows for greater flexibility in the labour markets of destination countries.[117]

What does the above imply for states' control of migration, according to transnationalist accounts? Following Sassen, states' desire to limit the movement of people across their territory clashes with their propensity to liberalise trade, services and finance, with the result that immigration restrictions cannot but fail to deliver the expected results.[118] In other words, the economic de-nationalisation caused by globalisation is incompatible with the political re-nationalisation encouraged by migration issues.[119] As far as state sovereignty is concerned, a process of transformation is taking place, in consequence of which sovereignty is being partly shifted to other entities, such as supranational institutions and the global capital market.[120]

---

[112] Sassen cited in Hollifield (1998, p. 607) and Dicken (2011, p. 515).
[113] Massey et al. (1998), cited in Cornelius and Rosenblum (2005, p. 102).
[114] Ibid.
[115] Boswell (2003), Hollifield (1998, p. 608), and Ghosh (2000, p. 9).
[116] Castles (2004a, p. 211).
[117] Talani (2015, p. 32) and Overbeek (2002, p. 3).
[118] Sassen (1996) and Léon and Overbeek (2015).
[119] Sassen (1996). See also Mittelman (2000, p. 73).
[120] Sassen (1996).

Similarly, Castles maintains that attempts to control the phenomenon of migration cannot succeed as long as immigration is a proxy for the wider tensions involved in the global divide between the North and the South, and until developed states do not commit to addressing the inequalities that cause it.[121] At the moment, coordinated and comprehensive action meant to reduce and, in the long term, nullify the divide is unlikely, although the academic seems to imply that such an outcome would be possible, if a drastic change in the North's perspectives and priorities occurred.[122]

Even more sceptical about states' role are Mittelman, Talani, and Overbeek, albeit with varying degrees. Indeed, they all argue that the dynamics of globalisation transcend national boundaries, not being defined by the latter's existence anymore, with the result that states lose their centrality in social relations.[123] In particular, globalisation transcends nation states by redistributing labour on a global level,[124] as it is an area's position in the global division of labour and power, and its forms of specialisation, that determine the conditions for in- and outflows.[125]

Nevertheless, for Mittelman, the role of variations in migration policies and their consequences should not be neglected, as state capacity is, yes, diminishing, but also continuing at the same time.[126] In particular, for the academic, 'the state's exclusive power to grant citizenship, order repatriation, and delimit the social and political rights of noncitizens in their territories', among other factors, may still play a very significant role.[127]

On the contrary, Talani and Overbeek seem more inclined to argue that, in the contemporary capitalist system, politics is being subordinated to economics. On one hand, businesses are able to leverage globalised production to gain more favourable economic policies from

---

[121] Castles (2004a).

[122] Castles (2004b, p. 865).

[123] Léon and Overbeek (2015, p. 40), Talani (2015, p. 30), and Mittelman (2000, pp. 58, 68).

[124] Mittelman (2000, pp. 58, 68) and Talani (2015, p. 31).

[125] Mittelman (2000, pp. 58, 68).

[126] Ibid., pp. 65–67.

[127] Ibid.

national governments, by threatening the displacement of their activities abroad.[128] On the other hand, 'the logic of global competition has made it almost impossible to completely hinder irregular migration', on which advanced economies strongly rely.[129] Finally, according to Overbeek and Léon, not only are states unable to control migration, but so are supranational institutions. Since the second half of the twentieth century, attempts at developing cooperation mechanisms on migration have been scarce, and typically regarded as failures, as exemplified by the 1990 UN Convention on the Protection of the Rights of all Migrant Workers and Members of their Families, ratified by no industrialised country.[130]

Overall, transnationalist theorists develop in-depth analyses that persuasively explain migration, especially at the macro-level. Yet, many of them have been criticised for sharing the neoclassical interpretation of economic mobility as a function of spatial disequilibria[131] taking place to balance the demand and supply of labour[132]: Globalisation, by exacerbating dualities, incentivises migration. While this neoclassical hypothesis has found vast consent, it has also been challenged for the scarce importance it attributes to migrants' agency,[133] so that transnationalist accounts may, at times, also risk resulting too deterministic. In addition to the above, the idea of a complete subordination of politics to economic forces has been regarded as problematic.[134] As a matter of fact, while it looks plausible to argue that states do not detain full control over migration, the effects that their actions and decisions regarding the economy, the structure of the labour market, education or foreign policy, can have on migratory flows, suggest a somewhat more active role for states,[135] than the one presented by some of the authors subscribing to the globalisation thesis. After all, Dicken himself argues that nation states are 'a most significant force in shaping the world economy', and that an enabling factor of

---

[128] Talani (2010, pp. 30, 40). Similarly, see Overbeek (1996, esp. pp. 62–64).

[129] Overbeek (1996, p. 79). Author's translation.

[130] Léon and Overbeek (2015, pp. 46–47).

[131] Hollifield (1998, p. 610).

[132] Borjas (1989) and de Haas (2011, p. 9).

[133] Tapinos (2000, p. 291).

[134] Hollifield (1998, p. 609).

[135] Czaika and de Haas (2013).

globalisation has been the gradual removal of political obstacles, by states themselves.[136]

To conclude, from the above theoretical discussion, it emerges that states' ability to control migration is regarded in significantly different fashion by different IPE traditions. If realists argue that states are able to control migration, provided that they want to do so, this assumption is challenged by both liberal institutionalist and transnationalist authors. Specifically, the former maintain that institutional and legal constraints play a key role in limiting states' potential to control the phenomenon, whereas the latter view globalisation processes as making migration largely structural and uncontrollable. Overall, while it is important to highlight that states may not only affect migratory flows through policies explicitly targeting them, the idea that the former may exert full control on the latter appears challenged by both theoretical arguments and contemporary events.

Having surveyed the theoretical framework that surrounds the question of whether states can control migration, in the next section I turn to the practical difficulties of migration policies, as specific tools of state action: How effective has the literature found them to be, in practice? What problems may influence their potential for success, and how may we operationalise them?

Migration Policies and Their 'Gaps'

Wayne Cornelius, in an influential paper written with Martin and Hollifield in 1994, argues that the literature has repeatedly noted a gap between the announced goals of immigration control policies and their eventual outcomes, suggesting that states' attempts to regulate inflows through policies are, in fact, rarely successful.[137] In delineating an explanation for the 'policy gap hypothesis', as this concept has come to be known, he distinguishes between gaps caused by the unintended consequences of policies and those generated by their inadequate implementation.[138]

---

[136] Dicken (2011, p. 171).
[137] Cornelius et al. (1994).
[138] Cornelius and Tsuda (2004).

Cornelius's contribution has been crucial to set in motion the research on the divergence between objectives and results of policies. Yet, it has also often been contested on the basis that it takes at face value politicians' stances on immigration and their pre-announced intentions,[139] without deconstructing their rhetoric, or investigating the hidden agendas that constitute the policy-making process.[140]

On the contrary, Mathias Czaika and Hein de Haas offer a promising categorisation, identifying three sets of possible contradictions in discursive, implementation and efficacy gaps.[141] Their framework permits a methodologically more accurate analysis, and I refer to it to organise immigration studies addressing policies' effectiveness from slightly different angles.

Starting with discursive gaps, these highlight the distance between political discourse and rhetoric on one hand, and the designed policies on the other hand.[142] The reasons why the two may not be identical not only lie in the fact that politicians' stated goals do not necessarily coincide with the actual ones,[143] but also in that the policy-making process necessarily entails compromises between the parties.[144] As a consequence, restraints stemming from socio-economic interests may also contribute to defining the magnitude of discursive gaps. Taking this argument further, some authors, as seen earlier, maintain that adopting contradictory measures meant to appease different interests should not be regarded as a failure, but as a deliberate reaction to unrealistic demands,[145] or even as a 'policy in itself'.[146] From this perspective, it should not therefore be surprising that, once promulgated, a policy meant to limit irregular migration is not as restrictive as originally intended, or is coupled with another, more inclusive measure. A clear example is the Italian Bossi-Fini law: Although, overall, it introduced restrictive norms on migration, it was then accompanied by one of the largest regularisations in Europe, thanks to which

---

[139] Guiraudon and Joppke (2001, p. 11).

[140] Boswell and Geddes (2011) and Bonjour (2011).

[141] Czaika and de Haas (2011).

[142] Czaika and de Haas (2013).

[143] Guiraudon and Joppke (2001, p. 11).

[144] Czaika and de Haas (2013, p. 494).

[145] Boswell and Geddes (2011, p. 48).

[146] Garcés-Mascareñas (2013, p. 6).

646,000 people (mainly domestic workers) were regularised between 2002 and 2003.[147]

Continuing with Czaika and de Haas's framework, after a certain strategy is adopted, it is possible for it to be affected by implementation gaps, that is to say differences between the provision on paper and its actual enforcement.[148] For Zolberg, this may occur to such an extent that 'the significance of the law will [...] depend largely on how the executive branch implements the legislation'.[149] Factors that can cause such gaps thus include practical constraints, the degree of discretion of agencies and agents at the enforcement level,[150] as well as courts' protection of foreigners' rights. Simultaneously, however, corruption[151] and the illegal side of the migration industry may also work against the smooth application of the measures, by hindering processes and increasing illegality.

Lastly, the study of efficacy gaps takes account of macro-structural dynamics to assess the relative effect of policies.[152] In other words, it considers the impact of an intervention as compared to that of external factors, among which changes in the political and economic situations in countries of origin, and contrasting measures in related fields (such as education, development, or foreign policy). It also assesses the presence of any potential substitution effects from a spatial, inter-temporal, categorical, and reverse-flow point of view,[153] resulting key to highlight the broader effects of a policy. In this context, unrealistic or inadequate policy goals (such as the expectation that temporary workers will leave the host country once their programs end), have the potential to broaden the distance between a policy and its effects.[154] Such deficiencies, in turn,

---

[147] ISTAT (2006), in Geddes (2008, p. 350).
[148] Czaika and de Haas (2013).
[149] Zolberg (1989, p. 272).
[150] Czaika and de Haas (2013, p. 496).
[151] Ibid.
[152] Czaika and de Haas (2013).
[153] de Haas (2011, p. 27) and Czaika and de Haas (2013, p. 497).
[154] Finotelli and Sciortino (2009, p. 127) and Cornelius and Tsuda (2004, p. 7).

may depend on the limited understanding of the dynamics that drive migration,[155] or on policies' propensity to 'fight the previous war'.[156]

To sum up, when inconsistencies are noted between the policies proposed in public debates, and the actual levels of immigration, a useful framework to explain such divergence is the 'policy gap' hypothesis. In particular, through Czaika and de Haas' interpretation of it, it is possible to distinguish between discursive, implementation, and efficacy gaps, thus obtaining greater details on the reasons behind the scarce coherence between the stated objectives and the eventual outcomes of the policies under review.

CONTROLLING MIGRATION THROUGH DETERRENCE?

We have seen so far that the realist confidence in states' ability to control migration is, to a notable extent, not perfectly persuasive. Rather, I have argued for a more limited role of the state, and proposed to follow Czaika and de Haas[157] and distinguish between discursive, implementation, and efficacy gaps to analyse the divergence between goals and outcomes of migration policies. With this in mind, I now turn to the more specific topic of deterrence, a strategy that has vastly been adopted in recent years by Western countries, in the hope of reducing irregular migration.

Among the scholars who analyse the relationship between deterrent measures and migratory dynamics, quite a few are sceptical of the effectiveness of the former. As an example, studying the indefinite detention of irregular migrants and asylum seekers in the United Kingdom, Robyn Sampson finds that such a practice does not yield any significant deterrent effects.[158] Instead, politicians' statements arguing for success generally overlook the influence of external factors, first of which political or economic developments in countries of origin and transit.[159] The scholar argues that lengthy detention neither encourages compliance with immigration law, nor leads to more swift resolutions of administrative procedures: Most often, either repatriations are carried out in the first

---

[155] Massey, cited in Guiraudon and Joppke (2001, p. 3) and Geddes (2003, p. 25).
[156] Sciortino (2000, p. 224).
[157] Czaika and de Haas (2013).
[158] Sampson (2015).
[159] Ibid., pp. 5–6.

year of custody, or they remain un-accomplishable.[160] Such findings are indeed replicated by a study on the French detention system, according to which, return rates in the country tend to drop after the 5th day of detention, and become marginal after the 25th day.[161] On the contrary, detention significantly disrupts migrants' integration possibilities in host societies, increasing the rates of depression, anxiety, and post-traumatic stress disorder.[162]

If detention is indefinite and inefficient in the UK, however, it has the opposite problem on the other side of the Atlantic: According to César Cuauhtémoc García Hernández, detention fails to deter potential migrants in the United States because of its temporal limitation, which is constrained to two months.[163] In addition, he argues, its excessive use (the United States detains 400,000 people each year) greatly undermines its power to stigmatise those who are detained, and the frequent presence of relatives in the country is a strong factor of attraction for potential migrants that is difficult to overcome.[164]

Criticism of deterrence also stems from quantitative studies of border controls, in both European and American contexts. Among others,[165] Eiko Thielemann, by conducting research on the impact of policies on forced migration in twenty OECD countries, concludes that structural pull factors generally have a greater effect than policy-related ones.[166] In particular, he finds deterrent measures based on access control not to be statistically significant, due to legislators' propensity to overestimate the degree of information possessed by migrants, and underestimate key structural factors, among which popular perceptions, social networks, and the availability of employment opportunities.[167] In addition, the academic maintains that, as most countries' immigration control measures tend to

---

[160] Ibid.

[161] Assfam, FTDA, La Cimade, Ordre de Malte, 'Centres et locaux de rétention' (2015), Rapport (2014), as cited in Lochak (2016, p. 4).

[162] Sampson (2015, pp. 5–6).

[163] García Hernández (2014, p. 1402).

[164] Ibid., pp. 1400–1401.

[165] See, for example, Gathmann (2008), and Hanson and Spilimbergo (1999), both cited in: Casarico et al. (2015, p. 687).

[166] Thielemann (2003, 2011).

[167] Thielemann (2003, p. 32).

converge, their effectiveness is often simply limited to a brief first-mover advantage.[168] Similarly, Emilio Reyneri argues that migrants are unlikely to be deterred by border controls as long as there is a prosperous underground economy offering them better prospects of employment than those available in their countries of origin,[169] a situation that characterises today several southern European countries, including Italy.[170]

Focusing on North America specifically, Cornelius studies the impact of the enhancement of border controls between the United States and Mexico from 1993 onwards, and argues that deterrence can only affect migrants' strategies, not their willingness to flee.[171] He finds that the predominant effects of the restrictive interventions consisted in the redistribution of entries to other parts of the frontier, the increase in permanent settlements, and the surge in the costs of the journey (both in terms of money and risks). From a social point of view, border controls also produced a rise in anti-immigration sentiments. Most significantly, no evidence of successful deterrence emerges from his study.[172] Instead, the propensity to migrate looks more dependent on the odds (as opposed to the difficulty) of successful entry, the cost of 'coyote fees', and the income gap between the United States and Mexico.[173]

The above studies provide helpful insights suggesting that border controls, as examples of deterrence, are not perfectly able to regulate migration. Yet, a number of controversies emerge with measuring the effectiveness of migration policies.

Thomas Espenshade, for instance, while discovering no evidence of a deterrent effect stemming from the enhanced border controls enacted after the US 1986 Immigration Reform and Control Act, still maintains that it would be misleading to define the intervention as ineffective.[174] Instead, he contends, border controls 'may still exert a broader though less visible deterrent influence', since they do probably deter at least some potential migrants, and, were they not in place, migration would be

---

[168] Ibid., p. 33.
[169] Reyneri (2003, p. 134).
[170] Talani (2018).
[171] Cornelius (2005) and Cornelius and Salehyan (2007).
[172] Ibid.
[173] Cornelius and Salehyan (2007).
[174] Espenshade (1994).

higher (thus recalling the realist argument mentioned above).[175] Continuing with Espenshade's words: 'Whenever public policy attempts to control an undesirable activity, governments will tolerate a residual level of the activity because the marginal cost of eliminating it altogether becomes prohibitively expensive and outweighs the extra benefit to society'.[176]

As the above points to, the interpretation of findings may at times imply a degree of subjectivity, being affected by scholars' theoretical position, and aggravated by the difficulty to identify what success or failure actually mean (as well as by the inaccuracy of data on irregular migration).

Furthermore, a study by Alberto Dávila and others on border and interior apprehensions in the United States between 1983 and 1997, partly overlapping in terms of timeframe with Cornelius's above-mentioned research, finds that increasing enforcement did have a deterrent effect, albeit one that only last for about two or three months, and then faded in the long term.[177]

This points to the fact that effectiveness may change through time, but also, more broadly, that the reasons for which a certain conduct is not undertaken are hard to establish with certainty. As an example, migrants' degree of knowledge regarding receiving countries' measures has frequently been found to be limited,[178] with the result that their perceptions may play a key role instead.[179]

The above considerations are key to keep in mind, and suggest a need to better understand the specific functioning of deterrence strategies. It is here that qualitative studies of deterrence may be of help, by contributing to the conceptualisation of the strategy's strengths and weaknesses.

An author who takes this approach is Ana López-Sala, whose work tackles deterrence through the investigation of the Spanish case. Two of her articles are of particular interest here. In the first one, the academic approaches 'dissuasion' as a strategy aimed at 'preventing emigration […], impeding […] unauthorised transborder crossing, and hindering irregular immigrants from settling'.[180] The key element at the core of such

---

[175] Ibid., p. 873.
[176] Ibid., p. 889.
[177] Dávila et al. (2002).
[178] Thielemann (2003) and Richardson (2010, p. 11).
[179] Ryo (2013, p. 576).
[180] López-Sala (2015, p. 522).

strategy is the 'discouragement' of irregular migration and, in that way, dissuasion is said to include elements of deterrence too.[181] López-Sala goes on to distinguish between different types of dissuasion, differentiating in particular between preventive, coercive and repressive policies, characterised in turn by different place of enforcement.[182] If preventive dissuasion measures take place in countries of origin, and are exemplified by information campaigns and economic support to such countries, coercive ones are located at states' frontiers, and include policies such as border controls and surveillance.[183] Finally, repressive measures are based in receiving countries, and represented by repatriations and detention.[184]

After an analysis of Spanish migration-management system, the academic briefly suggests that dissuasion may have worked to reduce irregular migration to specific routes, although it may have had a weaker effect if a wider area is taken into consideration, potentially triggering a shift in flows to other directions.[185] Interestingly, through López-Sala's interviews with institutional and NGO actors, diverging opinions regarding the effectiveness of the three types of dissuasion emerged. On one hand, consulted government representatives and law-enforcement agencies maintained that preventive dissuasion (as defined above) may only have limited success, when compared to migrants' motivations for leaving.[186] As a more effective solution, such actors tended to prefer repressive types of dissuasion, first of which repatriations.[187] On the contrary, NGO staff were generally more in favour of preventive measures, and especially of development programs, perceived as the only strategies able to reduce the risks involved in irregular journeys.[188]

In a later study with Dirk Godenau, López-Sala leaves aside the concept of dissuasion, focusing specifically on what she terms 'multi-layered deterrence', namely a process of chained layers of control that

---

[181] Ibid.

[182] Ibid. The author also distinguishes the three concepts by referring to the different degrees of intentionality and intensity of the measures.

[183] Ibid., pp. 522–523.

[184] Ibid.

[185] Ibid., pp. 529–531.

[186] Ibid.

[187] Ibid.

[188] Ibid.

extend both horizontally (across multiple locations) and vertically (as anchored to different governmental levels).[189] The goal of deterrence is presented as to reduce irregular migration, by: (a) Intercepting and (b) detaining undocumented aliens.[190] Here, deterrence is found as having some effectiveness, although caution is placed again on potential substitution effects.[191] Among different measures, technology is said to be most useful to reduce the uncertainties related to apprehension, but not sufficient to prevent migration by itself.[192] On the contrary, bilateral agreements with countries of origin (for example, on repatriations) are necessary to ensure costal patrols and repatriations, but not always easy to obtain.[193]

Altogether, López-Sala's insights capture nuanced aspects of the construction of dissuasion and deterrence, and conceptualise the strategies in detail. Simultaneously, however, two remarks should be made.

First, if López-Sala's discussion of 'dissuasion' emphasises the goal of discouraging migrants from migrating irregularly, it however neither dedicates significant attention to the effectiveness of such measure, nor perfectly extends to deterrence measures. Indeed, existing studies reveal an intrinsic difference between the two measures: While dissuasion employs both positive and negative incentives, deterrence only focuses on negative ones.[194]

Second, although her analysis of 'deterrence' investigates in slightly more detail the possible success of the strategy, it does not significantly highlight the idea of it as discouraging (rather than impeding) migration. The threat (of apprehension or punishment), however, has been recognised by previous studies as a key element of deterrent measures, which are supposed to reduce people's intentions to infringe the law *before* they actually do so.[195]

In sum, whereas some existing quantitative analyses of deterrence tend to regard the strategy as having rather limited effectiveness in reducing

---

[189] Godenau and López-Sala (2016, p. 4).

[190] Ibid.

[191] López-Sala (2015, p. 531) and Godenau and López-Sala (2016, p. 15).

[192] Godenau and López-Sala (2016, pp. 10, 11, 14).

[193] Ibid., p. 14.

[194] Freedman (2004, p. 104).

[195] See, for example, Zimring and Hawkins (1973, p. 91).

migration, qualitative ones do not always share such a viewpoint. At the same time, however, the former do not address the specific functioning of deterrence, and the latter place constrained attention to some of the key aspects of the strategy. As a result, states' capability to decrease potential migrants' willingness to flee, through the threat of sanctions, has many aspects that are still to be explored.

Conclusion

This chapter has discussed different international political economy understandings of states' ability to control migration. In particular, if realist scholars see states as being able to regulate the phenomenon, despite increasing globalisation, liberal institutionalist academics regard institutional and legal factors as significantly limiting such capability. Finally, scholars subscribing to the transnationalist view understand state sovereignty over migratory flows as having been greatly eroded by contemporary globalisation processes.

In this context, the 'policy gap' hypothesis has emerged as increasingly predominant, suggesting that substantial inconsistencies may be appreciated, between migration policies' stated goals, and their overall effectiveness. Although the thesis has been criticised for taking a simplistic approach to the matter, Czaika and de Haas's threefold interpretation of such gaps (as articulated between discursive, implementation, and efficacy ones) overcomes such deficiency, offering a more structured framework to understand policy effectiveness.

When looking specifically at measures of deterrence, some quantitative studies have argued for their rather limited role, although data availability and difficulties in defining success raise measurement challenges. On the contrary, Lopez-Sala's qualitative discussion of deterrence offers a nuanced understanding of the strategy, but also overlooks some of the key elements characterising threats based on it.

Overall, therefore, the question posed in the introduction of this book remains in search of a comprehensive answer: Are deterrence and criminalisation effective, in reducing irregular migration? Specifically, the following should be understood: Are all border-control measures based on deterrent principles? If not, what are the key elements at the basis of such strategy? Are there any limitations likely to affect its overall results? Finally, what are the specificities of criminalisation, as an example of deterrence?

To untangle the above, in the next chapter I discuss how deterrence and criminalisation may be conceptualised, in the field of international migration. Specifically, I do so by drawing from IPE, criminological and migration studies. While it may be unusual to rely on such different fields, criminological analyses significantly add to the other two disciplines, thanks to their long-standing attention to deterrence, as one of states' possible ways to promote compliance with the law. In the next chapter, I therefore combine studies from different fields, to discuss not only the functioning principles of deterrence, but also its potential weaknesses as a strategy to address migratory flows, and the specific meaning of criminalisation.

References

Bhagwati, J. 2003. Borders Beyond Control. *Foreign Affairs* 82 (1): 98–104.

Bonjour, Saskia. 2011. The Power and Morals of Policy Makers: Reassessing the Control Gap Debate. *IMR* 45 (1): 89–122.

Borjas, George J. 1989. Economic Theory and International Migration. *International Migration Review* 23 (3): 457–485.

Boswell, Christina. 2003. *European Migration Policies in Flux: Changing Patterns of Inclusion and Exclusion*. Oxford: Blackwell.

Boswell, Christina. 2007. Theorizing Migration Policy: Is There a Third Way? *International Migration Review* 41 (1): 75–100.

Boswell, Christine, and Geddes, Andrew. 2011. *Migration and Mobility in the European Union*. New York: Palgrave Macmillan.

Casarico, Alessandra, Giovanni Facchini, and Tommaso Frattini. 2015. Illegal Immigration: Policy Perspectives and Challenges. *CESifo Economic Studies* 61 (3): 673–700.

Castles, Stephen. 2004a. Why Migration Policies Fail. *Ethnic and Racial Studies* 27 (2): 205–227.

Castles, Stephen. 2004b. The Factors That Make and Unmake Migration Policies. *International Migration Review* 38 (3): 852–884.

Cornelius, Wayne A. 2005. Controlling 'Unwanted' Immigration: Lessons from the United States, 1993–2004. *Journal of Ethnic and Migration Studies* 31 (4): 775–794.

Cornelius, Wayne A., and Idean Salehyan. 2007. Does Border Enforcement Deter Unauthorised Immigration? The Case of Mexican Migration to the United States of America. *Regulation & Governance* 1: 139–153.

Cornelius, Wayne A., and Marc R. Rosenblum. 2005. Immigration and Politics. *Annual Review of Political Science* 8: 99–119.

Cornelius, Wayne A., Philip Martin, and James Hollifield. 1994. Introduction: The Ambivalent Quest for Immigration Control. In *Controlling Immigration: A Global Perspective*, ed. Wayne Cornelius, Philip Martin, and James Hollifield, 3–41. Stanford: Stanford University Press.

Cornelius, Wayne A., and Takeyuki Tsuda. 2004. Controlling Immigration: The Limits of Government Intervention. In *Controlling Immigration: A Global Perspective*, 2nd ed., ed. Wayne Cornelius, Takeyuki Tsuda, Philip Martin, and James Hollifield. Stanford: Stanford University Press.

Czaika, Mathias, and Hein de Haas. 2011. The Effectiveness of Immigration Policies: A Conceptual Review of Empirical Evidence. International Migration Institute Working Paper 33. Oxford: IMI.

Czaika, Mathias, and Hein de Haas. 2013. The Effectiveness of Immigration Policies. *Population and Development Review* 39 (3): 487–508.

Dávila, Alberto, José A. Pagán, and Gökçe. Soydemir. 2002. The Short-Term and Long-Term Deterrence Effects of INS Border and Interior Enforcement on Undocumented Immigration. *Journal of Economic Behavior and Organization* 49: 459–472.

de Haas, Hein. 2011. The Determinants of International Migration: Conceptualising Policy, Origin and Destination Effects. International Migration Institute Working Paper 32. Oxford: IMI.

Di Nicola, Andrea, and Giampaolo Musumeci. 2014. Confessioni di Un Trafficante di Uomini (Confessions of a People-Smuggler), Milano, Chiarelettere.

Dicken, Peter. 2011. *Global Shift: Mapping the Changing Contours of the World Economy*, 6th ed. New York: The Guilford Press.

Espenshade, Thomas J. 1994. Does the Threat of Border Apprehension Deter Undocumented US Immigration? *Population and Development Review* 20 (4): 871–892.

Finotelli, Claudia, and Giuseppe Sciortino. 2009. The Importance of Being Southern: The Making of Policies of Immigration Control in Italy. *European Journal of Migration and Law* 11: 119–138.

Freedman, Lawrence. 2004. *Deterrence*. Cambridge, UK; Malden, MA: Polity Press.

Freeman, Gary P. 1995. Modes of Immigration Politics in Liberal Democratic States. *International Migration Review* 29 (4): 881–902.

Freeman, Gary P. 1998. The Decline of Sovereignty? Politics and Immigration Restriction in Liberal States. In *Challenge to the Nation-State: Immigration in Western Europe and the United States*, ed. Christian Joppke, 86–108. Oxford: Oxford University Press.

Freeman, Gary P. 2011. Comparative Analysis of Immigration Politics: A Retrospective. *American Behavioural Scientist* 55 (12): 1541–1560.

Freeman, Gary P., and Alan E. Kessler. 2008. Political Economy and Migration Policy. *Journal of Ethnic and Migration Studies* 34 (4): 655–678.

Garcés-Mascareñas, Blanca. 2013. Reconsidering the 'Policy Gap': Policy Implementation and Outcomes in Spain. GRITim Working Paper No. 18. Barcelona: GRITim.

García Hernández, César Cuauhtémoc. 2014. Immigration Detention as Punishment. University of Denver Sturm College of Law, Legal Research Paper Series, Working Paper No. 13–41. 61 UCLA Law Review 1346.

Geddes, Andrew. 2003. *The Politics of Migration and Immigration in Europe*. London: Sage.

Geddes, Andrew. 2008. Il Rombo dei Cannoni? Immigration and the Centre-Right in Italy. *Journal of European Public Policy* 15 (3): 349–366.

Geddes, Andrew, and Oleg Korneev. 2015. The State and the Regulation of Migration. In *Handbook of the International Political Economy of Migration*, ed. Leila Simona Talani and Simon McMahon, 54–73. UK: Edward Elgar.

Ghosh, Bimal. 2000. Towards a New International Regime for Orderly Movements of People. In *Managing Migration: Time or a New International Regime?* ed. Bimal Ghosh, 6–26. Oxford: Oxford University Press.

Godenau, Dirk, and Ana López-Sala. 2016. Multi-layered Migration Deterrence and Technology in Spanish Maritime Border Management. *Journal of Borderlands Studies* 31: 1–19.

Guiraudon, Virginie. 2000. European Integration and Migration Policy: Vertical Policy-Making as Venue Shopping. *Journal of Common Market Studies* 38 (2): 251–271.

Guiraudon, Virginie. 2001. De-nationalizing Control: Analyzing State Responses to Constraints on Migration Control. In *Controlling a New Migration World*, ed. Virginie Guiraudon and Christian Joppke, 31–64. London: Routledge.

Guiraudon, Virginie, and Christian Joppke. 2001. Controlling a New Migration World. In *Controlling a New Migration World*, ed. Virginie Guiraudon and Christian Joppke, 1–27. London: Routledge.

Hollifield, James F. 1998. Migration, Trade, and the Nation-State: The Myth of Globalization. *UCLA Journal of International Law and Foreign Affairs* 3: 595–636.

Hollifield, James F. 2004. The Emerging Migration State. *International Migration Review* 38 (3): 885–912.

Hollifield, James F., Philip L. Martin, and Pia M. Orrentius. 2014. The Dilemmas of Immigration Control. In *Controlling Immigration: A Global Perspective*, 3rd ed., ed. James F. Hollifield, Philip L. Martin, and Pia M. Orrentius, 3–34. Stanford: Stanford University Press.

Hollifield, James F., and Tom K. Wong. 2014. The Politics of International Migration: How Can We 'Bring the State Back in'? In *Migration Theory: Talking Across Disciplines*, 3rd ed., ed. Caroline Brettel and James Hollifield, 227–288. New York: Routledge.

Joppke, Christian. 1998. Why Liberal States Accept Unwanted Immigration. *World Politics* 50 (2): 266–293.

Lahav, Gallya, and Virginie Guiraudon. 2006. Actors and Venues in Immigration Control: Closing the Gap Between Political Demands and Policy Outcomes. *West European Politics* 29 (2): 201–223.

León, Alba I., and Henk Overbeek. 2015. Neoliberal Globalisation, Transnational Migration and Global Governance. In *Handbook of the International Political Economy of Migration*, ed. Leila Simona Talani and Simon McMahon, 37–53. UK: Edward Elgar.

Lochak, Danièle. 2016. 'Pénalisation', in 'L'étranger et le Droit Pénal'. *AJ Pénal*, Janvier, 10–12. https://hal-univ-paris10.archives-ouvertes.fr/hal-01674295/document. Accessed 5 Jan 2019.

López-Sala, Ana. 2015. Exploring Dissuasion as a (Geo)Political Instrument in Irregular Migration Control at the Southern Spanish Maritime Border. *Geopolitics* 20 (3): 513–534.

Messina, Anthony M. 2007. *The Logics and Politics of Post-WWII Migration to Western Europe*. Cambridge: Cambridge University Press.

Meyers, Eytan. 2000. Theories of International Immigration Policy: A Comparative Analysis. *The International Migration Review* 34 (4): 1245–1282.

Migration Policy Centre team, and Peter Bosch. 2015. Towards a Pro-active European Labour Migration Policy: Concrete Measures for a Comprehensive Package. Migration Policy Centre, EUI.

Mittelman, James H. 2000. *The Globalization Syndrome: Transformation and Resistance*. Princeton: Princeton University Press.

Overbeek, Henk. 1996. L'Europe en quête d'une politique de migration : Les contraintes de la mondialisation et de la restructuration des marchés du travail. *Études Internationales* 27 (1): 53–80.

Overbeek, Henk. 2002. *Globalisation and Governance: Contradictions of Neo-Liberal Migration Management*. Hamburg Institute of International Economics. Discussion Paper Series.

Rame. 2019. Non ha i requisiti da profugo. Ma il giudice lo fa restare in Italia: "È integrato". *Il Giornale*, May 14. http://www.ilgiornale.it/news/cronache/non-ha-i-requisiti-profugo-giudice-fa-restare-italia-1694152.html. Accessed 5 Dec 2019.

Reyneri, Emilio. 2003. Immigration and the Underground Economy in New Receiving South European Countries: Manifold Negative Effects, Manifold Deep-Rooted Causes. *International Review of Sociology* 13 (1): 117–143. https://doi.org/10.1080/0390670032000087023.

Richardson, Roslyn. 2010. Sending a Message? Refugees and Australia's Deterrence Campaign. *Media International Australia* 135: 7–18.

Ryo, Emily. 2013. Deciding to Cross: Norms and Economics of Unauthorized Migration. *American Sociological Review* 78 (4): 574–603.

Samers, Michael. 2015. Migration Policies, Migration and Regional Integration in North America. In *Handbook of the International Political Economy of Migration*, ed. Leila Simona Talani and Simon McMahon, 373–398. UK: Edward Elgar.

Sampson, Robyn. 2015. Reframing Immigration Detention in Response to Irregular Migration: Does Detention Deter? International Detention Coalition Briefing Paper N. 1. Melbourne: IDC.

Sassen, Saskia. 1996. *Losing Control? Sovereignty in an Age of Globalization*. New York: Columbia University Press.

Sciortino, Giuseppe. 2000. Toward a Political Sociology of Entry Policies: Conceptual Problems and Theoretical Proposals. *Journal of Ethnic and Migration Studies* 26 (2): 213–228.

Talani, Leila Simona. 2010. *From Egypt to Europe: Globalisation and Migration Across the Mediterranean*. London: I.B. Tauris.

Talani, Leila Simona. 2015. International Migration: IPE Perspectives and the Impact of Globalisation. In *Handbook of the International Political Economy of Migration*, ed. Leila Simona Talani and Simon McMahon, 17–36. Cheltenham, UK: Edward Elgar.

Talani, Leila Simona. 2018. *The Political Economy of Italy in the Euro: Between Credibility and Competitiveness*. London: Palgrave Macmillan.

Talani, Leila Simona. 2021. *International Migration in the Globalisation Era*. London: Palgrave Macmillan.

Tapinos, Georges. 2000. Migration, Trade and Development: the European Union and the Maghreb Countries. In *Eldorado or Fortress? Migration in Southern Europe*, ed. Russel King, Gabriella Lazaridis, and Charalambos Tsardanidis. London: MacMillan Press LTD.

Teitelbaum, Michael S. 2002. The Role of the State in International Migration. *The Brown Journal of World Affairs* 8 (2): 157–167.

Thielemann, Eiko. 2003. 'Does Policy Matter? On Governments' Attempts to Control Unwanted Migration. IIIS Discussion Paper No. 09. https://www.tcd.ie/triss/assets/PDFs/iiis/iiisdp09.pdf. Accessed 13 December 2021.

Thielemann, Eiko. 2011. How Effective Are Migration and Non Migration Policies That Affect Forced Migration? LSE Migration Study Unit Working Paper No. 2011/14.

Wagstyl, Stefan. 2015. European Refugee Influx Leads to Temporary Housing Bonanza. *Financial Times*, August 7. http://www.ft.com/intl/cms/s/0/668b4bd0-3b75-11e5-bbd1-b37bc06f590c.html#axzz3jaKJ6u2x. Last accessed 25 Aug 2015.

Weiner, Myron. 1985. On International Migration and International Relations. *Population and Development Review* 11 (3): 441–455.

Weiner, Myron. 1995. *The Global Migration Crisis: Challenge to States and to Human Rights*. New York: Harper Collins College Publishers.

Weiner, Myron. 1996. Ethics, National Sovereignty and the Control of Immigration. *The International Migration Review* 30 (1, Special Issue: Ethics, Migration, and Global Stewardship): 171–197.

Zimring, Franklin, and Gordon Hawkins. 1973. *Deterrence: The Legal Threat in Crime Control*. Chicago: University of Chicago Press.

Zolberg, Aristide R. 1989. The Politics of Immigration 'Reform'. *Revue Française d'études Américaines* 41: 263–276.

CHAPTER 3

Deterrence and Criminalisation: Between Migration, IPE, and Criminology

INTRODUCTION

In recent years, politicians throughout Europe have increasingly emphasised the use of deterrence, as a way to prevent irregular migration. Among the various measures that governments may adopt to achieve such goal, deterrence is characterised by a focus on negative incentives, and operates on the basis of migrants' expected rationality.

From an IPE standpoint, deterrence is strongly related to the realist assumption that states are able to exert control over migratory flows. It is also strictly related to the *securitisation* of migration, namely the construction of migration as a threat, needing a security response[1]: Both the very usage of the concept, traditionally meant to discourage criminality and warfare-related actions, and the reliance on law-enforcement and military

---

[1] Huysmans (2000).

This chapter is based on Rosina, M. (2019), 'Globalisation and Irregular Migration: Does Deterrence Work?', in Talani, L. and Roccu, R. (eds.), *The Dark Side of Globalisation*, Palgrave Macmillan, 85–120, reproduced with permission of Palgrave Macmillan.

© The Author(s), under exclusive license to Springer Nature Switzerland AG 2022
M. Rosina, *The Criminalisation of Irregular Migration in Europe*, Politics of Citizenship and Migration,
https://doi.org/10.1007/978-3-030-90347-3_3

actors and means, contribute to creating an explicit connection between migration and security.

How does deterrence work? What elements make it more or less likely to succeed? Does it include any intrinsically problematic aspects? In this chapter, I delve into the functioning of deterrence strategies, adopting an interdisciplinary approach that draws from international political economy, criminology, and migration studies. In so doing, I first outline the key characteristics of the strategy, defining it and discussing the way in which it operates. Then, I highlight a number of weaknesses that I hypothesise may hinder its potential success. Finally, I examine the specific policy of the criminalisation of migration, as an example of deterrence, to understand its meaning and purpose.

THE FUNCTIONING OF DETERRENCE

Definition

The word 'deterrence', from the latin '*deterrere*', to frighten from or away, refers in its broadest sense to a strategy to 'discourage and turn aside or restrain by fear'.[2] From a criminological perspective, it is one of the multiple techniques available to legislators to increase individual compliance with the law.[3] While compliance may be generated by instrumental reasons, or by the normative belief in the legitimacy and morality of the law, deterrence only aims to affect the former by inserting or reinforcing negative incentives.[4] Its key characteristic is that observance of the law shall not be induced through the actual imposition of the penalty, but, rather, through its threat.[5]

The fundamental concept underlying the notion of deterrence is the utilitarian idea that people engage with actions that procure them more pleasure than pain,[6] or, in rational choice terms, that people balance the costs and benefits deriving from a certain conduct, then acting in pursuit

---

[2] Oxford English Dictionary (2016).

[3] von Hirsch et al. (1999, pp. 3–4).

[4] Ibid.

[5] Zimring and Hawkins (1973, p. 91).

[6] Bentham, cited in Andenaes (1968) and Beccaria, cited in Onwudiwe et al. (2005, p. 235).

of utility maximisation.[7] As a consequence, legislators should, in order to increase compliance, alter the legal costs connected with the unwanted behaviour, by introducing or amending sanctions, so that the fear of the expected punishment predominates over the desire for potential gains.[8] This, in turn, implies an interactive relationship between two parties, the threatening agent (the state), and the threatened audience (the public at large and potential offenders specifically).[9]

In light of the above, I refer to deterrence as a strategy meant to discourage irregular migration to a country, through potential migrants' fear of negative consequences.[10]

Building upon the discussion in the previous chapter, from an IPE perspective, the very idea that migration may be reduced or controlled through deterrence is strictly embedded in the realist emphasis on national sovereignty,[11] and understanding that states have the ability to regulate the phenomenon, if they want to.

Specifically, deterrence can be distinguished from dissuasion and containment: While the former aims to affect its targets' rationality through both positive and negative incentives,[12] the latter intends to effectively stop ('contain') people emigrating.[13] To better explain the difference, let me bring some examples. To begin with, fences and push-backs are not necessarily deterrents, if they physically prevent people from crossing borders. In contrast, the criminalisation and detention of irregular migrants, their exclusion from social benefit schemes, and the imposition of sanctions on employers of irregular foreign work are all cases of deterrence, in that they do not stop migrants, but rather attempt to affect their assessment of the costs involved. Finally, border patrols and repatriations may have a double function of containment and deterrence: Taking the latter as an example, increasing the number of repatriations acts as a deterrent when it causes a person not to cross a border, in fear of being returned. However, when a migrant is actually repatriated, that is

---

[7] Akers (1990, p. 654) and Nagin (2013, p. 9).

[8] Andenaes (1968, p. 79).

[9] Zimring and Hawkins (1973, p. 91).

[10] Based on von Hirsch et al. (1999).

[11] Briskman (2008, p. 128).

[12] Freedman (2004, p. 104).

[13] Hassan (2000, p. 185).

not deterrence anymore, but enforcement of the law. Indeed, in the latter case, deterrence has failed.

It may be useful to distinguish between general and specific deterrence: While the former applies to the whole society, and is based on the threat of a potential penalty, the latter is directed against individuals who have already offended, and exploits the restraining effect of the experience of punishment.[14] Another way to think about the difference is that special deterrence applies in the instances in which general deterrence has failed. In terms of migration, Hassan defines general deterrence as the set of measures taken in a country of immigration to discourage people from going there irregularly, and specific deterrence as the set of measures aimed at discouraging them from remaining there.[15]

It is also interesting to note that deterrence may not only be interpreted as a strategy to increase compliance, but also as a justification for punishment itself. As a matter of fact, scholars have long debated the moral and philosophical purposes of criminal punishment, suggesting retribution, rehabilitation, incapacitation, and deterrence itself, as key aspects. More specifically, if retribution refers to the idea that offenders deserve to be sanctioned for their actions, as a way to '[pay] their debt to society' (as exemplified by the *lex talionis*'s 'an eye for an eye and a tooth for a tooth'), rehabilitation stresses that punishment should enable offenders to reintegrate into society (for example, through training programs), so that the deeper causes of crime may be treated.[16] Incapacitation, on the contrary, aims to prevent people from carrying out crimes, by holding them in custody.[17] Finally, deterrence is said to make punishment acceptable insofar as it can discourage further offences.[18] In this way, deterrence represents not only a strategy for states to reduce unwanted behaviours, but also one of the justifications for imposing sanctions in the first place.

Having defined deterrence and explained the reasoning at its basis, I now focus on how it works. In other words, what are the key elements

---

[14] Andenaes (1968, p. 78).
[15] Hassan (2000, p. 185).
[16] Banks and Pagliarella (2004, pp. 110, 116–117).
[17] Ibid., p. 117.
[18] Ibid., p. 104.

on which the strategy rests? The starting point to address such discussion concerns the legal (or formal) costs embedded in the foreseen sanctions.

Legal Costs: Certainty and Severity

The basic assumption of deterrence implies that, for the strategy to work, there must be legal (or formal) costs associated to the unwanted conduct. In particular, Cesare Beccaria, the very first philosopher of deterrence, identifies the certainty, severity, and celerity of a punishment as the key factors shaping the overall effectiveness of a deterrent threat.[19] Despite some early findings that the success of deterrence is inversely proportional to the waiting time between the commission of a crime and the administration of its penalty,[20] the study of celerity has not been continued in later analyses of deterrence, due to its ambivalent theoretical construction and lack of evidence.[21] On the other hand, however, the principles of certainty and severity of the threat have constituted the foundation of nearly all subsequent investigations on deterrence.

Of the two, Beccaria hypothesised, perhaps contrary to common perceptions, that 'the certainty of incurring a modest sanction has a greater impact than the fear of incurring a more terrible one, but with hope of impunity'.[22] Indeed, absent detection and apprehension, there is no possibility of conviction or punishment[23]; and several empirical studies have now confirmed that the biggest restraint to the commission of an offence is the infallibility of the punishment (both in terms of apprehension and conviction).[24] As an example, a key element to raise the perception of certainty has been identified by Daniel Nagin in the presence of police officers acting in their role of sentinels, to the extent that a 10% increase in police deployment would produce a 3% reduction in total crime.[25]

---

[19] Beccaria (1764).
[20] Robinson and Darley (2004, p. 193).
[21] Nagin (2013, p. 10).
[22] Beccaria (1764).
[23] Nagin (2013).
[24] See Nagin (2013) and von Hirsch et al. (1999, p. 6).
[25] Nagin (2013, p. 45).

On the contrary, the effectiveness of increasing the severity of a punishment is more controversial, with many scholars sceptical about its success.[26] Indeed, the effects of altering 'how stringently the offender is punished, once he is caught and convicted',[27] appear to show little consistency: While a more severe sanction may sometimes reduce non-compliance, it may also prove ineffective, occasionally even counterproductive.[28] As an illustration, it has been argued that lengthening a short prison sentence may contribute to enhancing deterrence, but also that doing the same with sentences that are already long may not yield any benefit in terms of crime reduction.[29] These diverging effects may be explained by people's propensity not to perceive costs as proportional to punishment (that is to say, the cost of spending two years in prison is generally not considered as twice that of being imprisoned for one year).[30] Moreover, above and below certain thresholds, individuals appear to be immune to changes in the severity of the threatened punishment: On one hand, if the costs associated to a sanction are perceived as very low, any additional increment will not increase them enough to have an impact on the decision-making process. On the other hand, if the costs are already perceived as excessively high, any additional increment will not be relevant, as incurring the sanction would be detrimental anyway.[31] Finally, and perhaps quite paradoxically, increasing the severity of a penalty might result undermining the certainty of incurring it, as courts may be less willing to implement sanctions they deem too severe or unjust.[32] It is therefore vital that the severity of the punishment is proportional to the act performed.

The above criminological findings are corroborated by several migration studies: In their model of multi-layered deterrence, Godenau and Lopéz-Sala note that the deterrent effect of migration control strategies against experienced migrants is proportional to the ratio of failed to total

---

[26] See, for example, Nagin (2013) and von Hirsch et al. (1999).

[27] von Hirsch et al. (1999, p. 6).

[28] Ross (1982), cited in Nagin (2013, p. 20).

[29] Nagin (2013, p. 39).

[30] Nagin (1998, p. 21).

[31] von Hirsch et al. (1999).

[32] Nagin (2013, p. 20).

attempts.[33] In other words, the closer the number of instances in which the individual is either intercepted or repatriated is to the total number of attempted border crossings, the more deterred she will be.[34] Carling and Hernández-Carretero similarly find that the deterrent effect of repatriations can only be significant if migrants perceive it to be certain that they will be repatriated in absence of valid documents.[35] In their words: 'As long as prospective migrants see the outcome of their migration attempt as a question of luck', deterrence is likely to only have a limited effect.[36]

However, it is questionable whether the current situation in Europe is able to offer certainty: Large numbers of expulsion orders have been issued against unauthorised migrants, but then proven to be impossible to enforce.[37] This is usually due to the lack of either financial resources or the necessary agreements with third countries, and results reinforcing the perception that sanctions can be easily avoided.[38] Likewise, the practice of Italian officers not to register some of the apprehended undocumented migrants so that the latter could travel farther north and apply for asylum in another country[39] (which would not be possible, were they registered in Italy, under the Dublin Regulation) is a further example of inconsistent implementation. This undermines the very idea of certainty of apprehension and conviction, depicting instead the enforcement of immigration provisions as sporadic and subject to luck.

The rather controversial effect of severity is also supported by migration analyses, several of which report that strengthening penalties and worsening reception conditions do not succeed in deterring people from travelling to the West. A study by Cornelius and Salehyan on the propensity to emigrate from Mexican communities to the United States in the mid-2000s, for example, finds that, while the probability of successful entry does affect migrants' decisions, the perceived adversity of border patrols and the dangers of border crossing do not have a statistically

---

[33] Godenau and López-Sala (2016, p. 4).

[34] Ibid.

[35] Carling and Hernández-Carretero (2008, p. 56).

[36] Ibid.

[37] Carling and Hernández-Carretero (2011, pp. 47–48).

[38] Ibid.

[39] See Fargues and Bonfanti (2014, p. 13).

significant effect, simply inducing people to devise more accurate 'evasion strategies'.[40]

A further interesting contribution is the research conducted by Mieke Kox on the specific deterrent effect of detention in the Netherlands, that is the ability of the practice to induce detained migrants to leave the country.[41] Through interviews, the author concludes that detention did not affect the willingness of most migrants in custody to leave the Netherlands. Instead, it only managed to influence a minority's intentions to leave (and either go back, or to another European country), while, at the same time, strengthening a few people's decision to remain.[42] In particular, Kox notes that deterrence has a somewhat stronger effect on people who migrated for reasons other than security, who did not request international protection, and who neither have relatives in the country of destination nor health issues.[43] Those who had acceptable work opportunities in their own countries were also less deterred by the threat of being detained.[44] The acceptance of rough conditions may be partly explained by the fact that migrants are not scared by deprivations and an inhospitable atmosphere, especially at the beginning of their journey, for they view them as characterising an only temporary phase.[45]

Finally, to support the thesis that too much severity may undermine certainty, Bhagwati reports that both Swiss and German courts have, in multiple instances, found evidence of employers hiring irregular foreign work, but decided not to sanction them severely, considering the punishment excessive, and that the situation in the United States has been rather similar.[46] In similar fashion, Boswell stresses that such sanctions have often been imposed in weaker forms than envisaged by the law, either because of an actual difficulty to determine employers' degree of responsibility, or out of sympathy with the usually small firms involved.[47]

---

[40] Cornelius and Salehyan (2007).
[41] Kox (2011).
[42] Kox (2011, pp. 89–95).
[43] Ibid.
[44] Ibid.
[45] Talani (2018).
[46] Bhagwati (2003).
[47] Boswell (2011, p. 18).

Bounded Rationality and Perceptions

As seen in the previous section, criminology stresses the role of certainty, rather than that of severity, as a deterrent of crime, and studies in the field of migration seem to corroborate the hypothesis. Yet, the idea of men being extremely logical human beings, comparing the costs and benefits of their actions in a context of perfect information, seems rather unconvincing. Indeed, although deterrence is based on the utilitarian idea that people act in pursuit of pleasure and in escape of pain, the criminological literature has soon distanced itself from the assumption that men are perfectly rational. Rather, it considers this as an unrealistic expectation, and acknowledges that individual behaviour can deviate from the *homo oeconomicus* model,[48] opting therefore for the concept of 'bounded rationality', in which choices are affected by imperfect or scarce information, structural constraints, moral principles, and 'non-rational' influences specific to each individual.[49] As a matter of fact, finding evidence of extremely rational decisions may be the result of respondents' rational reconstruction of events,[50] unconsciously performed either in expectation of less severe penalties, to please the interviewer, or to avoid moral culpability.[51]

In the same vein, Richard Ned Lebow and Carol Bohmer question three assumptions that they consider at the basis of deterrence measures in migration: That individuals make decisions on the basis of a cost–benefit analysis, that this calculus can be altered from the outside, and that people are free not to act if the expected costs exceed the expected benefits.[52] Instead, they argue, not only is individual assessment of costs not necessarily rational, and its calculation frequently 'opaque to outsiders', but people can also feel compelled to act even if the anticipated costs are higher than the gains.[53]

To take account of people's bounded rationality, criminologists have, since the 1960s, focused not so much on the actual legal costs involved in

---

[48] Loughran et al. (2016, p. 107).

[49] Akers (1990, p. 661).

[50] Cromwell et al. (1991), cited in Jacobs (2010, p. 424).

[51] von Hirsch et al. (1999, p. 35) and Sykes, Gresham M. and Matza, David, 'Techniques of neutralization', in: McLaughlin and Muncie (2013, p. 250).

[52] Bohmer and Lebow (2015).

[53] Ibid.

sanctions, but rather on how they are *perceived* by the individuals whose decision-making process they are expected to affect, so that personal assessments of reality can be integrated into the analysis.[54] Indeed, '[t]here can be no direct relationship between sanctions and criminal action; the two must be linked through the intervening variable of subjective perceptions of the risks and rewards of committing an offense'.[55] This implies that a necessary condition for deterrence to work is that the potential offender is aware that a certain conduct is proscribed, and that she may be sanctioned for it.[56] As a result, the threatening agent should not only increase the certainty and (sometimes) severity of a penalty, but also make sure that potential offenders are aware of such changes, and that their assessment of costs and benefits are modified in response to them.[57] Overall, what is important to keep in mind is that the deterrent effect of a threat depends on the perceptions of the audience, rather than on the message as meant by the state.[58]

According to several academics, sanction risk perceptions are heavily affected by personal and peer experiences, respectively relevant for experienced offenders and novices.[59] On one hand, people who do not have experience with criminal offences tend to base their perceptions on information derived from friends, family members, other offenders and the media,[60] as well as to overestimate the certainty of apprehension and conviction.[61] On the other hand, 'career criminals' are expected to rely on their personal 'experiential effect' to estimate the likelihood of being caught, and to have a more accurate appreciation of it.[62]

---

[54] This has given birth to a literature denominated 'perceptual deterrence', developed since the 1970s, which has found vast consensus. See von Hirsch et al. (1999, p. 33).

[55] Decker et al. (1993, p. 135), cited in Jacobs (2010, p. 418).

[56] Zimring and Hawkins (1973, p. 142), Robinson and Darley (2004, p. 175), and Nagin (2013, p. 7).

[57] Zimring and Hawkins (1973, pp. 142, 147).

[58] Ibid., p. 157.

[59] Stafford and Warr (1993, p. 130) and Pogarsky et al. (2004, p. 364).

[60] Stafford and Warr (1993, p. 130).

[61] Robinson and Darley (2004, p. 184).

[62] Paternoster et al. (1982), cited in Nagin (2013, p. 60).

Such hypotheses look consistent with recent migration studies finding that most details regarding undocumented border crossing are transmitted to potential migrants through personal networks and smugglers.[63] They may also indicate experienced migrants' propensity to estimate detection in light of their own previous attempts at crossing irregularly, as suggested by Godenau and López-Sala.[64]

Social Costs: Stigmatisation and Social Control

In addition to studying the effects of legal costs and individual perceptions of them, criminologists highlight the role of the expected social costs associated to being sanctioned for the commission of an unlawful act, as powerful promoters of compliance with the law.[65]

In this context, social costs refer to the stigmatisation of the offender by the community,[66] following the former's apprehension or punishment,[67] and can be represented by disapproval, loss of status or exclusion from a group.[68] Their role has been found so important as to make some scholars argue that the relative deterrent effect of informal social costs may be more relevant than that of formal legal costs.[69] Indeed, the process of socialisation is often regarded as a strong tool to produce social control,[70] and it has been found that, because being labelled as deviants entails significant changes in a person's social participation and public image,[71] the more integrated people are, the more deterred they feel by the potential stigmatisation following public exposure for the commission of an unlawful act.[72] Strictly speaking, social costs are not always part of the deterrent threat as devised by governments.[73] Yet, they are a key aspect

---

[63] Thielemann (2011, p. 7) and Richardson (2010).
[64] Godenau and López-Sala (2016, p. 4).
[65] Nagin (2013).
[66] Zimring and Hawkins (1973).
[67] Nagin (2013).
[68] Zimring and Hawkins (1973, p. 191).
[69] Akers (1990, p. 675).
[70] Zimring and Hawkins (1973, p. 119).
[71] Becker (1963, pp. 31–32).
[72] Nagin (1998, p. 20).
[73] Zimring and Hawkins (1973, p. 174).

of punishment,[74] especially in cases of criminalisation, where stigmatisation plays a crucial role.[75] As a consequence, they cannot be separated from the study of the effects of deterrence.[76]

Following the labelling and social-control theories, the role of communities is important because social rules are not universal or absolute, but rather the product of specific social groups.[77] In the words of Howard Becker, '[d]eviance is not a quality that lies in behaviour itself, but in the interaction between the person who commits an act and those who respond to it', and a deviant act is therefore simply one that is defined as such by a group.[78] As a consequence, whereas the community that enforces a certain rule perceives those who break it as outsiders, it is also possible that, from the latter's point of view, those who have made the rules are the outsiders.[79] In particular, if the offender's entourage does not perceive an illicit conduct as especially blameworthy, or the enforcing actor as legitimate (either because it condemns with disproportionate punishment, processes in an unfair way, or overreaches its intervention area), the mechanism of stigmatisation may fail, and even have a counterproductive effect.[80] Moreover, if being detained becomes normal in a group, no stigma will be associated to it: For deterrence to be effective, the actual imposition of the punishment must be a relatively rare event.[81]

Social control theory is also useful to highlight the vicious cycle produced by deviance and its marginalisation. According to Becker, when an individual is labelled as deviant in a specific context, people tend to assume that they are more likely to engage with other types of offences as well, and often end up marginalising them, by making of the deviant characterisation the prevailing one.[82] However, because marginalisation implies exclusion from conventional groups, the process generates a self-fulfilling prophecy, in which the person labelled as deviant is incentivised

[74] Andenaes (1968, p. 81).
[75] Simester and von Hirsch (2011).
[76] Zimring and Hawkins (1973, p. 174).
[77] Becker (1963, p. 15).
[78] Ibid., p. 14.
[79] Ibid., p. 2.
[80] von Hirsch et al. (1999, p. 40).
[81] Nagin (2013).
[82] Becker (1963, pp. 33–34).

to move closer to groups at the margins of society, and to break other rules they did not intend to break.[83] As a consequence, deviant behaviour is reinforced by its marginalisation, and its reiteration becomes, in part, a consequence of the public reaction to non-compliance.[84]

The labelling theory has been criticised for being extremely relativistic,[85] and for focusing only on minor crimes, failing to explain more serious ones.[86] Yet, because irregular migration has often been depicted as a victimless crime (insofar as it does not harm an individual as, for instance, murder would),[87] it provides perfect application of the theory. Moreover, the accent on the question 'deviant for whom?' highlights the presence of multiple communities that may interpret deviance differently, and therefore greatly contributes to the analysis of potential migrants' assessment of social costs.

When transposing the analysis of social costs to migration policies, it may be useful to consider stigmatisation from the perspective of the two different groups with which migrants interact: The sending and receiving communities. Starting from the former, migrants' community in their home country is likely to have a great impact on the decision to migrate, especially by shaping the individual's perception of migration norms. However, existing studies suggest that irregular migration is not necessarily perceived as an immoral act, but most often simply as a way to achieve a better life. Carling and Hernández-Carretero, for example, find that, in Senegal, (irregular) emigration is a strategy meant not only to acquire economic prosperity, but also to ameliorate one's social status, to the extent that leaving the country is seen as the easiest way to success, and embodies a symbolic value of masculinity and honourability.[88] Simultaneously, Ryo argues that the US immigration system is not perceived as

---

[83] Ibid.

[84] Ibid., p. 35.

[85] Wellford (1975).

[86] John Muncie, in McLaughlin and Muncie (2001, p. 160).

[87] As argued by Guild (2010, p. 4): 'Leaving aside the issue of trafficking in human beings, an individual who irregularly crosses a border or stays on the territory of a state beyond his or her permitted period does not harm a specific individual. To the extent that harm is done at all, it is to the integrity of the state's border and immigration control laws'.

[88] Carling and Hernández-Carretero (2008).

fair by actual and potential Mexican migrants, who instead view it as illegitimate, arbitrary, and hypocrite, being biased in terms of both class and race of newcomers, offering better opportunities only to certain groups, and not acknowledging the structural need for migrants' labour.[89]

On the contrary, deterrence measures, and in particular detention, are likely to involve high social costs for migrants in receiving communities, by associating the former with dangerous people and criminals, and justifying the state's limited responsibility towards them.[90] Moreover, the fundamental mistrust at the basis of the asylum system in several Western countries risks criminalising asylum seekers in the eyes of the public, even before their claims are assessed, and, when the word 'detention' is used as a synonym of 'imprisonment' to refer to criminals' convictions (as is the case, for example, in the United Kingdom), the link between migrants and wrongdoers is further strengthened.[91]

It might be possible to hypothesise that the social costs stemming from the stigmatisation of migrants in receiving countries may be weakened by the fact that the former are not full members of the host society,[92] and softened by the help and support of local ethnic communities. Moreover, the extensive use of detention may undermine its stigmatising effect for, as seen above, punishment should be rare to produce deterrence.

Still, a vicious cycle between irregularity and marginalisation may emerge. Reyneri finds that migrants' widespread involvement in the underground economy in southern European countries contributes not only to their marginalisation and to the growth of negative attitudes among locals against newcomers, but also to the emergence of negative perceptions of locals among migrants themselves.[93] The author explains the phenomenon through the above-mentioned social-control theory, according to which an actor's conduct is to be attributed to his as much as to others' actions.[94] In other words, migrants' marginalisation causes

---

[89] Ryo (2015).
[90] Hassan (2000).
[91] Malmberg (2004, pp. 13–15).
[92] Stumpf (2006, p. 412).
[93] Reyneri (2003), cited in Talani (2010, p. 194).
[94] Becker (1963, p. 35).

a vicious cycle to start, leading to an increased attractiveness and necessity of further engagement with irregular activities,[95] and to an enhanced difficulty to be regularised (in both employment and residence status).[96]

In sum, an important aspect of deterrence is the expectation of incurring not only formal legal sanctions, such as detention or fines, but also informal social costs, in the form of stigmatisation. At the same time, however, stigmatisation may lead to further marginalisation, and prompt the person to get involved with further underground behaviour.

The Pitfalls of Deterrence

As seen thus far, legislators may increase deterrence by making the sanction associated to a certain act more certain and, at times, more severe. At the same time, they should pay attention to the social dynamics surrounding the relevant behaviour, to have a better understanding of the community's perception and assessment of it. Yet, even in cases of certain and severe penalties, and high social costs, threatened punishment may fail to bring about the expected deterrent effect, and non-compliance may occur. Indeed, the above may not be enough to explain the outcome of strategies aiming to deter irregular migration, and it appears that a number of problematic aspects are likely to emerge, some of which might be easier for states to contain, others less so. It is to these possible weaknesses—or 'pitfalls'—of deterrence that I now turn.

Do Costs Outweigh Benefits?

As seen above, the aim of deterrence is to alter a potential offender's cost-and-benefit analysis, by inserting or strengthening negative incentives. Yet, the focus on the cost-side of the decision-making equation alone, and lack of attention to the benefit-side, have been the object of criticism in the criminological literature.[97]

Indeed, because the full behavioural formula concurrently includes both positive and negative incentives, Akers argues that deterrence is not enough to explain decisions of compliance or non-compliance, and

---

[95] Talani (2010, p. 186).
[96] Reyneri (2003, p. 134).
[97] Loughran et al. (2016, p. 90).

suggests instead to maintain a broader approach, including the study of positive inducements.[98] Taking this argument further, Loughran and others suggest that rewards may even have a slightly stronger effect than sanctions, although recognising that both elements are important.[99] They also note that rewards shall not only be monetary, but may also involve social benefits, such as increased social status.[100] Even James Wilson, one of the main thinkers of deterrence, maintains that the wisest solution is to simultaneously raise the benefits of compliance and the costs of non-compliance.[101]

Yet, recalling the origin of the word deterrence as a strategy meant to frighten away,[102] it looks evident that it can only involve attempts aimed at affecting the cost-side of the target's decision-making. It is thus possible to argue that the basic feature of deterrence, namely its focus on negative incentives, is also its most problematic one for, by neglecting positive inducements, it may not be able to significantly affect people's decisions.

When it comes to (irregular) migration, deterrence may affect two types of costs. First, it can affect the costs of the journey, which can be both monetary (including smuggling fees, transportation, and accommodation expenses), and non-monetary (such as the bearing of abuse and hardships along the trip). Second, it can shape the costs of living in, and adapting to, the new environment (including finding a job, and integrating in a new community). Such costs may be reduced through the help of facilitating factors, such as social networks, which can provide social capital to reduce migrants' transition and adaptation costs.[103] In any case, it is these two kinds of costs that deterrence measures can alter, by introducing or strengthening legal sanctions for unauthorised crossing or residence, and by leveraging the related negative social implications.

In introducing negative incentives, however, the deterrence paradigm neglects the structural factors at the root of migratory flows. Indeed, several authors emphasise the need to address the drivers of flight, rather

[98] Akers (1990).
[99] Loughran et al. (2016, p. 107).
[100] Ibid., p. 91.
[101] Wilson in McLaughlin and Muncie (2013, p. 355).
[102] Oxford English Dictionary (2016).
[103] See, for instance, Neumayer (2004, pp. 164–165) and Borjas (1989, p. 470).

than raising the costs of reaching the West. Holzer and colleagues, for instance, argue that deterrent policies can be assumed not to have any relevant impact if the conditions that generate migration in a neighbouring country 'reach a critical level'.[104] Similarly, Castles adds that even though, in the short term, addressing the causes of economic and forced migration may result in a surge in outflows from the interested country (as predicted by the migration hump hypothesis),[105] in the long term these would stabilise.[106] Furthermore, Cornelius and Salehyan maintain that taking a labour-market approach would be more effective than introducing stricter border controls,[107] and Talani similarly writes that reducing the rewards deriving from the easy access to employment in the underground economy of countries like Italy would greatly reduce unauthorised migration.[108]

Overall, the intrinsic characteristic of deterrence as a strategy that only introduces negative incentives appears highly problematic. From a rationalist viewpoint, if potential migrants perceive the situation that originates the desire to migrate as costlier than the hardships to be shouldered along the journey and to integrate in the new environment, then, deterrence is likely to fail. Beyond the above, however, the deterrence paradigm neglects the structural factors at the root of migratory flows, and it is indeed the disregard of such aspects that contributes to the next pitfall, namely the generation of substitution effects.

Availability of Alternatives

While the aim of deterrence is to discourage non-compliance with the law, individual deterrent measures can only target a specific conduct among several that may lead to the same objective. If, on one hand, this suggests that the more alternatives a person has, the more deterrence is likely to succeed[109]; on the other hand, it also implies that the decrease in a specific offence caused by a new or strengthened deterrent measure

---

[104] Holzer et al. (2000, p. 1205).
[105] Martin and Taylor (2001).
[106] Castles (2004, p. 2).
[107] Cornelius and Salehyan (2007).
[108] Talani (2018).
[109] Zimring and Hawkins (1973, p. 135).

may generate an unexpected negative outcome: The increase in other substitute offences.[110]

Indeed, following Zimring and Hawkins, the strength of the motivation for committing a crime does not only depend on the intensity of the need or desire the individual hopes to satisfy, but also on the availability of alternative means to reach the same goal.[111] As a result, a greater number of alternatives usually leads to a higher effectiveness of deterrence.[112] However, Steven Levitt argues, while raising the arrest rate for a certain behaviour may generate a positive effect (the drop in that offence), it may also produce a counterproductive outcome (the rise in other crimes).[113] In other words, if governments sanctioned robberies with firearms more heavily, the effect may be a reduction in such activities, but also a rise in robberies with knives.[114]

Likewise, in the field of migration, deterrence measures may produce substitution effects, reducing arrivals through a specific route, but, at the same time, increasing those occurring through other ones. De Haas studies such dynamics and identifies four types of potential substitution effects: Spatial, inter-temporal, categorical, and reverse-flow.[115] The first two produce a diversion of migration towards other countries or moments: While the first may be exemplified by the increase in African migration to Greece in the early 2000s, following the intensification of border patrols along the Spanish and Italian routes[116]; the second is illustrated by the Dutch decision to promote the independence of Suriname in 1975 and preparation of a ban on migration, which only made arrivals spur in the short term, due to Surinamese people's fear of not being able to enter the Netherlands later on.[117] By contrast, the categorical substitution effect can be illustrated by the contemporary propensity of migrants to apply for asylum even if they are not escaping persecution, in light of the fact that the opportunities for low-skilled workers to legally enter

---

[110] Levitt (1998, p. 361).
[111] Zimring and Hawkins (1973, p. 135).
[112] Ibid.
[113] Levitt (1998, p. 361).
[114] Nagin (1998, p. 18).
[115] de Haas (2011, p. 27).
[116] Fargues and Bonfanti (2014, p. 5).
[117] de Haas (2015).

and remain in the EU are limited. Finally, an example of the reverse-flow substitution effect may be seen in the decision of many guest workers from Turkey and Morocco to permanently settle in northern Europe after the 1973 immigration limitations, rather than returning home and not knowing whether they would be able to re-enter in the following years.[118]

The notion of substitution effects should be employed with caution, as it assumes that potential migrants are thoroughly aware of the legislation in place, and carefully weigh the possible alternatives against each other. Moreover, it implies the existence of a fixed number of would-be migrants, who are then pulled towards different directions, spatially or otherwise, according to the convenience of different countries' immigration provisions.[119] Finally, it should be understood whether would-be migrants perceive changes in sanctions as route-specific, or as generally applicable to the European region as a whole.[120]

Despite such warnings, the concept of substitution effects is key, insofar as it highlights the broader impact of deterrent policies. It is also especially relevant in the Schengen area, in view of the absence of internal barriers between participating states.

Multiple Audiences: The Political Dimension

As a policy meant to preserve the external boundaries of a state, deterrence is inherently embedded in the political context, the dynamics of which may bear significant implications for the eventual developments and outcomes of the strategy. As criminological studies tend to avoid taking into consideration the environment in which deterrence measures are formulated, I rely in this section on insight deriving from the discipline of international relations.

As an important theorist of deterrence in the field of international relations, Lawrence Freedman analyses how the political context may affect the articulation of deterrence threats, and identifies in the presence of multiple audiences a major potential problem of deterrence.[121] The actual target of deterrence strategies, he argues, is often ambiguous: Deterrent

---

[118] de Haas (2011, p. 27) and Bhagwati (2003, p. 99).
[119] Boswell (2003, p. 36).
[120] Cf. Nagin (1998, p. 19).
[121] Freedman (2004, pp. 47–52).

messages and threats are meant not only for the notional audience—potential migrants in our case—but also for the threatening country's domestic public, and for its international allies.[122]

In this context, the danger is for a policy to become so preoccupied with the concerns of those who are closer and louder, that it increasingly neglects the requirements of those who are further away and quieter.[123] Yet, the more deterrence is driven by internal considerations, rather than by external ones, the wider the gap between how the strategy should be undertaken, and how it actually is, becomes.[124] As a consequence, Freedman concludes, when the deterrent message is not targeted at the notional audience, but at other ones, strategies of deterrence risk resulting incoherent.[125] Certainly, the above may be a natural consequence of the presence of multiple forces attempting to influence policy-makers, and represent the latter's need to formulate strategies that simultaneously address practical and political matters, both factors that some scholars argue are unavoidable in democratic regimes.[126] Nevertheless, in strict terms of effectiveness of deterrence, attributing prevalence to the domestic audience is problematic.

In contemporary Europe, where the electoral value of immigration is high, concerns about the domestic public play a key role. Indeed, the politicisation of immigration, the rise of populist radical right parties, as well as the insecurities and frustrations generated by the recent economic and financial crises, all contributed to the increase in the political salience of immigration-related matters. They also raised the electoral incentive for policy-makers to adopt highly visible and symbolic measures with an expected immediate effect, rather than committing to long-term goals, in the hope of appeasing public hostility by showing commitment and of creating, at least, an '*appearance* of control'.[127] While such a choice

---

[122] Ibid.

[123] Ibid., p. 50.

[124] Morgan (1985, p. 136), cited in Freedman (2004, p. 51).

[125] Freedman (2004).

[126] Ibid., p. 52.

[127] Massey et al. (1998, p. 288, emphasis in original), cited in de Haas (2006), Rudolph (2003), in Hollifield (2004, p. 903), Weiner (1995, p. 198), and Cornelius and Tsuda (2004, p. 42).

may look beneficial in terms of political gains and stability, the aforementioned discussion shows that, if the interests of the indirect target audiences are considered by the threatening actor to be more important than those of the direct one, the deterrent strategy is likely to be affected by inconsistencies.[128]

In addition to the domestic public, a further possible secondary audience for measures of deterrence is composed by the governments of the countries of origin and transit, which, in the case of migration to Europe, are mainly represented by states on the Southern and Eastern shores of the Mediterranean, as well as in Africa and the Middle East more in general. Indeed, third countries are key interlocutors on issues of migration: Following Weiner, it is possible to argue not only that migration policy can affect international relations between states, but also that the latter can shape the former.[129] In particular, the situation in which a country allows or promotes emigration to a country that restricts entry is extremely delicate.[130] It is indeed not rare to find third countries leveraging the control they can exercise over departures, with the purpose of obtaining enhanced economic or other kind of concessions. Colonel Gaddafi represents a case in point: He attempted to exploit the threat of rising migratory flows from Libya in multiple occasions, by first warning that, unless the EU had provided Libya with 5bn euro a year, it would have turned 'black' as a result of uncontrolled migration,[131] and then that, if his regime fell, 'thousands of people from Libya [would] *invade* Europe'.[132] Likewise, in the context of the EU-Turkey Statement, the latter country stated in August 2016 that it would not have felt bound by the Statement anymore, unless the European Union had lifted the visa requirement for Turkish citizens.[133] While the position was later softened to provide for the continuation of the agreed measures,[134] it looks evident

---

[128] Freedman (2004, p. 51).

[129] Weiner (1985, pp. 447–448).

[130] Ibid.

[131] Squires (2010).

[132] Fargues and Bonfanti (2014, p. 6, emphasis added).

[133] See http://www.aljazeera.com/news/2016/08/turkey-eu-refugee-deal-visa-free-travel-160816101936490.html.

[134] Maurice (2016).

that third countries are key interlocutors on issues of migration, of which deterrence messages have to take stock.

Overall, deterrence is weakened by 'threats issued to impress audiences other than the notional target'.[135] In migration just like in international relations, different audiences have contrasting interests and varying degrees of influence on governments' decision-making. Yet, if the interests of secondary audiences are allowed to prevail over those of the primary audience, the overall message risks resulting contradictory and inconsistent.

Information and Communication

Deterrence is about persuading a target population that the costs involved in a certain behaviour are going to be greater than the benefits to be potentially gained from it. Thus, the transmission of information from the threatening agent to the threatened audience makes up for a key part of the strategy. While criminological studies assessing the effectiveness of sanctions take this problem into account by analysing citizens' perceptions of certainty and severity, rather than their official rates, political strategies based on deterrence may still fail because of a lack of attention to the communicative elements.

Certainly, a necessary condition for deterrence to work is that the potential offender is aware that a certain conduct is proscribed, and that they may be sanctioned for it.[136] In fact, however, although perfect information was ruled out of the deterrence paradigm from the very early stages, even the possession of at least some accurate information is in itself problematic. After all, the probabilities of the outcomes of most human decisions are hard to know objectively.[137] An interesting study suggests that, often, people tend to assume that the law is just how they think it should be, making it particularly hard to raise awareness of a sanction in a community that is not persuaded of the illegality and immorality of the related criminalised behaviour.[138]

---

[135] Freedman (2004, p. 49).

[136] Zimring and Hawkins (1973, p. 142), Robinson and Darley (2004, p. 175), and Nagin (2013, p. 7).

[137] Camerer and Weber (1992), cited in Loughran et al. (2011).

[138] Robinson and Darley (2004, pp. 176–177).

In migration even more than in criminology, the problem of limited accurate knowledge of sanctions is extremely relevant. Thielemann, among others, argues that the amount of information possessed by potential newcomers is systematically overestimated by receiving governments.[139] Indeed, although the development of communications has made it easier to gain access to information, it should not be assumed that the degree of exposure to the media enjoyed in the West is available throughout the world,[140] and it has been shown that personal networks and smugglers play a key role in providing information to potential migrants.[141]

Still, both networks and smugglers may carry inaccurate information. On one hand, previous migrants generally have quite a general understanding of migration policies, and tend to conceal the hardships endured in their migration experience.[142] On the other hand, smugglers, whose role as providers of information seems to have now replaced that of personal networks,[143] are more likely to be more knowledgeable on the topic. Yet, they engage with migrants in order to make a profit, and have therefore an incentive to oversell the probability of success.[144] Additionally, news accounts reveal that even facilitators' knowledge may be dubious at times: In Libya, for example, some smugglers were reported in 2013 as not aware first of the existence, and then of the suspension, of the Italian search and rescue operation Mare Nostrum.[145]

To raise awareness of the hardships and consequences of irregular migration, receiving states could invest in information campaigns in countries of departure. Persuading an audience, however, involves key challenges. To begin with, a message can be interpreted by different audiences in different ways, as its meaning cannot be fixed by the sender.[146] In the field of migration in particular, the transnationality of the threat, and

---

[139] Thielemann (2003).

[140] See Richardson (2010, p. 11).

[141] Thielemann (2011, p. 7), Richardson (2010), and IOM (2011, p. 14).

[142] Richardson (2010), Carling and Hernández-Carretero (2008, p. 8), and Talani (2010, p. 193).

[143] Richardson (2010, p. 12).

[144] See, for example, Kingsley (2015b).

[145] See Kingsley (2015a) and Bohmer and Lebow (2015).

[146] Richardson (2010, p. 8).

the consequent cultural distance between sender and receiver, contribute to increasing the gap between the message as intended by the former, and as interpreted by the latter. As a consequence, if strategies of deterrence are ethnocentric and fail to account for the due differences, the communication may result ineffective or counterproductive. Adding to that, the Bayesian updating model suggests that when people are presented with new information, they do not completely abandon prior beliefs, but rather update them.[147] The degree of adjustment depends on the amount of prior knowledge, so that individuals with less prior information tend to change their judgements more readily, and vice versa.[148] Finally, according to the cognitive bias hypothesis, people have a systematic bias in favour of information that is consistent with prior knowledge, which makes it easier for them to ignore, deny, or discount messages that are inconsistent with original beliefs, to the extent that they could even result reinforcing contrary opinions.[149]

As a matter of fact, Carling and Hernández-Carretero, through the study of Senegalese migration to the Canary Islands in the late 2000s, argue that would-be migrants are not unaware of the risks involved in migrating with pirogues, as Western governments portray them.[150] Instead, they employ specific techniques to reduce the perceived hardships, either by stressing their experience and ability to overcome the dangers involved, or by adopting strategies to minimise risk (from practical expedients such as avoiding two people from the same family travelling together, to more psychological ones, like seeking spiritual protection).[151] Moreover, the authors of the study also highlight potential migrants' propensity to (1) avoid contrasting information, by refraining from asking questions, or thinking about it, (2) discredit the sources of such information, by deeming governmental organisation untrustworthy, campaigns as imprecise and partisan, and testimonials as bribed, and (3) dismiss conflicting information as insignificant to their case.[152] These are

---

[147] Pogarsky et al. (2004, p. 346) and Nagin (2013, p. 61).

[148] Nagin (2013, p. 61).

[149] Lebow (2007, pp. 72–75) and Stein (2009, p. 63).

[150] Carling and Hernández-Carretero (2008). For similar findings, see Richardson (2010, p. 15).

[151] Carling and Hernández-Carretero (2008, p. 11).

[152] Ibid., pp. 9–10.

meaningful findings, which support the argument for the presence of a cognitive bias in favour of information consistent with one's mind-set.

It should be highlighted that Carling and Hernández-Carretero's analysis refers to migrants' awareness of the risks involved in irregular migration, rather than to their knowledge of the policies and sanctions in place. Yet, their study is useful in that it shows the difficulties of getting a message across, and empirically demonstrates that communicating the hardships involved in undocumented border crossing may not be sufficient to deter people.

To sum up, in the above sections I have argued that key elements for the analysis of deterrence in international migration are not only the legal costs of the sanctions imposed, dependent on the certainty and, to a lesser extent, severity of the threat, but also the social costs, identifiable in the expected stigmatisation of migrants following their apprehension and punishment. Further, I have highlighted four major problems that may impact deterrence threats' ability to deliver, namely: The neglect of positive incentives and of the structural drivers of migration; the generation of substitution effects (towards other entry routes or categories); the presence of multiple audiences and prevalence of the interests of secondary ones; and, finally, the difficulties involved in the communication of the threat, in terms not only of how much information is possessed by potential migrants, but also of how they engage with it. In this context, how does the criminalisation of migration emerge as an instance of deterrence? Specifically, how does it work? In the remainder of the chapter, I turn to such aspects, to present the use of the criminal law to address migration, as a specific tool of deterrence.

CRIMINALISING MIGRATION IN EUROPE

The criminalisation of a certain conduct is a good avenue to study deterrence for, by introducing criminal sanctions where previously they had been civil or administrative, legislators both set stricter punishment, and a moral condemnation of the unwanted behaviour. Indeed, deterrence is a fundamental function of the criminal law.

Following Simester and von Hirsch, the criminal law is a tool meant to affect people's behaviour through both instrumental and normative mechanisms.[153] By criminalising a certain conduct, states communicate in fact two different messages: That individuals should not get involved with that action because it is morally wrong, and that, if they do, they will be subjected to unpleasant consequences.[154]

On one hand, the criminal law is therefore intrinsically different from the civil law in that it possesses a moral voice, which can convene censure and be used as an additional reason to make people refrain from committing an offence.[155] The censure generated is particularly powerful and authoritative because it is made in the name of society and represents the community's assessment of a certain behaviour.[156]

On the other hand, however, by threatening and actually imposing sanctions, criminalisation also offers a safety check with which to instrumentally disincentivise individuals from undertaking a certain behaviour.[157] This mechanism represents deterrence, which therefore becomes both the goal of the issued threat, and the justification for the consequences of criminalisation.[158] In Simester and von Hirsch's words:

> by criminalising the activity of φing, the state declares that φing is morally wrongful; it instructs citizens not to φ; it warns them that, if they φ, they are liable to be convicted and punished [...]; and, further, the state undertakes that, on proof of D's φing, it will impose an appropriate measure of punishment, within the specified range, that reflects the blameworthiness of D's conduct.[159]

The study of criminalisation in the context of irregular migration has developed in recent years, although in two different directions.[160] On one hand, crimmigration law scholars investigate the progressive criminalisation of immigration law, stressing that the latter increasingly resembles

---

[153] Simester and von Hirsch (2011).
[154] Ibid., p. 6.
[155] Ibid., p. 12.
[156] Ibid., p. 13.
[157] Ibid., pp. 14–15.
[158] Ibid., p. 6.
[159] Ibid., p. 6.
[160] Provera (2015, pp. 2–3).

the criminal law in terms of substance of the law, enforcement, and procedures.[161] In doing so, they focus on the legitimacy of the use of criminal law, purposefully omitting to investigate its effectiveness.

On the other hand, a second group of academics, often based in Europe, tends to be more concerned about understanding the deterrent effect of criminalisation. At the same time, however, they also frequently intend the term in a much broader way, most often embracing, in addition to the use of criminal law for migration purposes, the administrative detention of migrants.[162] It is also common to find studies that include surveillance, police actions, and immigration discourse as tools of criminalisation, or conflate the term with the criminalisation of facilitators of irregular migration or with the issue of migrants' criminality.[163]

As Provera argues, however, maintaining the distinction between the criminal and administrative law is important not only because of the great symbolic power of the former, but also because of the different standards of proof required to sanction individuals.[164] Detaining migrants, however symbolically very powerful, would be better defined by the term 'administrative criminalisation',[165] as it is, in fact, controlled by a specifically distinct branch of the law.

In light of the above, this study adopts on a narrower interpretation of the concept of criminalisation of irregular migration, referring to it as the use of the criminal law to sanction immigration-related offences and, in particular, foreigners' irregular entry and stay. In doing so, it leaves aside the broader interpretation and, specifically, detention practices.

Understood in this way, the criminalisation of migration has emerged as a key trend in Western Europe since the mid-1980s.[166] According to Jacobson, residing in a country without the proper documentation is still considered as an administrative infraction by most legislations.[167] Yet, the European Union Agency for Fundamental Rights (FRA) reveals that all but two member states actually sanctioned irregular entry and/or stay

---

[161] Stumpf (2006, pp. 381–390). See also Hastie and Crepeau (2014).

[162] See Jacobson (2016) and Mitsilegas (2015).

[163] See Mitsilegas (2015), Palidda (2011a, b), and Provera (2015, p. 3).

[164] Provera (2015, p. 3).

[165] Campesi (2013).

[166] Provera (2015, p. 1).

[167] European Migration Network, cited in Jacobson (2016, p. 2).

with a fine or imprisonment in 2014,[168] thus making the investigation of the effectiveness and consequences of such measures of key contemporary importance.

Criminalisation is also a key example of the securitisation of migration. Traditionally targeted at domestic rule-breakers, it draws heavily from the internal security and crime-control language, as well as contributes to it, thus strengthening the link between migrants, criminality, and security. By sanctioning third country nationals with criminal fines or prison sentences, it further creates and reinforces the perception of foreigners as dangerous individuals and criminals.

In this context, France and Italy are considered here as key cases to analyse the above dynamics. Indeed, not only do both countries sanction irregular entry and stay with severe, although significantly different, penalties, and have made of deterrence their explicit goal, but they have also both witnessed changes in legislation in recent years, thus making it possible to assess whether and how such variations affected migratory flows. Additionally, the criminalisation of migration has been politicised greatly in Italy, but only marginally in France, hence offering an avenue to analyse the extent to which the different political context may have led to different outcomes.

In the remainder of the book, I therefore look at the criminalisation of third country nationals' irregular entry and stay in France and Italy, since the early 2000s, to investigate whether any policy gaps affected their results and, if so, how they may be understood.

Conclusion

In this chapter, I have combined criminological, migration, and international political economy studies, to discuss the way in which deterrence works in the context of human mobility. At its basic level, deterrence is about persuading potential migrants that the costs associated with irregular migration and its consequences are going to be greater than the benefits potentially gainable from it.[169] In this sense, it is based on the assumption that individuals act rationally, and on the realist view that

---

[168] FRA (2014, pp. 4–5).
[169] McLaughlin and Muncie (2013, p. 130).

migration is controllable by states. Importantly, deterrence also relies on the potentiality of the threat, rather than on its actual enforcement.

Criminological studies of deterrence highlight the relevance of a number of factors, in building a successful threat. To begin with, the certainty of a sanction is often found more effective than its severity at ensuring compliance. Second, because people are not perfectly rational, perceptual deterrence studies stress the role of people's *perceptions* of certainty or severity, in increasing effectiveness. Third, as highlighted by the labelling and social-control theories, costs may not only be of legal nature, but of social one too (including stigmatisation and marginalisation), with the result that potential migrants' communities may thus be key actors in discouraging them to enter (or remain in) a country irregularly.

Studying deterrence in the specific field of international migration, I have hypothesised that the strategy may be affected by several weaknesses, which are likely to undermine its overall effectiveness. First, by focusing on negative incentives, and neglecting the structural determinants of migration, deterrence is unable to comprehensively address the phenomenon. Second, even though it might succeed in decreasing migration to a specific route at a certain moment, it can simultaneously produce unintended substitution effects towards other countries or categories of migration, therefore nullifying the overall result. Third, if the domestic or international audiences gain more relevance in the eyes of governments than potential migrants, the threat is likely to be shaped in an inconsistent and ineffective way. Finally, migrants may not have accurate information about the foreseen sanctions, or they may discount its relevance.

In this context, the criminalisation of irregular migration, that is to say the adoption of the criminal law to address migration infractions, emerges as a key example of deterrence, thanks to its double aim to simultaneously increase the legal and social costs of migration. Indeed, this happens through penalties foreseeing fines or imprisonment, but also through the moral condemnation implied by the criminal law itself. The investigation of the cases of France and Italy will permit to appreciate how the above-mentioned weaknesses of deterrence impacted criminalisation in two key EU member states, both of which experienced high irregular migration in the last decades, but had significantly different degrees of politicisation of the norm itself.

I will return to the functioning and weaknesses of deterrence in later chapters, by first discussing the evolution and effectiveness of the criminalisation of migration in Italy (Chapter 4) and France (Chapter 5), and then evaluating the importance of the four pitfalls of deterrence in the two countries (Chapter 6).

References

Akers, Ronald L. 1990. Rational Choice, Deterrence, and Social Learning Theory in Criminology: The Path Not Taken. *Journal of Criminal Law and Criminology* 81 (3): 653–676.

Andenaes, Johannes. 1968. Does Punishment Deter Crime? *Criminal Law Quarterly* 11: 76–93.

Banks, Cyndi, and Irene Pagliarella. 2004. *Criminal Justice Ethics: Theory and Practice*. Thousand Oaks: Sage.

Beccaria, Cesare. (1973) 1764. *Dei Delitti e delle Pene*. Milan: Letteratura Italiana Einaudi.

Becker, Howard S. 1963. *Outsiders: Studies in the Sociology of Deviance*. New York: The Free Press.

Bhagwati, J. 2003. Borders Beyond Control. *Foreign Affairs* 82 (1): 98–104.

Bohmer, Carol, and Ned Lebow. 2015. Why Deterrence Won't Solve the Mediterranean Migrant Crisis. *Telegraph*, April 21. http://www.telegraph.co.uk/news/worldnews/europe/11551401/Why-deterrence-wont-solve-the-Mediterranean-migrant-crisis.html. Accessed 25 Aug 2015.

Borjas, George J. 1989. Economic Theory and International Migration. *International Migration Review* 23 (3): 457–485.

Boswell, Christina. 2003. *European Migration Policies in Flux: Changing Patterns of Inclusion and Exclusion*. Oxford: Blackwell.

Boswell, Christina. 2011. Migration Control and Narratives of Steering. *British Journal of Politics and International Relations* 13: 12–25.

Briskman, Linda. 2008. Detention as Deterrence. In *Asylum Seekers: International Perspectives on Interdiction and Deterrence*, ed. Alperhan Babacan and Linda Briskman, 128–142. Newcastle: Cambridge Scholars Publishing.

Campesi, Giuseppe. 2013. *La detenzione amministrativa degli stranieri in Italia: storia, diritto, politica*. Università di Bari "Aldo Moro".

Carling, Jørgen, and Maria Hernández-Carretero. 2008. Kamikaze Migrants? Understanding and Tackling High-Risk Migration from Africa. Paper presented at the Narratives of Migration Management and Cooperation with Countries of Origin and Transit, Sussex Centre for Migration Research, University of Sussex.

Carling, Jørgen, and María Hernández-Carretero. 2011. Protecting Europe and Protecting Migrants? Strategies for Managing Unauthorised Migration from Africa. *British Journal of Politics and International Relations* 13: 42–58.

Castles, Stephen. 2004. The Factors That Make and Unmake Migration Policies. *International Migration Review* 38 (3): 852–884.

Cornelius, Wayne A., and Takeyuki Tsuda. 2004. Controlling Immigration: The Limits of Government Intervention. In *Controlling Immigration: A Global Perspective*, 2nd ed., ed. Wayne Cornelius, Takeyuki Tsuda, Philip Martin, and James Hollifield. Stanford: Stanford University Press.

Cornelius, Wayne A., and Idean Salehyan. 2007. Does Border Enforcement Deter Unauthorised Immigration? The Case of Mexican Migration to the United States of America. *Regulation & Governance* 1: 139–153.

de Haas, Hein. 2006. Turning the Tide? Why 'Development Instead of Migration' Policies Are Bound to Fail. IMI Working Paper.

de Haas, Hein. 2011. The Determinants of International Migration: Conceptualising Policy, Origin and Destination Effects. International Migration Institute Working Paper 32, IMI, Oxford.

de Haas, Hein. 2015. Borders Beyond Control? January 7. Hein de Haas Blog. http://heindehaas.blogspot.co.uk/2015/01/borders-beyond-control.html. Accessed 25 Aug 2015.

Fargues, Philippe, and Sara Bonfanti. 2014. When the Best Option Is a Leaky Boat: Why Migrants Risk Their Lives Crossing the Mediterranean and What Europe Is Doing About It. Migration Policy Centre, EUI.

FRA. 2014. *Criminalisation of Migrants in an Irregular Situation and of Persons Engaging with Them*. Vienna: FRA.

Freedman, Lawrence. 2004. *Deterrence*. Cambridge, UK and Malden, MA: Polity Press.

Godenau, Dirk, and Ana López-Sala. 2016. Multi-Layered Migration Deterrence and Technology in Spanish Maritime Border Management. *Journal of Borderlands Studies* 31: 1–19.

Guild, Elspeth. 2010. Criminalisation of Migration in Europe: Human Rights Implications. Council of Europe Issues Paper, CommDH/IssuePaper (2010)1, Council of Europe, Strasbourg. https://rm.coe.int/16806da917. Accessed 4 Nov 2021.

Hassan, Lisa. 2000. Deterrence Measures and the Preservation of Asylum in the United Kingdom and United States. *Journal of Refugee Studies* 13 (2): 184–204.

Hastie, Bethany, and Francois Crepeau. 2014. Criminalising Irregular Migration: The Failure of the Deterrence Model and the Need for a Human Rights-Based Framework. *Journal of Immigration, Asylum and Nationality Law* 28 (3): 213–236.

Hollifield, James F. 2004. The Emerging Migration State. *International Migration Review* 38 (3): 885–912.

Holzer, Thomas, Gerald Schneider, and Thomas Widmer. 2000. The Impact of Legislative Deterrence Measures on the Number of Asylum Applications in Switzerland (1986–1995). *The International Migration Review* 34 (4): 1182–1216.

Huysmans, Jef. 2000. The European Union and the Securitization of Migration. *Journal of Cutaneous Medicine and Surgery: Incorporating Medical and Surgical Dermatology* 38 (5): 751–777.

IOM. 2011. Egypt After January 25: Survey of Youth Migration Intentions. Cairo. http://www.migration4development.org/sites/default/files/iom_2011_egypt_after_january_25_survey_of_youth_migration_intentions.pdf. Accessed 31 May 2019.

Jacobs, Bruce A. 2010. Deterrence and Deterrability. *Criminology* 48 (2): 417–441.

Jacobson, Hyla. 2016. The Criminalization of Irregular Migration in Italy. Mediterranean Migration Mosaic, May 13.

Kingsley, Patrick. 2015a. Libya's People Smugglers: Inside the Trade That Sells Refugees Hopes of a Better Life. *Guardian*, April 24. http://www.theguardian.com/world/2015/apr/24/libyas-people-smugglers-how-will-they-catch-us-theyll-soon-move-on. Accessed 25 Aug 2015.

Kingsley, Patrick. 2015b. People Smugglers Using Facebook to Lure Migrants Into "Italy Trips". *Guardian*, May 8. http://www.theguardian.com/world/2015/may/08/people-smugglers-using-facebook-to-lure-migrants-into-italy-trips. Accessed 25 Aug 2015.

Kox, Mieke. 2011. *Leaving Detention? A Study on the Influence of Immigration Detention on Migrants' Decision-Making Processes Regarding Return*. The Hague: IOM.

Lebow, Richard Ned. 2007. Cognitive Closure and Crisis Politics. In *Coercion, Cooperation, and Ethics in International Relations*, ed. Richard Ned Lebow. New York, NY: Routledge.

Levitt, Steven D. 1998. Why Do Increased Arrest Rates Appear to Reduce Crime: Deterrence, Incapacitation, or Measurement Error? *Economic Inquiry* 36: 353–372.

Loughran, Thomas A., Raymond Paternoster, Alex R. Piquero, and Greg Pogarsky. 2011. On Ambiguity in Perceptions of Risk: Implications for Criminal Decision Making and Deterrence. *Criminology* 49 (4): 1029–1061.

Loughran, Thomas A., Ray Paternoster, Aaron Chalfin, and Theodore Wilson. 2016. Can Rational Choice Be Considered a General Theory of Crime? Evidence from Individual-Level Panel Data. *Criminology* 51 (1): 86–112.

Malmberg, Mari. 2004. Control and Deterrence: Discourses of Detention of Asylum-Seekers. Sussex Centre for Migration Research, Sussex Migration Working Paper No. 20.
Martin, Philip L., and J. Edward Taylor. 2001. Managing Migration: The Role of Economic Policies. In *Global Migrants Global Refugees: Problems and Solutions*, ed. Aristide R. Zolberg and Peter M. Benda, 95–120. Oxford: Berghahn Books.
Maurice, Eric. 2016. Turkey Sends EU Mixed Message on Migration. *EUObserver*, September 4. https://euobserver.com/foreign/134896. Accessed 30 Apr 2017.
McLaughlin, Eugene, and John Muncie, eds. 2001. *The SAGE Dictionary of Criminology*. London: Sage.
McLaughlin, Eugene, and John Muncie, eds. 2013. *Criminological Perspective: Essential Readings*, 3rd ed. London: Sage.
Mitsilegas, Valsamis. 2015. *The Criminalisation of Migration in Europe: Challenges for Human Rights and the Rule of Law*. London: Springer.
Nagin, Daniel S. 1998. Criminal Deterrence Research at the Outset of the Twenty-First Century. *Crime and Justice* 23: 1–42.
Nagin, Daniel S. 2013. Deterrence in the Twenty-First Century: A Review of the Evidence. Carnegie Mellon University Research Showcase @ CMU, Working Paper, Heinz College Research.
Neumayer, Eric. 2004. Asylum Destination Choice: What Makes Some West European Countries More Attractive Than Others? *European Union Politics* 5 (2): 155–180.
Onwudiwe, Ihekwoaba D., Jonathan Odo, and Emmanuel C. Onyeozili. 2005. Deterrence Theory. In *Encyclopedia of Prisons & Correctional Facilities*, vol. 2, ed. M. Bosworth, 234–237. Thousand Oaks, CA: Sage. https://doi.org/10.4135/9781412952514.n91.
Oxford English Dictionary. 2016. Deter. OED. http://www.oed.com/view/Entry/51209?rskey=ns6vnz&result=1&isAdvanced=false#eid. Accessed 18 Jan 2017.
Palidda, Salvatore. 2011a. Introduction. In *Racial Criminalization of Migrants in the 21st Century*, ed. Salvatore Palidda, 1–22. Farnham: Ashgate.
Palidda, Salvatore. 2011b. A Review of Principal European Countries. In *Racial Criminalization of Migrants in the 21st Century*, ed. Salvatore Palidda, 23–30. Farnham: Ashgate.
Pogarsky, Greg, Alex R. Piquero, and Ray Paternoster. 2004. Modeling Change in Perceptions About Sanction Threats: The Neglected Linkage in Deterrence Theory. *Journal of Quantitative Criminology* 20 (4): 343–369.
Provera, Mark. 2015. The Criminalisation of Irregular Migration in the European Union. CEPS Paper in Liberty and Security in Europe, No. 80/February 2015.

Reyneri, Emilio. 2003. Immigration and the Underground Economy in New Receiving South European Countries: Manifold Negative Effects, Manifold Deep-Rooted Causes. *International Review of Sociology* 13 (1): 117–143. https://doi.org/10.1080/0390670032000087023.

Richardson, Roslyn. 2010. Sending a Message? Refugees and Australia's Deterrence Campaign. *Media International Australia* 135: 7–18.

Robinson, Paul H., and John M. Darley. 2004. Does Criminal Law Deter? A Behavioural Science Investigation. *Oxford Journal of Legal Studies* 24 (2): 173–205.

Ryo, Emily. 2015. Less Enforcement, More Compliance: Rethinking Unauthorized Migration. *UCLA Law Review* 62: 622–670.

Simester, A.P., and Andreas von Hirsch. 2011. *Crimes, Harms, and Wrongs: On the Principles of Criminalisation*. Oxford: Hart Publishing.

Squires, Nick. 2010. Gaddafi: Europe Will 'Turn Black' Unless EU Pays Libya £4bn a Year. *The Telegraph*, August 31. http://www.telegraph.co.uk/news/worldnews/africaandindianocean/libya/7973649/Gaddafi-Europe-will-turn-black-unless-EU-pays-Libya-4bn-a-year.html. Accessed 30 Apr 2017.

Stafford, Mark C., and Mark Warr. 1993. A Reconceptualisation of General and Specific Deterrence. *Journal of Research in Crime and Delinquency* 30 (2): 123–135.

Stein, Janice Gross. 2009. Rational Deterrence Against Irrational Adversaries? No Common Knowledge. In *Complex Deterrence*, ed. Patrick M. Morgan, 58–84. Chicago: University of Chicago Press.

Stumpf, Juliet. 2006. The Crimmigration Crisis: Immigrants Crime and Sovereign Power. *American University Law Review* 56 (2): 367–419.

Talani, Leila Simona. 2010. *From Egypt to Europe: Globalisation and Migration Across the Mediterranean*. London: I.B. Tauris.

Talani, Leila Simona. 2018. *The Political Economy of Italy in the Euro: Between Credibility and Competitiveness*. London: Palgrave Macmillan.

Thielemann, Eiko. 2003. 'Does Policy Matter? On Governments' Attempts to Control Unwanted Migration. IIIS Discussion Paper No. 09. https://www.tcd.ie/triss/assets/PDFs/iiis/iiisdp09.pdf. Accessed 13 Dec 2021.

Thielemann, Eiko. 2011. How Effective Are Migration and Non Migration Policies That Affect Forced Migration? LSE Migration Study Unit Working Paper No. 2011/14.

von Hirsch, Andrew, Anthony E. Bottoms, Elisabeth Burney, and P.-O. Wikström. 1999. *Criminal Deterrence and Sentence Severity: An Analysis of Recent Research* (University of Cambridge Institute of Criminology). Oxford, UK: Hart Publishing.

Weiner, Myron. 1985. On International Migration and International Relations. *Population and Development Review* 11 (3): 441–455.

Weiner, Myron. 1995. *The Global Migration Crisis: Challenge to States and to Human Rights*. New York: Harper Collins College Publishers.
Wellford, Charles. 1975. Labelling Theory and Criminology: An Assessment. *Social Problems* 22 (3): 332–345.
Zimring, Franklin, and Gordon Hawkins. 1973. *Deterrence: The Legal Threat in Crime Control*. Chicago: University of Chicago Press.

CHAPTER 4

Italy: Tough Rhetoric, Underground Economy, and Counterproductive Consequences

Introduction

Italy has a rather short history of immigration, when compared to other EU member states. If in the late 1940s, 'collective migration [was] inexistent', with only about 500–600 entries per year,[1] until the 1970s the country still saw large migratory out-flows, rather than inflows.[2] The situation changed following the economic boom starting in the early 1960s, and the growing restrictions posed by traditional immigration countries in northern Europe (including France), in the aftermath of the oil crisis.[3] In 1973, immigration figures outnumbered emigration ones for the first time, although it would take until the 1990s to see immigrants represent 1% or more of the population.[4] In this context, migration issues did not take long to be politicised in the mid-1980s, in parallel with trends in the rest of Europe.[5]

---

[1] Group Salaires et questions sociales (1950, IT, pp. 1–2). Data refer to 1948 and 1949.

[2] Fasani (2010, p. 167).

[3] Ibid.

[4] Ibid.

[5] Delvino and Spencer (2014, p. 4).

Following the absence of a proper migration policy over all those years,[6] the 1990 Martelli Law[7] was the first legislative act to extensively address migration (partly upon incentive by other EU member states, so that the country could join the Schengen agreement).[8] The law was subsequently integrated with additional provisions under the 1998 Turco-Napolitano law,[9] which created the Consolidated Text of Provisions Governing Immigration and the Status of Aliens (*Testo Unico sull'Immigrazione*, TUI)[10] on which most of the migration legislation is based today. Since the creation of the TUI, several laws have come to modify the system, including the 2002 Bossi-Fini law,[11] 2009 Security Package,[12] the 2018 and 2019 Security Decrees,[13] and the 2020 changes to them.[14]

In this context, irregular migration was criminalised only recently, when the 2009 legislative decree introduced the crime of irregular entry and stay, to be punished by monetary sanction or expulsion, in what became article 10-bis of the TUI.[15]

Today, more than ten years after the introduction of article 10-bis, its role in deterring would-be migrants has been contested, especially in light of sustained migratory pressure and of the worsening congestion of the Italian judicial system.[16] This was to the extent that, in 2014, the Italian Parliament delegated the Government to transform the crimes of irregular entry and stay into administrative offences.[17] Despite such concerns,

---

[6] Group salaires et questions sociales (1950, IT, p. 1).

[7] Law No. 39/1990.

[8] Paoli (2018) and di Robilant (1990).

[9] Law No. 40/1998.

[10] '*Testo unico delle disposizioni concernenti la disciplina dell'immigrazione e norme sulla condizione dello straniero*', Legislative Decree No. 286/1998.

[11] Law No. 189/2002.

[12] Law No. 94/2009.

[13] Decree law No. 113/2018 and Decree law No. 53/2019.

[14] Law No. 173/2020.

[15] Law No. 94/2009, art. 1.

[16] See Report Committee on Migration (2014, p. 9) and Chamber of Deputies (14/01/2016, pp. 61–62).

[17] Law 67/2014, Art. 2, para. 3, lett. b.

however, the reform was not pursued by the Renzi Government and, at the time of writing, the norm is still in place.

Italian developments raise key questions: Was criminalisation affected by 'policy gaps'? What were the elements responsible for its failure, and what have its consequences been? Finally, why was it not repealed, and who benefitted from it? In this chapter, I discuss these and related issues, to analyse the effects and consequences of the norm on migratory flows to Italy.

The discussion is organised as follows: First, I investigate the possible presence of discursive gaps, looking at the rhetoric surrounding the introduction of the norm and its adoption. Second, I use original datasets to analyse the extent to which the norm has been applied, and possible implementation gaps. Third, I apply the five evaluation criteria presented in the introduction, to assess whether efficacy gaps occurred.

From Rhetoric to Paper: Discursive Gaps

Before the formal introduction of the crime of irregular entry and stay, the right-wing fringes of the Italian Parliament had long advocated for the criminalisation of irregular migration as a solution to unauthorised inflows of migrants. In 1999, for example, the far-right-sponsored Fini-Landi Bill proposed to punish irregular entry with criminal sanctions and expulsion.[18] In 2002 again, there were widespread expectations that the new centre-right government would adopt such measure,[19] and in 2008, the electoral manifesto of La Destra re-proposed the introduction of the crime.[20]

While none of these initiatives was successful, shortly after winning the 2008 general elections, the governing coalition including *Il Popolo della Libertà* (PdL) and Lega Nord (LN) proposed a 'Security Package' bill. This classified irregular entry as a criminal offence, to be punished with imprisonment ranging from six months to four years.[21] Imprisonment had not been foreseen in previous proposals of criminalisation: While La Destra's manifesto had supported immediate expulsion, the

---

[18] Law Proposal N. 5808 (1999).

[19] Campesi (2013).

[20] Vesci (2008).

[21] Law Proposal N. 733/2008, art. 9, para. 1.

Landi-Fini bill had preferred house arrest, on the basis that the latter would not only reduce the pressure on the Italian prison system, but also (interestingly) avoid conflating migrants with criminals.[22]

The 2008 Government's proposal was received with broad condemnation. Among others, the UN High Commissioner for Human Rights publicly stigmatised Italy's decision on humanitarian grounds, and the Secretary of the Pontifical Council for Migrants and Itinerant People criticised the norm for subjecting people to imprisonment for what was in fact an administrative infraction.[23] Adding to that, practical constraints made the measure potentially troublesome: In particular, the state of Italian prisons, steadily over-populated since the early 2000s,[24] was especially concerning.[25]

At the policy-making level, the proposed norm caused dissent not only among the opposition, but also within the governing coalition itself. The strongest hit was suffered following the withdrawal of support to the provision by then Prime Minister (PM) Berlusconi, declared on the same day the bill was presented to the Italian Senate (3 June 2008).[26] Indeed, after the PM's statement, differences emerged, with the bill-proponent and Minister of Interior Maroni (LN) supportive of imprisonment, the Defence Minister La Russa (LN) showing a preference for immediate expulsion, and MP Ghedini (PdL) suggesting to make the penalty dependent on the dangerousness of the individual involved.[27]

As a result, in November 2008, the Government itself proposed an amendment to the bill, substituting imprisonment with a monetary sanction ranging from €5000 to €10,000, and envisaging the possibility for justices of the peace (JPs) to substitute the fine with an expulsion order. While accepting to amend the severity of the norm, however, the

---

[22] Law Proposal N. 5808 (1999).

[23] See http://www.repubblica.it/2008/05/sezioni/cronaca/sicurezza-politica-6/vaticano-2giu/vaticano-2giu.html?ref=search.

[24] According to ISTAT, prisons reached 151% of their capacity in 2010. The trend is consistent for all years since 2000 except 2006, when an amnesty was granted (see https://www.istat.it/it/files/2015/03/detenuti-2015-1.pdf).

[25] Interviewees 20 and 26.

[26] See http://www.antigone.it/component/content/article/76-archivio/1940-immigrazione-berlusconi-frena-qclandestinit-sia-solo-aggravanteq-la-repubblica-030608.

[27] See http://www.ilsole24ore.com/art/SoleOnLine4/Italia/2008/06/prostitute-espulsione.shtml?uuid=ad0ac57a-3396-11dd-b639-00000e251029.

cabinet also extended it to both foreigners' irregular entry *and* stay, thus substantially enlarging its scope. Contestation by civil society persisted, including petitions by Catholic associations,[28] lawyers,[29] and intellectuals,[30] and even a hunger strike,[31] but protests did not lead to further amendments, and the Security Package including the norm was eventually promulgated on 24 July 2009.

In introducing criminalisation, the Government had two objectives. First, it hoped to take advantage of the deterrent effect of the criminal law to reduce annual irregular entries[32]: When presenting the bill that eventually resulted in the 2009 security package, the Government explicitly stated that the crime of irregular migration was meant to reduce annual unauthorised arrivals by 10%, from 54,500 in 2007, to an expected 49,050 in later years.[33] Likewise, much of the discussion regarding the repeal of article 10-bis reverted around the issue of its deterrent effect, or lack thereof (see section "Efficacy (II): Criminal Law and Landings").[34]

Second, the Government aimed to facilitate and increase immediate expulsions, by eluding the Return Directive and thus avoiding the concession of a delay for voluntary returns.[35] Indeed, criminalisation was intended to exploit an exception foreseen by the Directive itself, according to which Member States can decide not to apply the Directive to third country nationals who 'are subject to return as a criminal law sanction or as a consequence of a criminal law sanction'.[36] Through this exception, the delay for voluntary return could have been avoided, and forced expulsions be used extensively.[37]

---

[28] Fondazione Migrantes et al. (2009).

[29] Caputo et al. (2009).

[30] Camilleri et al. (2009).

[31] See http://www.inviatospeciale.com/giornale/2009/05/da-oggi-litalia-si-avvia-ad-essere-un-po-fascista/.

[32] Law Proposal N. 733 (2008) and Report Committee on Migration (2014, p. 9).

[33] Law Proposal N. 733 (2008, p. 9).

[34] See, for example, Corriere del Mezzogiorno (2014) and Chamber of Deputies, 15/01/2014, p. 14.

[35] Chamber of Deputies, 15/10/2008. See also interviewee 30, Jacobson (2016, pp. 22–33), and Provera (2015, p. 20).

[36] Directive 2008/115/EC, art. 2, para. 2 lett. B.

[37] Interviewee 30.

Legal Aspects and Judicial Controversies

Following the adoption of the law, despite critiques and campaigns[38] against it, no changes affected the norm, which is still in place and the text of which has not seen any alterations. Yet, this does not mean that the norm has not been the object of discussion and analysis by Italian and EU courts, and that its interpretation has not changed.

By October 2009, two justices of the peace had already remitted the issue of its compatibility with national and supranational law to the Italian Constitutional Court. On one hand, by allowing magistrates to substitute the fine with immediate expulsion on a systematic basis, art. 10-bis was said to be in contrast with the objectives of the EU Return Directive,[39] which foresees the granting of a period for voluntary departure as the standard procedure.[40] On the other hand, the norm was deemed incompatible with Italian constitutional law, for it sanctioned a personal condition, rather than an actual infraction, and infringed the principle of reasonableness, by prescribing parallel administrative and criminal procedures with the same goal.[41]

In July 2010, the Italian Constitutional Court rejected the issue of constitutionality.[42] In the Court's opinion, as far as the EU Directive was concerned, the term for conforming to it had not yet expired, and, in any case, it would have been other provisions of the TUI to be in contrast with it, rather than art. 10-bis.[43] Furthermore, the norm did not sanction the status of a person but rather a specific behaviour, namely irregularly entering or staying in the country.[44] Finally, the Court highlighted that the choice of the best instruments to manage migratory flows was a matter

---

[38] The most recent of which being the 'Ero straniero' campaign (see https://www.nev.it/nev/2019/04/11/ero-straniero-la-campagna-contro-la-bossi-fini-arriva-alla-camera/).

[39] JP of Lecco by order of 1 October 2009 and JP of Turin by order of 6 October 2009, as reported in: Constitutional Court, Sentence 250/2010.

[40] Directive 2008/115/EC, art. 7.

[41] JP of Lecco by order of 1 October 2009 and JP of Turin by order of 6 October 2009, as reported in: Constitutional Court, Sentence 250/2010, Caputo et al. (2009), and Ferrero (2010).

[42] Constitutional Court, sentence n. 250/2010.

[43] Ibid., section 9.

[44] Ibid., section 6.2.

for the Italian Parliament to decide,[45] and that, although doubts on the deterrent effect of the norm were plausible, it was not the Court's role to deliberate on the effectiveness of the criminal law.[46] In doing so, it showed unwillingness to be linked to the politicised debate surrounding the norm.

Following the decision of the Constitutional Court, the issue of the compatibility of the norm with EU law was raised to the European Court of Justice (ECJ) in 2011. In December of the following year, the Court upheld the compatibility of the norm with the EU Directive, arguing that criminally sentencing a monetary sanction for irregular stay is compatible with EU law, as it does not impede a return decision (it would be different if the criminal fines were *an alternative to* expulsion, as later confirmed in Case C-38/14, Zaizoune).[47] Importantly however, the ECJ also stressed that immediate expulsion under national criminal law is only possible under specific circumstances, namely following a risk of absconding, to be assessed on a case-by-case basis,[48] therefore nullifying the Italian Government's hope to use such provision in an extensive way.

A final judicial development that should be highlighted concerns irregular migration and Search and Rescue (SAR) operations. Indeed, 2014 saw increasing sea arrivals to Italy, which forced courts to address the debate on the applicability of art. 10-bis to migrants who were rescued by SAR missions. If on one hand, they were entering irregularly, and therefore subjectable to sanctions, on the other, they did not autonomously choose to land in Italy, but were brought there following maritime rescues. In a **2016 ruling**, the Supreme Court (*Corte di Cassazione*) resolved the dispute by concluding that these persons should not be held responsible for entering irregularly, since it is not possible to assume that the peril to their lives which resulted in the SAR operation was intentional.[49] The ruling substantially reduced the scope of applicability of the

---

[45] Ibid., section 6.3.
[46] Ibid., section 10.
[47] Case C-430/11, Sagor.
[48] Ibid., paras. 40–42.
[49] Supreme Court of Cassation, sentence 53691/2016, section 2.1.

norm, since most of the irregular migrants landing in Italy at the time were being rescued at sea.[50]

Overall, despite widespread criticism, high courts did not consider the crime of irregular entry and stay as infringing existing legislation, and the law remained unchanged. Such rulings, however, did not prevent local actors from finding strategies to absolve, nonsuit or dismiss cases, as will clearly emerge from the following pages.[51]

Demanded Repeal and Political Implications

Demands for the repeal of the norm did not only come from legal circles, but from political ones too. Indeed, four years after the introduction of the norm, Five Star Movement (*Movimento Cinque Stelle*, M5S) MPs Andrea Cioffi and Maurizio Buccarella proposed to delegate the Government to decriminalise migration.[52] The initiative was met with significant scepticism, especially by the leaders of the Movement, who viewed the proposal as easily contestable by voters.[53] Yet, on 2 April 2014, Law 67/2014 was approved and, through it, the Parliament delegated the Government to transform the crime of irregular entry and stay into an administrative offence within eighteen months.[54]

According to Delvino and Spencer, the Parliament's decision was based on three main considerations.[55] First, the ineffectiveness of criminalisation in reducing irregular landings or encouraging persons in irregular situations to leave the country, was emerging.[56] Indeed, the Fiorella Commission for the revision of the penal system had stressed in 2012

---

[50] The sentences of the Supreme Courts are mandatory for the case they refer to, but not for other ones. On average however, these are regarded as important precedents for judges in other cases, who therefore generally abide by the Courts' rulings. See: http://www.cortedicassazione.it/corte-di-cassazione/it/funzioni_della_corte.page.

[51] See, for example, JP Civitanova Marche, 2011, as reported in Supreme Court of Cassation, Sentence n. 35588, 2013, JP Turin, Sentence 314/2011, JP Rome, Sentence 16/06/2011, and JP Monza, Sentence 2560/2012.

[52] Amendment 1.0.1/3, available at: http://leg17.senato.it/japp/bgt/showdoc/17/Emendc/734317/714470/index.html.

[53] Grillo and Casaleggio (2013).

[54] Law 67/2014, art. 2, c.3, lett. b.

[55] Delvino and Spencer (2014, p. 35).

[56] Ibid.

that criminalisation of migration 'is a criminal norm [that is] completely ineffective and symbolic, which foresees irrational sanctions'.[57] Second, several voices had contested the detrimental effects the norm had on courts' workload.[58] Third, the criminal pursuit of the survivors of the 2013 deadly shipwreck near Lampedusa (which caused 368 deaths)[59] had generated public 'outrage'.[60] Even then-PM Enrico Letta stated that '[a]s prime minister I felt ashamed in view of such zeal: It is an enormous human tragedy'.[61]

After long deliberation, and after the expiration of the Parliament's proxy, in December 2015 the Government presented a legislative decree to follow up decriminalisation. The decree was proposed by the Minister of Justice Andrea Orlando, sponsored by the national Anti-Mafia and Anti-Terrorism Public Prosecutor Franco Roberti, and vastly supported by the House of Representative's Justice Committee (*Commissione Giustizia della Camera*).[62]

Even though, as a legislative decree, the bill did not need further parliamentary approval, decriminalisation was eventually not completed. Instead, it soon sparked very politicised reactions from a variety of actors. First of all, the centre-right *Forza Italia* (FI), which was not part of the governing coalition but had been so at the time of the adoption of the norm, was rather sceptical. As an example, the president of the FI Senate group Paolo Romani warned about what he called the 'inevitable negative consequences' of decriminalisation, and the leader of the FI House of Representatives group Renato Brunetta argued that Italy 'would risk imploding'.[63] Several politicians associated with the Northern League also strongly opposed the reform: Both the president of the Lombardy region Roberto Maroni, and the Senate vice president Roberto Calderoli, referred to decriminalisation as 'madness', and the LN leader Matteo

---

[57] Commissione Fiorella (2012), section 3.3.

[58] Democratic Party Senator Felice Casson, as cited in Delvino and Spencer (2014, pp. 35–36).

[59] Live Sicilia (2013).

[60] Delvino and Spencer (2014, p. 35).

[61] Live Sicilia (2013). Author's translation.

[62] See https://www.repubblica.it/politica/2016/01/08/news/immigrati_clandestini_il_decreto_del_governo_per_cancellare_il_reato-130812428/.

[63] Repubblica (2016). Author's translation.

Salvini threatened retaliation and even a referendum on the matter.[64] Opposition to decriminalisation seemed largely of political, rather than substantial, nature.

Political were also the considerations of many in the governing coalition, and particularly of then-Minister of the Interior Angelino Alfano, considering the potential repercussions that the choice would have had in electoral terms. On one hand, the Minister was concerned about the potential inflammation of rhetoric that decriminalisation could imply. In particular, he did acknowledge the null deterrent effect of criminalisation, and that its repeal would in fact have been reasonable.[65] Yet, he also appeared to perceive the high sensitivity of migration issues as being easily instrumentalised, with the result that the wrong message would be received by the public, and that the appeal of far-right discourse would risk increasing.[66] In his words, 'the people would not understand'.[67]

On the other hand, it must also be stressed that Alfano had himself been a cabinet member as Minister of Justice at the time of the introduction of the norm in 2009. Consequently, his adversity to decriminalisation might partly be linked to the wish to look coherent in his voters' eyes. According to the coordinator of the Immigration Forum of the Democratic Party (*Partito Democratico*, PD) Marco Pacciotti, Alfano's New Centre-Right (*Nuovo Centro Destra*, NCD) party was also split between far-right sympathisers and more central MPs,[68] which may have constrained the politician's choices.

Eventually, the political considerations presented by Alfano seemed to resonate more with the cabinet, and especially with Prime Minister Matteo Renzi, than the judicial and efficiency arguments supported by Orlando, with the result that the norm was left untouched, and has been so to the present day. This has been the case despite loud criticism of the norm by several actors, including key public figures such as the Head of the National Police Alessandro Pansa, the National Anti-Mafia and Anti-Terrorism Prosecutor Franco Roberti, and the former President of

---

[64] Ibid.
[65] Repubblica (2016), Baer (2016), and Bei (2016).
[66] Ibid.
[67] Baer (2016) and Bei (2016).
[68] Interviewee 17.

the Italian Supreme Court Giovanni Canzio.[69] According to Pacciotti, the political majority was not sufficiently stable for the Government to abrogate the crime of irregular migration.[70]

Requests for decriminalisation continued in later years too, for example with the 2017 popular campaign 'Ero Straniero' ('I was a foreigner'), which gathered over 90,000 signatures and has been discussed in the Italian Parliament since 2019.[71] The bill started being debated under the Lega-Five Star Movement governing coalition, whose immigration policy went however in a radically different direction (for example with the power given to the Minister of the Interior to close ports or limit transit in national waters, and the removal of humanitarian protection).[72] Moreover, having proposed to criminalise migrant rescuers and to sanction up to €5500 per saved migrant (although this was, eventually, not done),[73] the Government was unlikely to support the decriminalisation of migration. The bill is still under discussion at the time of writing, although no decision has been adopted on it yet.[74]

Overall, as detailed thus far, a very politicised debate over the criminalisation of irregular migration emerged, providing evidence of the first kind of policy gaps: A significant distance between the tough rhetoric (initially proposing imprisonment), and the actual measure (foreseeing a fine), with the latter being somewhat watered down (albeit not in scope). This is further supported by the fact that, a month after passing the law on criminalisation, the Government approved an amnesty expected to regularise over 500,000 domestic workers and carers.[75] As will be argued later in the book, such counterbalancing act suggests that the 2009 Security Package

---

[69] Pansa (2016), Baer (2016), and Canzio (2016).

[70] Interviewee 17.

[71] See https://erostraniero.radicali.it/iter-in-parlamento/, https://www.huffingtonpost.it/entry/ce-unitalia-che-non-si-arrende-alla-paura-del-diverso_it_5d04e8e3e4b0dc17ef0a874a.

[72] Decree law 53/2019 and Decree law 113/2018.

[73] See https://www.ilfattoquotidiano.it/2019/05/29/dl-sicurezza-bis-la-nuova-versione-tolte-multe-per-soccorso-migranti-via-pene-per-resistenza-passiva-durante-manifestazioni/5199622/.

[74] For more on the status of the discussions, see: https://www.camera.it/leg18/126?tab=1&leg=18&idDocumento=13&sede=&tipo=.

[75] Ferrero (2010, p. 62). Although, eventually, only 294,000 applications were registered (Pasquinelli 2009).

was more concerned with pursuing visible measures to restrict entry, than with the actual factors causing or incentivising migration, in a strategy that can be interpreted as a case of deliberate malintegration.[76] This is also supported by the analysis of the discussion concerning the norm's repeal, which gave precedence to unwanted political consequences, rather than considerations on a rational use of resources.

If the criminalisation of irregular entry and stay is to remain in the Italian legislation, it is important to assess its implementation, in order to understand how widely the norm is applied. This is indeed the object of the next section, in which I discuss criminalisation in practice, relying on data from ISTAT and the Ministry of Justice, as well as interviews with key stakeholders.

From Paper to Enforcement: Implementation Gaps

Having analysed the rhetorical dimension related to the criminalisation of migration, I now turn to its implementation. To do so, I first provide an overview of how the norm is applied, and of how it relates to the overall Italian migration-reception system. Then, I look at the numbers of criminalisation, to evaluate its implementation.

The Unfolding of Criminal Trials in the Italian Reception System

The Italian reception system has been modified several times in recent years, leading to a complex network of provisions. For the purposes of this discussion, I highlight the main procedures migrants go through, then explaining in greater detail how the criminal ones unfold.

First reception is the initial stage. Traditionally, migrants landed in Italy following a search and rescue operation were received in first aid and reception centres (*Centri di primo soccorso e accoglienza*, CPSA).[77] As the name indicates, these provided temporary reception, waiting for the persons to be relocated to other facilities throughout the country.[78] Following the adoption of the EU Agenda on Migration, in 2015 and

---

[76] For more on this, see section "Utility: Who Benefits from Criminalisation?"
[77] MoI (2015).
[78] Parliamentary commission on the reception system (2017, p. 52).

2016, 'hotspots' were introduced, to increase the efficiency of the identification process, by channelling arrivals to specific ports and promptly separating migrants according to their situations.[79] In hotspots, migrants receive medical screening and are interviewed by immigration service officials (and sometimes also by the police, Frontex or Europol, for information on smuggling or trafficking networks).[80]

A significant change brought about by hotspots relates to the percentage of people identified after landing: Until 2015, it was not uncommon for Italian police officers not to register migrants and let them travel to central and northern Europe, where they would apply for asylum.[81] Yet, with the introduction of hotspots in 2015–2016, such possibility was taken away, with the result that today, virtually all apprehended migrants are registered with photographic evidence following their landing.[82]

Following this initial phase, migrants are transferred to other centres, according to their specific situation.[83] On one hand, those who show willingness to request asylum are either sent to asylum seeker reception centres,[84] or to specific hubs if they qualify for relocation to other EU member states (since 2015).[85]

On the other hand, those who do not request international protection, or who do not qualify for it, are considered 'irregular'. The same is true for foreigners found irregularly staying in the country, and it is at this point that procedures on criminalisation start, in parallel with administrative processes aiming to achieve expulsion. In particular, the procedure described in article 10-bis may start in one of two situations: (a) When

[79] Ibid., pp. 69–70.

[80] Italian Roadmap (2015, p. 6) and Interviewees 34 and 37.

[81] Fargues and Bonfanti (2014), p. 13 and interviewees 13 and 15.

[82] Parliamentary commission on the reception system (2017, p. 58) and interviewees 15, 16 and 17.

[83] Sometimes, the centres' functions are mixed, so it can happen to find tasks of first reception, assistance for asylum seekers and detention, in the same facilities (Parliamentary commission on the reception system 2017, p. 53).

[84] The name of the centres, and their specificities, have changed with time. See MoI (2015), Parliamentary commission on the reception system (2017), EMN (2019), https://www.interno.gov.it/it/temi/immigrazione-e-asilo/sistema-accoglienza-sul-territorio.

[85] COM (2015) 240.

a foreigner has entered the country in an irregular way (without a valid passport and visa) and/or outside of border crossing points; (b) when a foreigner is in the country without proper documentation (either because this is absent, or expired).[86]

Briefly discussing the administrative pathway, before delving into the criminal one, this involves issuing expulsion orders.[87] In some cases, this implies not only the obligation to leave the country, but also an entry ban for three to five years, and registration in the Schengen Information System database.[88] If there is no immediate threat or risk of absconding, the persons can request a delay for voluntary departure (between 7 and 30 days).[89] On the contrary, if the risk of absconding is present, but expulsion is not immediately possible, they may be held in permanence and return centres (*Centri di permanenza e rimpatrio*, CPR).[90] The length of time migrants could remain in CPRs has changed over time, from up to 18 months in 2011, to maximum 90 days in 2014.[91] In practice, however, the limited number of available places in CPRs, coupled with the difficulties in returning migrants,[92] means that the majority of people are ordered to leave the country on their own.[93] This, of course, creates a high number of persons who are supposed to leave the country but who neither do so independently, nor can be returned by the state itself,[94] and therefore remain in a state of limbo.

In parallel to the administrative process leading to expulsion, article 10-bis requires that a criminal procedure is also started. This represents

---

[86] Legislative Decree 286/1998, art. 4.

[87] Parliamentary commission on the reception system (2017, p. 77).

[88] Ibid., pp. 78, 93.

[89] Ibid., p. 79.

[90] Ibid.

[91] The maximum length of detention was further increased to 180 days in 2018, and then reduced to 120 days in 2020. See Legislative Decree 286/1998, art. 14, para. 5; Decree Law 113/2018, art. 2; Law 173/2020; https://www.antigone.it/tredicesimo-rapporto-sulle-condizioni-di-detenzione/01-detenzione-amministrativa/.

[92] See section "Efficacy (I): Criminal Law and Expulsions."

[93] Parliamentary commission on the reception system (2017, p. 80).

[94] Ibid.

the criminalisation of migration and, importantly, is compulsory for prosecutors to pursue[95]: Once a situation of irregularity is discovered, both the administrative and criminal procedures start.

The criminal pathway starts with officers of the judicial police registering the offence of irregular entry or stay, and sending the crime report (*notizia di reato*) to the competent public prosecution.[96] There, public prosecutors register the infraction and decide whether to (1) request the dismissal of the case, (2) initiate legal action, or (3) pursue further investigations.[97]

When first registered, files are generally clustered by apprehension or landing,[98] with some of them including over a hundred people.[99] As the criminal procedure unfolds, however, names on the files are taken away depending on whether they regard minors (whose cases go to the competent Minors' Tribunal), asylum seekers (whose cases are suspended), or other special circumstances.[100] Likewise, if the crime of irregular entry or stay is contested together with other offences, these are unbundled and sent to the competent judges.[101] Eventually, despite the clustered origin of initial files, rulings are individual.[102]

Importantly, as the norm does not foresee arrest, during all these developments, migrants are not held in custody, but free to move.[103] This implies that before the cases go to trial, indicted foreigners need to be traced, so that they can be informed of the charges against them.[104]

---

[95] According to art. 112 of the Italian Constitution, '[t]he public prosecutor has an obligation to pursue criminal proceedings'.

[96] Interviewee 40.

[97] Interviewee 40 and Simone (2019, p. 210).

[98] Interviewees 32 and 40.

[99] See, for example, JP Agrigento, Settlement 12/11/2016.

[100] Interviewees 32 and 46, legislative decree 286/1998, art. 10*bis*. paragraph 6. If the asylum claim is rejected, the criminal trial should start again, although this does not seem to often be the case (interviewee 47).

[101] Interviewee 29.

[102] Interviewee 46.

[103] Interviewee 14.

[104] Interviewee 32.

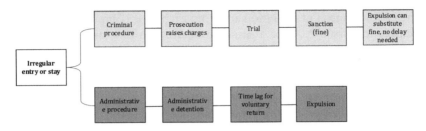

Fig. 4.1 Criminal and administrative procedures for irregular entry or stay in Italy (since 2009) (*Source* Author's elaboration. Criminal procedures in light grey; administrative ones in dark grey)

On the day of trial, it is not mandatory for the accused to attend, and there is evidence that this is indeed rarely the case.[105] Hearings are therefore generally attended by the justice of the peace, public prosecutor, (possibly) the police officers who apprehended or identified the migrant, a clerk of the court, and, if needed, an interpreter.[106] The role of police officers is to confirm the identity and apprehension of the person under trial[107] (though their statement can be replaced by a reading of the crime report upon agreement of the parts).[108] In the last stage of the trial, following a varying number of hearings (depending on the difficulty of the case), the JP issues a sentence to either (1) acquit the person, (2) sanction a fine of between 5000 and 10,000 euro, or (3) substitute the latter with an expulsion order.[109]

Thus, when a migrant is apprehended irregularly entering or staying, a double procedure starts: The administrative one leading to expulsion, and the criminal one leading to a fine, or its possible substitution with expulsion (Fig. 4.1).

How extensive is the use of the criminal procedure to pursue irregular entry or stay, in Italy? How many trials take place and how often are people sanctioned? What nationality do indicted people usually hold? These questions are the object of the next section, which investigates

---

[105] Interviewees 32, 41, 46, 54, 55.
[106] Interviewees 46, 54.
[107] Interviewee 32.
[108] Interviewee 55.
[109] Interviewee 32 and Law decree 286/1998, art. 16, para. 1.

the actual implementation of the norm, based on previously unreleased official data.

The Numbers of Criminalisation

Among the Italian cities interested by migrant landings, Siracusa is in charge of one of the ports with the highest numbers of arrivals in recent years: Augusta.[110] Yet, when discussing the criminalisation of migrants with one of the city's justices of the peace (JP), she reported that, despite being operative for 16 years, she had no more than three or four cases in which article 10-bis was pursued, as in all other instances the public prosecution had demanded the dismissal of the cases.[111] To what degree can this be representative of the wider national trend?

Several studies concerned with the implementation of criminalisation of irregular migration report an all but homogenous application of the norm, with several courts having autonomously stopped enforcing it.[112] Most of them, however, fail to explain to what extent this is the case, hindered in their analysis by either the lack of data, its inaccuracy, or both.

To overcome such shortcomings, I rely on previously unreleased figures, obtained primarily from the Italian National Statistical Institute (ISTAT) and the Italian Ministry of Justice (MoJ), to assess how extensive the implementation of the norm has been in the last ten years.[113] Data from ISTAT concern years 2009–2016 (sometimes 2009–2015), whereas those from the MoJ cover years up to 2017 (and a couple of months of 2018).

In the following sections, I first discuss the extent to which trials are initiated or dismissed, then present how likely it is for them to result in condemnations, and finally compare the number of convictions to overall cases. I conclude the discussion on implementation by looking at specific regional dynamics. What emerges is a picture consistent with previous studies, but significantly more detailed, depicting the implementation of the norm on criminalisation as very fragmented, and highlighting the role

---

[110] See MoI (2016, 2017, 2018).

[111] Interviewee 39.

[112] Ambrosini and Guariso (2016), di Martino et al. (2013), and Savio (2016).

[113] ISTAT (2018) and MoJ (2018). Unless specified, data on initiated trials, dismissals, and condemnations relating to art. 10-bis TUI either originate from ISTAT (2018), or represent the author's elaboration upon the dataset, received in May 2018.

of the judiciary (in particular that of public prosecutors and justices of the peace) as crucial.

Initiated Trials

Despite a rapid increase in the number of trials for irregular migration in the early years of the norm, the trend was soon reverted (Fig. 4.2). Indeed, initiated trials saw a significant upward trend right after the introduction of the crime of irregular migration, which led processes to more than double from 4612 in 2009, to 10,109 in 2010.[114] The following years already experienced a diminishing pattern however, gradual at first (down to 9063 in 2011, and 8720 in 2012), and then steeper (reaching 6760 in 2013, and less than 4300 in 2014 and 2015). On the contrary, dismissed cases increased continuously since the very beginning: From representing a third of initiated trials in 2009, they reached three times the latter's number in 2015. Interestingly, this happened even though all years from 2011 onwards experienced higher sea landings than 2010.[115]

Overall, with time, the number of instances in which charges were raised became increasingly divergent with the total number of known

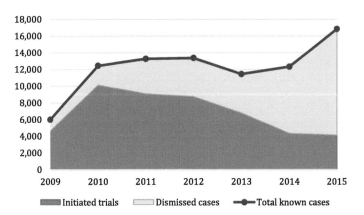

Fig. 4.2 Total known cases in relation to art. 10-bis TUI, as divided by initiated trials and dismissed cases, 2009–2015 (*Source* Author's elaboration on ISTAT [2018])

---

[114] Of note: Processes may include more than one person.
[115] Ismu (2014); Fig. 4.7.

cases: Whereas in 2009, 77% of known cases were taken to trial, this percentage was 59% in 2013, and fell to 24% in 2015.[116]

In this context, some regional differences emerge. On one hand, the provinces of Rome, Milan and Imperia registered the most cases of irregular entry or stay (3052, 2619, and 2441 respectively, in 2009–2015).[117] If the concentration in Imperia is likely to be linked to the presence of high numbers of migrants waiting to cross into France,[118] the peaks in Rome and Milan might be explained by their role as key railway nodes (potentially attracting migrants aiming to travel further North), and high concentration of migrants (both regularly residing, and in reception centres).[119]

On the contrary, Sicily had rather low numbers of cases (no higher than 320 in any of its nine provinces, in the same timespan). Although Sicily's low figures may seem surprising, two factors need to be considered. First, the available data does not allow to appreciate the number of persons accused in each case (as they are clustered). Second, the low figures seem to confirm the hypothesis that most cases either concern irregular stay, or autonomous irregular entry, rather than landings through SAR operations. Indeed, the general disapplication of article 10-bis to people saved by maritime operations has been confirmed by the First Director of the Central Directorate for Immigration and Border Police of the Italian National Police, Tommaso Palumbo.[120] More specifically, an official of the Interservice Group for Contrasting Irregular Migration (CIGIC) at the Siracusa Public Prosecution explained that, if the boats are rescued in international waters, it is considered that the SAR operation could have taken the migrants to other countries too, and the norm is not applied.[121] Should the rescue occur in Italian national waters, instead, the infraction would be seen as voluntary, and criminally pursued.[122]

---

[116] Elaboration based on ISTAT (2018).

[117] ISTAT (2018).

[118] See for example: https://www.ilsecoloxix.it/imperia/2015/09/30/news/ventimiglia-la-polizia-sgombera-il-presidio-dei-migranti-dalla-scogliera-1.32083897?refresh_ce.

[119] See Caritas and Migrantes (2019, p. 2) and MoI (2019).

[120] Interviewee 18.

[121] Interviewee 37. Similarly, anonymous interviewee informed on the facts.

[122] Ibid.

Condemnations

Having seen a decreasing trend in the initiation of trials over time, the extent to which people were condemned for irregular migration, or absolved, is now to be discussed. To do so, I first analyse the overall trends in condemnations, and then focus on the characteristics of the people who were convicted.

General trends

As for initiated trials, condemnations for art. 10-bis saw an initial increase in 2009–2012, which was however short-lived (Fig. 4.3). According to ISTAT, sentences of guilt for irregular migration saw a steep increase from 62 in 2009 to 12,646 in 2012, but then dropped to 8112 in 2014, and to 2952 in 2016. MoJ data confirms the downward trend following 2012, and there is evidence that convictions kept falling in subsequent years too, down to roughly 1500 in 2017, and to less than 200 in the first four months of 2018, in what emerges as a significant reduction in the use of the norm.

The trend parallels the inverted-U shape line seen for initiated trials. Although condemnations were more numerous than trials following

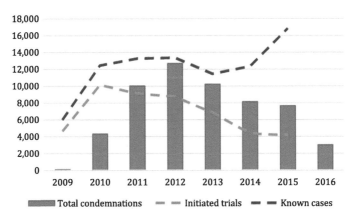

Fig. 4.3 Condemnations of foreigners for article 10-bis TUI, 2009–2016 (*Source* Author's elaboration on ISTAT [2018])

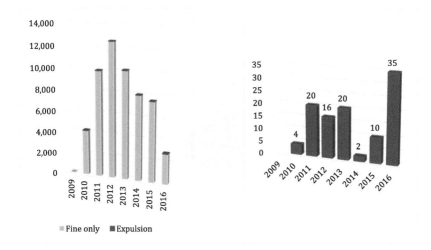

Fig. 4.4 Condemnations of foreigners for article 10-bis by type of sanction, 2009–2016 (*Source* Author's elaboration on ISTAT [2018])

2011, this appears consistent with the fact that while the latter often cluster several migrants together, sentences of guilt are individual.[123]

Concerning the type of sanctions given by courts, only few convictions resulted in expulsion, with most of them leading instead to fines (Fig. 4.4). In relative terms, expulsions averaged 0.3% of total condemnations for article 10-bis between 2010 and 2016, with a low in 2014 of 0.02% and a 'high' in 2016 of 1.2%.[124] In absolute terms too, returns for article 10-bis were extremely limited: Starting from zero in 2009, and only 4 in 2010, they then slightly increased to 20 in 2011 and 2013, only to fall back again to 2 in 2014. Curiously, given the lower number of initiated trials in 2014–2015, ISTAT data show repatriations as slightly increasing following 2014 (up to 35 in 2016). I hypothesise this to be the result of the sanction being used in cases in which persons were charged for other offences too. Indeed, local actors in both Imperia (Liguria) and Agrigento (Sicily) reported that, in recent years, the standard in their

---

[123] Interviewee 46.
[124] Elaboration based on Istat (2018).

offices has been to dismiss cases related to article 10-bis, unless contested in connection with other offences.[125]

Interestingly, while MoJ figures are lower than ISTAT ones regarding the number of monetary sanctions issued (with the former representing less than one fourth of the latter in 2015), they are higher concerning expulsions (with a peak in 2014 of over 150 expulsions registered by the MoJ, against only 2 by ISTAT). Such divergences are likely to be caused by delays in registering cases and backlog issues, as acknowledged by the MoJ database itself.[126] In the absence of clarifying evidence,[127] I continue the analysis based on data provided by ISTAT, in view of the more granular details that the database entails (including sex, age, nationality, and length of trials), but also refer to MoJ data when relevant.

How high were the fines, in practice? On average, they amounted to €4597 in 2009–2018, which is below the minimum €5000 foreseen by the law. They were constantly below such threshold between 2010 and 2014 (included).[128]

As for the length of trials, this was rather low in the first years (only one month in 2009, and four in 2010). The mean interval between the date of infraction and that of the final sentence however increased soon afterwards, doubling in 2011 to eight months, exceeding one year in 2013, and overcoming two years in 2016.[129] The delay is likely to have been caused by backlogs, but also by local actors' decision not to proceed. As an example, one of the deputy public prosecutors in Genoa revealed that files were often intentionally set aside, waiting for them to expire, to avoid costly and useless processes.[130] The spiralling number of cases that expired in 2014 and 2015 throughout the country (1657 and 3349, respectively)[131] suggests that the same may have been happening in other public prosecution offices too.

---

[125] Interviewees 41, 32.

[126] MoJ (2018).

[127] Databases obtained from local authorities do not perfectly overlap with either ISTAT or MoJ data.

[128] MoJ (2018).

[129] ISTAT (2018).

[130] Interviewee 47.

[131] ISTAT (2018).

Finally, examining regional differences, the Court of Appeal districts with the highest numbers of condemnations in 2009–2016 were Milan (7841 condemnations), Bologna (6759), Venezia (6042) and Turin (5497).[132] In Sicily, the Palermo Court of Appeal district issued 3365 condemnations in years 2009–2016.

Who Is Being Condemned?

So far, I have discussed data on the institutional aspects related to criminalisation, showing the role of local actors in determining how extensively the norm is applied. Who are the persons being indicted? Are women or special nationalities over-represented? Are minors ever condemned? In this section, I address these and related questions.

Looking at gender first, men represent the majority of the people condemned for irregular entry or stay: Summing up data for years 2009–2016, they accounted for over 88% of overall convictions. At first sight, this seems consistent with the fact that the greatest part of irregular migrants in the country are male.[133]

A closer look, however, reveals a slightly different picture. Indeed, comparing the percentage of women condemned for article 10-bis, with that of women apprehended as irregularly staying, it emerges that the former was consistently higher than the latter between 2010 and 2016.[134] This was particularly significant in 2011 and 2014, when the proportion of women condemned for unauthorised entry or stay was almost double that of females apprehended as irregularly staying.[135] As will be discussed in the next chapters, a similar trend is present in France,[136] raising questions on the reasons for women's over-condemnation for irregular migration infractions.

Concerning the age of convicted foreigners, most condemnations involve adults. As a judge of the Minors' Court in Palermo reported, while article 10-bis could, in theory, be applied to persons over 14 years

---

[132] ISTAT (2018).

[133] See Eurostat migr_eipre.

[134] The comparison has not been carried out with landings in this case, as the main sources used for landings (Ismu 2014; MoI 2017, 2018) do not provide data disaggregated by gender.

[135] Specifically, the proportion of women condemned for art. 10-bis was 1.7 and 1.9 times that of women found irregularly staying in 2011 and 2014, respectively.

[136] See Chapter 6, section "Implementation Gaps" and Fig. 6.4.

old, according to several magistrates it should be disapplied in the case of minors. This leads most courts to register infractions for art. 10-bis for minors too, but then dismiss them.[137] ISTAT data largely confirms the statement, showing that, although certain cities do register minors for irregular entry or stay, only very few of them are actually condemned (two in 2011 and 2014 each, one in 2016, and none in 2009, 2012, 2013, 2015).[138]

Finally, the most common nationalities of the people condemned for article 10-bis are Tunisian and Moroccan, each group having over 7000 convicted people in the timespan 2009–2016. The two countries are then followed by Nigeria, Afghanistan, China, Pakistan, and Senegal, with over 3000 nationals condemned each. Finally, Albania and Bangladesh had over 2000, and Egypt more than 1900 citizens found guilty. Overall, half of the top 10 countries are African ones (Tunisia, Morocco, Nigeria, Senegal, Egypt), four are Asian (Afghanistan, China, Pakistan, Bangladesh), and one is European (Albania).

Putting Data into Perspective
How do the above numbers fare, compared to data on the total number of irregular migrants? According to Colombo (writing in 2010), the number of initiated trials was rather high compared to the overall number of irregular migrants apprehended on Italian territory in 2009: Considering that in August–December 2009, 13,068 people were indicted for the crime, and that 52,823 were apprehended as irregular in 2009, it is possible, he argues, to average roughly 4000 apprehensions per month, which would lead to 62-to-64 foreigners being indicted every 100 apprehended.[139]

The analysis is interesting, but could be slightly amended. First, as there is evidence of initiated trials involving different number of people, ranging from one,[140] to over a hundred,[141] it seems more productive to use individual condemnations as a benchmark. Second, to account for asylum seekers' requests of international protection, the figure should

---

[137] Interviewee 40.
[138] Istat (2018).
[139] Colombo (2011, pp. 303–304).
[140] JP Rome, Sentence 16/06/2011.
[141] JP Agrigento, Sentence 195/2016.

detract asylum applicants, to whom charges do not apply. Third, the number of irregular landings should also be accounted for since, at least in early years, people rescued through SAR operations were also indicted for the crime, and since not all landings take place through SAR.

In light of the above, the yearly number of people subjectable to article 10-bis is made up of third country nationals (TNCs) who are found irregularly present and/or who landed in Italy, minus the number of asylum applicants. The figure ranged from 45,343 in 2009 to 90,841 in 2016, through a low of 25,262 in 2012 and a peak of 130,775 in 2014 (Fig. 4.5).[142] This compares with yearly condemnations ranging, as seen above, between 60 and 12,700, in an inverted U-shaped curve peaking in 2012 (see section "General Trends").

Interestingly, an inverted U-shaped curve also describes the proportion of condemnations over the total number of foreigners subjectable

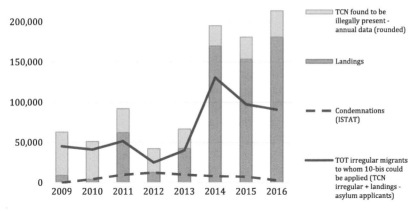

Fig. 4.5 Third country nationals subjectable to article 10-bis, 2009–2016 (*Source* Author's elaboration based on Ismu [2014], MoI [2017, 2018], EUROSTAT [migr_eipre and migr_asylapp], ISTAT [2018])[143]

---

[142] Data on landings are from Ismu (2014), MoI (2017, 2018); data on TCNs irregularly present from EUROSTAT migr_eipre; data on asylum applicants from EUROSTAT migr_asyapp; data on condemnations from ISTAT (2018).

[143] The number of TCNs subjectable to art. 10-bis is calculated as follows: TNCs in irregular situation, either through entry or stay, minus those who requested international protection.

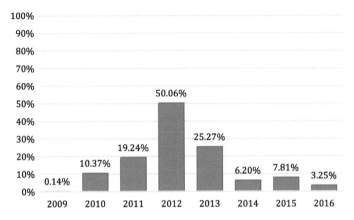

Fig. 4.6 Proportion of condemnations for article 10-bis, over total number of TNCs subjectable to the norm, 2009–2016 (*Source* Author's elaboration based on Ismu [2014], MoI [2017, 2018], EUROSTAT [migr_eipre and migr_asylapp], ISTAT [2018])

to article 10-bis (Fig. 4.6). The trend starts with a very low 0.14% in 2009, which is in stark contrast with Colombo's argument, but also partly understandable given that the law was only adopted in the middle of the year, and recalling the initial concerns about the constitutionality of the norm (see section "Legal Aspects and Judicial Controversies").[144] The proportion of condemned migrants then increased to 10 and 20% in 2010 and 2011 respectively, to reach 50% in 2012. It should be noted that this higher percentage is not only the result of the peak in convictions, but also, and most significantly, of the low in people subjectable to 10-bis: While sentences of guilt increased by almost a third (27%) from 2011 to 2012, migrants to whom the norm could be applied halved (−51%) in the same timeframe. Indeed, with the rise of flows since 2013, the proportion fell back to 25% that year, to then steeply drop to less than 10% in the following three years, to a low of 3% in 2016.

Overall, the above analysis shows an inverted U-shaped curve as characterising both initiated trials, condemnations, and the proportion of convictions over the total number of potentially pursuable people. In particular, 2014 emerges as decisive point in leading towards a significant

---

[144] See Public Prosecutor Turin (2009).

disapplication of the norm. Recalling that 2014 was also the year in which the Parliament delegated the Government to decriminalise migration, it may be possible that the act contributed to decreasing the implementation of the norm. Although decriminalisation was eventually not pursued, in that year the number of dismissed cases overtook that of criminal proceedings, and the proportion of condemned people fell to single-digit figures, pointing to a substantial change in trend.

The sections above also show, however, significant variations in local dynamics, for example with regard to offence registration and condemnations. Who are the key actors responsible for such contrasting patterns? The issue is addressed in the following section.

The Role of Local Actors
Police officers, deputy public prosecutors, and justices of the peace all play a key role in determining the degree of implementation of the norm.

To begin with, the police may decide whether to report a situation of irregularity or let it go. An interview with a police official revealed that it is not uncommon for officers to decide whether to do so based on practical considerations.[145] In particular, the possession by the apprehended person of a passport is a key determinant, since it implies a higher likelihood of expulsion, and therefore makes public officers more willing to opt for the administrative process leading to returns, rather than initiating criminal proceedings.[146]

Public prosecutors also retain significant leeway, being able to decide whether to pursue criminal charges or not. Despite their legal obligation to follow up reported cases of criminal infractions, evidence points to several methods used by prosecutors to limit or avoid such duty, including asking for mass dismissal of the files,[147] encouraging a rational use of resources by JPs and police officers,[148] and letting files expire.[149]

As an example, a deputy honorary prosecutor in Imperia reported that her decision on whether or not to demand the dismissal of cases is usually based on the context: If the person has only been in the country for about

---

[145] Interviewee 14.
[146] Ibid.
[147] Ziniti (2016).
[148] Interviewee 39 and Public Prosecutor Siracusa (2017).
[149] Interviewee 47.

10–20 days, is leaving Italy, or if it is his/her first time being registered, then her favoured pathway is to request the dismissal of the case.[150]

Enforcement may also change through time. In Genoa, while before 2014, initiated trials were higher than dismissed cases, the situation suddenly reversed in 2015, with over 1600 requested dismissals.[151] Indeed, a local prosecutor recalled that, in the early years of the norm, enforcement was strict: Not only did PPs pursue cases, but migrants were held in the police offices until their hearings, and JPs often sanctioned expulsion (though these were not necessarily carried out, due to high costs).[152] Following a meeting between the prosecution and JPs, however, the general orientation started to change, and dismissals to prevail.[153]

As the above highlights, it is not only policy officers and prosecutors that play a key role, but justices of the peace too. Some JPs have started very early on to concede the dismissal of the cases, or to avoid condemning migrants when only the crime of irregular entry or stay is contested.

The case of justices of the peace in Agrigento (under whose jurisdiction falls Lampedusa) is significant here as, like in Genoa's public prosecution, their decisions saw a significant change through time. Indeed, despite significant critiques from the local public prosecutor,[154] and even after the 2014 law, JPs in the city tended to pursue and condemn cases for article 10-bis, often sanctioning €5000 fines,[155] in a trend that significantly diverged from the overall tendency to dismiss cases. As explained by a local justice of the peace, while the public prosecution generally asked for the dismissal or acquittal of cases, dismissal meant a de facto repeal of article 10-bis, which had not been done by the legislator and, in the interviewee's view, should not have been the responsibility of local JPs.[156] It was only around 2016 that, following consultations with other JPs, the orientation of the interviewee eventually verted towards either dismissing

---

[150] Interviewee 46.
[151] ISTAT (2018).
[152] Interviewee 47.
[153] Ibid.
[154] Ziniti (2016).
[155] Interviewee 32; also confirmed by other sources.
[156] Interviewee 32.

new cases, or absolving existing ones on the basis of a tenuity of the offence, except in cases in which other crimes have been committed.[157]

Overall, the above clearly shows how the discretion of individual law-enforcement agents, prosecutors and justices of the peace has been crucial to alter the implementation of the measure. This is to the extent that, despite rulings by both the Italian Constitutional Court and the ECJ supporting the legality of the norm, local actors have found ways to avoid enforcing it, especially following the Parliament's request to repeal the norm. Legal considerations, together with concerns about efficiency, have been the major factors leading to such choices, and are therefore responsible for the resulting gap between the norm on paper, and its application.

From Enforcement to Overall Outcomes: Efficacy Gaps

As seen above, the objective of criminalisation was twofold: Increasing expulsions, and reducing annual irregular entries by 10%, namely from 54,500 in 2007 to an expected 49,050 in later years.[158] Have such goals been achieved? In the next sections, I discuss the effectiveness of the measure, by first addressing its efficacy in meeting its targets, and then analysing its efficiency, coherence, sustainability, and utility.

Efficacy (I): Criminal Law and Expulsions

Focusing on returns first, from the analysis carried out above, it is possible to appreciate that criminalisation did not widely result in expulsion sentences. In the vast majority of the cases in which the accused was found guilty, the sentence foresaw a fine, rather than expulsion. Indeed, the latter were consistently below 20 per year in 2009–2016 (except for the last year, in which they reached 35), and never accounted for more than 1.2% of overall condemnations for irregular entry or stay in 2010–2016 (see Fig. 4.4).[159] Even considering MoJ data, expulsions were still extremely low: Less than 1% of overall convictions in all years between

---

[157] Ibid.
[158] Law Proposal N. 733 (2008).
[159] Based on Istat (2018).

2009 and 2018, except for 2014 and 2015, when they accounted for 7% and 2%, respectively.[160]

The low number of expulsions is to be explained by two factors: The practical difficulties involved in enforcing returns, even when deriving from a criminal sentence, and the evolution of ECJ jurisprudence on the norm.

Looking at practical constraints first, criminal trials are unable to overcome the difficulties associated with returning migrants. To begin with, justices of the peace can only sanction expulsion when there are no obstacles to its immediate implementation, that is to say if the indicted person is identified, possesses travel documents, and there is an available means of transport.[161] This, however, is rarely the case, and made more unlikely by the fact that migrants are often difficult to identify,[162] and that many of the processes are celebrated *in absentia*.[163] Therefore, even if JPs wanted to, they could not sanction expulsion most of the times.[164] Moreover, if the above conditions were met, the administrative pathway (which, as seen, starts in parallel to the criminal one) would carry out the expulsion before the decision of the justice of the peace, thus making the latter redundant.[165] Finally, expulsions of criminal nature do not make returns any easier when agreements with third countries are lacking.

The First Director in the Immigration and Border Control Directorate of the National Police, Tommaso Palumbo, clearly stated that there is no direct relationship between the crime of irregular entry or stay and returns.[166] In his view, the relationship is however present in the cases in which migrants are held in prison for some time (for other crimes), since during that period it is possible for officials to carry out verification and identification procedures, which then increase the likelihood of returns.[167] Yet, this opens the door to related problems, including

---

[160] Based on MoJ (2018).
[161] Interviewee 39 and Legislative Decree 286/1998, art. 14 c. 1.
[162] Interviewees 41, 46, 17, 30, 39.
[163] Interviewees 54, 55, 32, 41, 46.
[164] Ibid.
[165] Savio (2016) and interviewee 30. That this is often the preferred option was confirmed by a senior police official (interviewee 14).
[166] Interviewee 18.
[167] Ibid.

possible further marginalisation and lack of efficient use of resources (as will be discussed in the case of France). Indeed, according to a cultural mediator in Alessandria, who previously worked in prisons, some penal institutions do have workshops to train detainees (e.g., to become carpenters), but knowing that, after the expiration of the prison term, migrants will either go back to irregularity or be returned, such programs are useless, and imprisonment only creates further criminality.[168]

Shifting now to the role of the ECJ, the Court nullified the Italian Government's aim to carry out immediate expulsions through criminalisation, in its 2012 Sagor sentence. In it, the ECJ maintained that the risk of absconding that justifies forced return following criminal processes cannot be broadly assumed, but needs to be demonstrated for each individual case.[169] In this way, the goal of increasing returns was made hard to achieve from a legal perspective too.

Overall, the original intent to have systematic and immediate expulsions failed, due to both practical difficulties in enforcing returns, and supranational judicial restraints.

Efficacy (II): Criminal Law and Landings

Focusing now on the second goal, has criminalisation managed to reduce arrivals? I first discuss the main viewpoints that emerged from political debates and interviewed stakeholders, to then proceed to analyse relevant data.

Whether or not criminalisation had a deterrent effect, has been greatly debated, in the context of the attempted repeal of the norm. On one hand, multiple politicians were sceptical of this. In supporting decriminalisation in a 2015 parliamentary audition, for instance, Minister of Justice Orlando repeatedly stressed that the norm had no deterrent effect, and therefore its repeal would 'bear no consequences on policies to fight irregular immigration'.[170] Even Interior Minister Alfano, while arguing against decriminalisation, recognised that with migration being criminalised, in 2011, irregular arrivals still increased to 62,000, implying that the norm

---

[168] Interviewee 22.
[169] Case C-430/11, Sagor, paras. 40–42.
[170] Senate, 08/07/2015, pp. 7, 17–18. Author's translation.

did not manage to deter migration.[171] In fact, he himself later acknowledged the lack of effectiveness of the measure when, asked whether it had been a mistake to support the norm under the Northern League's pressures, he answered: 'It was an attempt of dissuasion. There are norms that end up being successful, and others that don't'.[172]

On the other hand, several actors feared that the repeal of the norm would incentivise migration. Beppe Grillo and Gianroberto Casaleggio, for instance, criticised decriminalisation, viewing it as 'an invitation to emigrants from Africa and the Middle East to embark for Italy',[173] thus suggesting it would represent a pull factor. Similarly, while discussing the 2014 law in Parliament, Senator Marin from FI-PdL insisted on a psychological and pragmatic deterrent effect of article 10-bis, mentioning that the number of migrants arriving to Italy had shrunk since 2009 (whether this actually occurred, will be discussed in the next sub-sections).[174]

When it comes to the stakeholders interviewed by the author (including political and institutional figures), the majority did not perceive criminalisation as having had a deterrent effect. Respondents holding such views include politicians such as former Prime Minister and Commission President Romano Prodi,[175] former Minister of Social Affairs Livia Turco,[176] Undersecretary of the Interior on Migration Matters Domenico Manzione,[177] and the coordinator of PD's immigration Forum Marco Pacciotti,[178] as well as Sant'Egidio migration responsible Daniela Pompei.[179] Interestingly, of similar views were also institutional actors, such as police officers including first director Tommaso Palumbo (and others),[180] as well as public prosecutors and judges across the country.[181]

---

[171] Corriere del Mezzogiorno (2014).

[172] Bei (2016).

[173] Grillo and Casaleggio (2013).

[174] Chamber of Deputies, 15/01/2014, p. 14.

[175] Interviewee 20.

[176] Interviewee 26.

[177] Interviewee 30.

[178] Interviewee 17.

[179] Interviewee 16.

[180] Interviewees 18 and 14.

[181] Interviewees 34, 38, 41, 46, 32, 29, 31.

Returns were perceived by some of the interviewees as potentially more effective in terms of deterrence.[182] It should however be specified whether forced returns are interpreted as deterring others' departure, or simply as putting an end to a person's stay in the country, as only in the former case they would qualify as having an effective deterrent effect. In the experience of cultural mediator Osman, returns influence the surrounding society, but the extensiveness of the effects depends on the situation of the countries of origin: In his view, in countries like Egypt, where migrants would have no direct risk of death, deterrence may have an effect, but in states like Mali and Nigeria, nothing would change.[183]

Going back to criminalisation, the above shows that, despite diverging statements by centre-right and M5S politicians, most interviewed stakeholders, including both policy-makers and local actors, are sceptical concerning the norm's deterrent effect.

Looking at the issue from a systematic perspective, what can be derived? In the following sub-sections, I delve into the efficacy of criminalisation in deterring irregular migration. Specifically, starting by outlining the overall patterns characterising migratory flows to Italy since 2005, I then discuss three aspects: (1) Migrants' perspectives and motivations for leaving (based on the questionnaires carried out in 2017–2018); (2) the micro-level aspects influencing the degree of success of the norm (based on the criminological criteria discussed in Chapter 3); and (3) the macro-level international political and economic factors affecting the flows.

Irregular Migration in Italy

Irregular migration trends to Italy in the last fifteen years have been fluctuating (Fig. 4.7). Considering sea landings first, after being roughly 20,000 per year in 2005–2007, they saw an increase to 36,951 in 2008, which however soon dropped to 4406 in 2010.[184] This figure was the lowest in the 2005–2017 period, but was soon followed by a renewed intensification of arrivals, with 62,692 landings in 2011. In 2014–2016, figures went up again, amounting to over 150,000 per year, and to

---

[182] Interviewees 15, 31, 41.
[183] Interviewee 22.
[184] Ismu (2014) and MoI (2017, 2018).

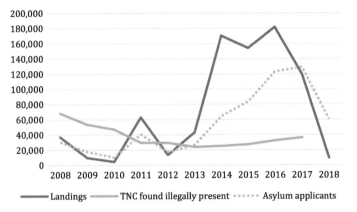

Fig. 4.7 Comparison of landings, TCNs found irregularly staying, and asylum applicants, 2008–2018 (*Source* Author's elaboration on Eurostat [migr_eipre, migr_asyappctza], ISMU [2014] and MoI [2017, 2018])

181,436 in 2016. Following the peak, 2017 and 2018 experienced lower arrivals, down to 23,370 in the latter year.

In parallel to the above, third country nationals found irregularly present decreased from 68,175 in 2008 to 23,945 in 2013.[185] After that year, they gradually increased to 36,230 in 2017, to then shrink to 26,780 in 2018.

Interestingly, despite the high visibility and political salience of sea arrivals, for a long time entering irregularly has not been the main way to be an irregular migrant in the country. Until at least 2006, the vast majority of migrants having an irregular status in fact entered Italy legally, and then overstayed the expiration of their visa.[186] Overstayers represented 60–75% of irregularly present foreigners between 2000 and 2006, when landings accounted for only between 4 and 17%, respectively.[187] Although the trend is likely to have changed following 2015, the above points to the relevance of discursive constructions of migration in shaping the public debate.

---

[185] Eurostat migr_eipre.
[186] MoI (2007, p. 383).
[187] Ibid.

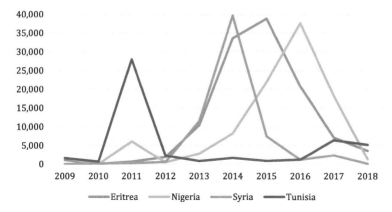

Fig. 4.8 Detection of irregular border crossings on the Central Mediterranean route by nationality, 2009–2018 (top four nationalities in overall period) (*Source* Author's elaboration on Frontex [2019])

Going back to landings, the main nationalities changed over time (Fig. 4.8). If 2011 saw a peak of Tunisians (following the country's uprisings), since 2013, Syrians and Eritreans accounted for the highest number of arrivals (before decreasing in 2015 and 2016, respectively), and 2016 saw an increase in Nigerians.[188] Such fluctuations appear strongly related to developments in countries of origin and of transit, and in particular to the worsening Syrian crisis.[189]

Indeed, together with the increase in landings, the number of asylum applications also saw an upward trend in the last decade, passing from 26,620 in 2015, to 128,855 in 2017.[190] The trend can be attributed to the general increase in the flows, but also to a rise in specific nationalities. Indeed, Nigerians accounted for the highest number of requests of international protection in Italy, in the whole 2013–2017 period (peaking at 27,105 first-time asylum applicants in 2016).[191] Syrian requests for protection also increased in 2016–2017 (although to a more limited

---

[188] Frontex (2019).
[189] See section "Macro-Level: International Political and Economic Determinants of Flows".
[190] Eurostat migr_asyappctza.
[191] Ibid.

extent). Notably, both nationalities had rather high recognition rates,[192] which further underlines the strong relationship between landings and developments in countries of origin.[193] Finally, the increase in the number of asylum requests in 2016–2017 might also be related to a more systematic enforcement of migration provisions, particularly following the introduction of hotspots.[194]

As the above highlights, in the years following the introduction of the crime of irregular migration, flows to Italy were in fact higher, peaking at over 180,000 in 2016, despite the possibility of incurring criminal sanctions. The criminalisation of migration does not thus appear to have had a significant impact on foreigners' willingness to migrate to Italy, even without authorisation. To delve deeper into the efficacy of criminalisation, in the next pages I first look at migrants' motivations for going to Italy, and then investigate the efficacy of the norm from a micro- and macro-level perspective.

Migrants' Viewpoint

So far, I have discussed figures related to migration and its criminalisation. But how do migrants themselves perceive criminalisation? Why did they migrate to Italy? To gain better appreciation of their own viewpoint, I discuss the results of the questionnaire conducted with third country nationals in Italy.[195]

Briefly introducing respondents' generalities, over half (55%) were between 20 and 30 years old, with 33% over 31. The vast majority (88%) were male, with about 12% female respondents (which is roughly in line with UNHCR data, according to which women represent just under 10% of arrivals to Italy).[196] While about a fourth of respondents said to have only attended primary school (24%), the relative majority (42%) also went to secondary school. 22% attended university, and 12% did not have any formal education. Finally, 69% of respondents were not married.

---

[192] See MPI (2018).

[193] For more on this, see section "Macro-Level: International Political and Economic Determinants of Flows."

[194] Parliamentary commission on the reception system (2017, p. 58).

[195] Questionnaires were carried out in Autumn 2017 and Spring 2018, on a total of 99 respondents. All percentages related to the survey refer to the number of valid answers.

[196] UNHCR (2019), based on data from January 2018 to July 2019.

The survey captured a recent snapshot, with two thirds of respondents having entered the country in either 2016 or 2017. In particular, three fourths of them did not have a job contract before arriving in the receiving country, and 35% paid to enter Italy. Based on the answers provided, entry payments varied notably, ranging from €500 for a person from Guinea, to €12,000 for one from Pakistan. Amounts varied greatly even among respondents from the same countries: For example, if Guineans reported sums from €500 to €8000, Pakistanis declared having paid from €1000 to €12,000, and Bangladeshis from €6400 to €12,800. The divergence may be caused by actual differences, but is also likely to reflect answers focused on different segments of the journey (e.g., accounting for the whole journey from Guinea to Italy, or only for that from Lybia to Italy). Indeed, the literature stresses that migrant journeys are often the result of several steps, rather than a coherent and pre-determined plan.[197]

In light of the above, why did migrants choose to go to Italy, rather than another country? The first finding of the survey is that personal contacts, especially among fellow countrymen, appear to have been important in geographically directing migrants' journeys to Italy (although not always necessary). Family ties were given as a justification for choosing Italy over other countries in 19% of cases (Fig. 4.9). 31% of respondents also said they had family or friends in the receiving country, and 57% of those who reported knowing someone (not just family or friends) mentioned that the latter shared their nationality.

The findings support social network theories, linking migration to transnational networks. Yet, even taking into account that Italy may not be the target destination country for all respondents (who may therefore have contacts in other European countries), results also show that other explanatory factors need to be considered. Indeed, 61% of respondents moved to Italy even without having any contacts there. Geographical proximity is likely to be a key factor (which was indeed mentioned by 13% of respondents), together with the facilitation of entry by criminal organisations. The presence of a social security system was also mentioned by some as an important factor (18%).

In this context, immigration rules emerge as having played a role for approximately one in seven respondents (14%). Importantly, however, while legislation on migration might, in some cases, contribute to

---

[197] Talani et al. (forthcoming).

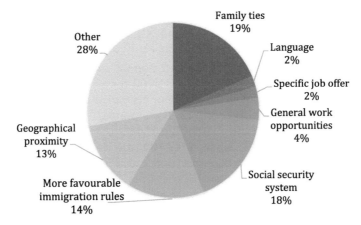

Fig. 4.9 Reasons for choosing Italy rather than other countries (*Source* Author's elaboration)

Fig. 4.10 Reasons for choosing Italy rather than other countries: Definition of 'Other' as added by respondents (*Source* Author's elaboration)

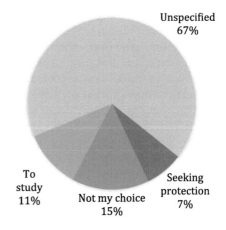

directing some of the migrants to a country more than others, it should be understood to what extent this is their choice, or smugglers'. Indeed, substantial numbers of surveyed persons (28%) indicated 'other reasons' as being prevalent (Fig. 4.10). Of these, 7% expressed the need for protection as the main motivation, whereas 15% said that going to Italy (rather than another country) was 'not their choice'. The latter answer is important, and may be understood in view of the stories recounting how migrants were in some cases forced to embark on boats crossing the Mediterranean Sea,[198] but also as pointing to the relevance of smugglers in determining the specific migratory routes to be pursued. Perhaps, the response might also be interpreted with reference to the fact that some respondents reported having been saved, suggesting that search and rescue operations were responsible for taking them to Italy (rather than, for example, Malta).

Finally, 6% of interviewees reported having chosen Italy for (either general or specific) work opportunities. The reported lower relevance of such motivations, compared to the role of networks for instance, may be the result of two factors. First, several of the respondents were in the process of seeking asylum, which may have made them wary of showing economic motivations. Second, the figure may indicate respondents' willingness not to remain in Italy, but travel further north.

Overall, from the questionnaire, family ties appear to have played an important role in shaping respondents' journeys. Immigration laws, on the other hand, were considered relevant in 14% of cases, although the role of smugglers should be critically evaluated. To these topics I will return shortly, to examine how information was obtained.[199] More broadly, to shed light on the specific role of criminalisation in affecting migration to Italy, in the next sections I discuss the efficacy of the norm, from both a micro- and a macro-level perspective.

Micro-Level: Criminological Criteria
Perceived Legal Costs

As seen in Chapter 3, the key criterion for deterrence to work, in a criminological context, is not so much the severity of the sanction, but rather its certainty. This may be an important factor to explain the lack of deterrence in the case of Italy: As seen in Fig. 4.6, certainty of being apprehended and sanctioned is limited, with an average of 15.29% of non-compliant foreigners being condemned each year in 2009–2016. The highest proportion of condemnations was reached in 2012, but was never repeated since then, and fell to single-digit figures since 2014. Moreover, according to a report by EMN Italy, internal controls in 2010 concerned only about 1 in 10 irregular migrants actually present in the

---

[198] Interviewee 24.
[199] See section "Information and Communication Issues".

country[200]: While foreigners apprehended with unauthorised status were about 47,000,[201] their likely overall number was about half a million.[202] If this was the case, the proportion of migrants condemned for article 10-bis in that year, instead of being about 10% of the total non-compliants, would be closer to 0.9%. Finally, given migrants' status, collecting the sanctioned fines is also complicated (see section "Efficiency: The Cost of Criminalisation"). Indeed, several interviewees, including the head public prosecutor of Siracusa, mentioned limited prosecution as being partly responsible for the lack of deterrence.[203]

Not only is certainty low, but migrants are often aware of this. According to cultural mediator Osman, they frequently regard France as implementing migration measures more systematically than Italy, and perceive being irregular in Italy as not dangerous.[204] Likewise, a judge in the Court of Rome recalled obtaining recordings of Romanian and Jordanian citizens who were aware of the fact that, in Italy, criminal sanctions are not always applied.[205] Beyond singular episodes, a survey carried out in 2011 in Egypt by the IOM found that Italy is indeed perceived as a rather 'easy' country to migrate to irregularly: In a context in which 70% of respondents thought *regular* migration was not simple, and the same proportion did not possess a valid passport, Italy was said to be the easiest country to migrate to, in an *irregular* way.[206]

Finally, in the view of some of the interviewees, the lack of severity of the norm also played a role in not obtaining higher deterrence.[207] According to the judge of the Court of Rome, however, as sanctions in migrants' countries of origin are far worse than a €5–10,000 fine, not even prison would play a significant role in deterring them from coming.[208] Considering that several of those countries (such as

---

[200] EMN (2012, p. 84).

[201] Eurostat mig_eipre.

[202] EMN (2012, p. 79).

[203] Interviewees 38 and 41.

[204] Interviewee 22.

[205] Interviewee 29.

[206] IOM (2011, pp. 17–20).

[207] Interviewees 18, 29, 38.

[208] Interviewee 29; for a similar argument see interviewee 17.

Nigeria, Somalia, and Egypt) foresee capital punishment,[209] anything less is unlikely to be considered a significant deterrent (and even then, it may be disputable whether death sentences have a deterrent effect).[210] Of course, what this implies is, in fact, that even strengthening the severity of the norm would not achieve its goal, given the liberal democratic nature of European states, which suggests that criminalisation is in fact unable to deter migration.

Information and Communication Issues
As already highlighted, the key factors for deterrence to work are not so much the actual legal costs involved in sanctions, but rather how these are perceived by migrants themselves. In this way, the role of information emerges as crucial, from two perspectives: The knowledge of foreseen sanctions, and the latter's ability to persuade potential migrants not to move irregularly. As discussed in the next paragraphs, both are problematic with regard to the criminalisation of migration in Italy.

To begin with, knowledge of criminalisation was found to be extremely limited, from the questionnaires conducted by the author. Indeed, the survey revealed that while 41% of respondents stated they knew the consequences of entering Italy irregularly, 74% were unaware of the documents required to regularly go to Italy, and 66% were unable to identify any of the specific sanctions foreseen by the country, among a provided list (Fig. 4.11). Of the persons who did indicate a particular sanction, the vast majority were aware of the possibility of being returned, and fewer of being detained. Significantly, only 2% of respondents were aware of the possibility of incurring a fine for entering or staying irregularly in the country.

Remarkably, most people did not know about the above-mentioned measures before leaving (Fig. 4.12). Most respondents (57%) reported learning about them only once in the receiving country, whereas about a third (35%) had heard about them before migrating, and 8% obtained information in transit countries. Although the survey is not statistically representative, the results are telling as, quite evidently, ignorance of criminalisation can neither have a substantial effect on one's decision to migrate, nor help achieve the goal of reducing irregular flows.

---

[209] See: https://www.bbc.co.uk/news/world-45835584.
[210] See for example Andenaes (1968, p. 82).

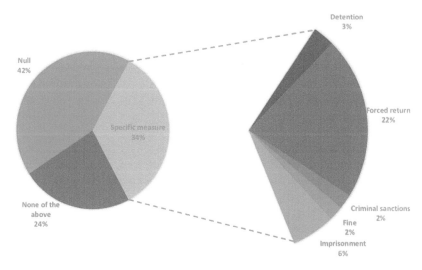

Fig. 4.11 Knowledge of the sanctions for unauthorised migration in Italy (*Source* Author's elaboration)

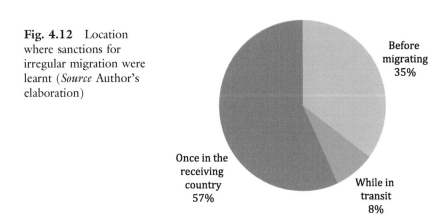

Fig. 4.12 Location where sanctions for irregular migration were learnt (*Source* Author's elaboration)

The lack of information among migrants is also stressed by interviewees. Persons involved with migrants noted for instance their limited understanding of documents provided by Italian authorities,[211] exemplifying how information is all but straightforward and well-understood for migrants themselves. Beyond these specific cases, the above supports existing literature arguing that migrants' degree of information is significantly overestimated.[212]

Of a different viewpoint was the cultural mediator in Alessandria, who reported that the migrants coming from Libya with whom he interacted were generally well-informed, including on the crime of irregular entry or stay.[213] In his view, however, they were equally well-informed regarding the possible getaways, such as declaring to be under 18, and knowing where to go to find employment even without a permit to work.[214]

Importantly, a significant part of the information possessed by migrants is obtained through personal networks. Looking at the answers provided by surveyed migrants, most information was said to have been passed on through personal networks, either in the form of family or friends (39%), or other people (19%) (Fig. 4.13). Indeed, according to a journalist active in Ventimiglia, information is often gathered through WhatsApp or Facebook groups, but thus also generally outdated.[215] On the contrary, public authorities in countries of origin and destination, and embassies, represented together the source of information for 13% of respondents, just above the internet, which was the main information means for 12% of respondents. NGOs were reported as being relevant to that end for only 7% of respondents, and newspapers or magazines for a meagre 2%.

In this context, it is interesting to report the findings of a survey conducted by the IOM in March–April 2011 as, while it is significantly antecedent to the one discussed above, it was carried out in Egypt, and can therefore be complementary, by presenting the views of persons who have not yet migrated. In particular, out of the 750 Egyptians between 15 and 29 years old who were interviewed, 75% of respondents reported having learnt information regarding migration through family or friends, and 86% declared knowing someone who had left Egypt.[216] As migration

---

[211] Interviewees 12, 21.
[212] See, for example, Thielemann (2011, p. 7) and Richardson (2010).
[213] Interviewee 22.
[214] Ibid.
[215] Interviewee 13.
[216] IOM (2011, pp. 14, 16).

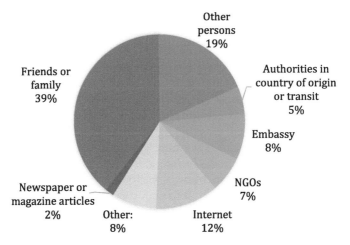

Fig. 4.13 How migrants learnt about sanctions for irregular migration (*Source* Author's elaboration)

was at that point still potential, rather than actual, it is possible to interpret the over-representation of personal contacts, when compared to the author's survey, in light of the specificities of the situation: Interviewed people may have heard about migration, rather than actively sought information about it and diversified sources. Moreover, although the survey did not foresee a category of 'other persons',[217] which could include smugglers, it is possible to think that the 'friends/relatives' category also reflects knowledge acquired through them. Either way, the IOM survey confirms the finding that personal networks are crucial in supporting migration projects.

Overall, the above shows a substantial lack of information among migrants. Moreover, both the author's and IOM's surveys confirm previous literature findings that personal contacts play a key role as providers of information, and recall the importance of not assuming that the internet is accessible to, and used by, all.

In order to contrast misinformation among migrants, Italy has been involved in several information campaigns. A particularly significant one

---

[217] The categories presented by the survey are: Friends/relatives, internet, governments, general readings, media, embassies, other. IOM (2011, p. 28).

is Aware Migrants, a 2016 campaign targeting North and Western African countries.[218] The website of the campaign includes a section explaining how to migrate regularly to selected European countries, sometimes outlining the potential sanctions foreseen. In particular, in the Italian subsection, expulsion and possible detention are mentioned, but no reference is made to the criminal sanctions possibly applicable.[219] The fact that the website does not mention the fine seems indicative of the country's actual little reliance on criminalisation as a tool of deterrence, in recent years.

In any case, the effects of such campaigns should not be taken for granted. As an example, according to an official of the European Commission, the greatest effect among potential migrants was that of the CNN video showing young men being sold at an auction,[220] rather than that of information campaigns.[221] Likewise, as recalled by a person involved with migrants in Genoa, although she provided them with links to online platforms containing information regarding migratory routes, suggesting that they share them with families and friends in countries of origin, several of the migrants did not even look at them.[222] Instead, the message sent home was one of freedom and lack of arbitrary rule.[223] In the interviewee's experience, even though, with time, migrants in Europe realised the hardships and asked their relatives to only come once they had secured a permit and a job (so that they could help them enter legally), this message was not believed by people in their home countries.[224] The neglect of challenging conditions might also be a result of such situations being perceived as temporary,[225] perhaps even necessary, to achieve the final goal of remaining in Europe.

---

[218] IOM and MoI (date unknown).

[219] See: https://awaremigrants.org/regular-channels-enter-italy, updated August 4, 2016, accessed 31 May 2019.

[220] See: https://edition.cnn.com/videos/world/2017/11/13/libya-migrant-slave-auction-lon-orig-md-ejk.cnn.

[221] Interviewee 15. Similarly on the great effect of the video: Interviewee 22.

[222] Interviewee 12.

[223] Ibid.

[224] Ibid.

[225] Talani (2018).

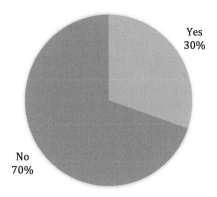

Fig. 4.14 Did knowing about the consequences of irregular migration affect your decision on how to migrate? (*Source* Author's elaboration)

Even more tellingly, the survey carried out by the author shows that even those migrants who knew about the measures sanctioning unauthorised entry or stay, were not particularly affected by them. In 70% of cases, knowledge of the consequences of irregular migration did not affect potential migrants' decision on how to leave their country (Fig. 4.14). The finding is particularly relevant as, using migrants' voices, it shows that even when sanctions are introduced, they do not necessarily affect potential migrants' choices, leading us to question the suitability of focusing on a cost-and-benefit approach to understand complex phenomena such as migration.

In conclusion, from the above section, information emerges as substantially problematic. It is not only often scarce and inaccurate, but also hard to trust for migrants themselves.

Marginalisation and Stigmatisation

Following the analysis of legal costs and information, the criminological aspect left to be discussed is the role of social costs. As discussed in previous chapters, the potential stigma associated to irregular migration, either in countries of origin or in countries of destination, could have a 'social-control' effect.

Interestingly, cultural mediator Osman reported that in many of the migrants' views, being in prison for migration-related offences was not particularly worrisome, but rather equalled being in detention camps.[226]

---

[226] Interviewee 22.

Indeed, some of them (especially Moroccans and Egyptians) asked him to tell their families they were in detention camps rather than in prison, as the latter knew it is normal for migrants reaching Europe to be detained or to live in the shadows for some time.[227] The message sent home was therefore a positive one—partly not to scare their relatives, partly because it was not seen as extra-ordinary.[228] Similarly, in the experience of a person involved with migrants in Genoa, the latter do not have the perception of committing a crime.[229] The above suggests that little stigma may be associated to detention and prison by actual and potential migrants in countries of origin, where such measures may instead be increasingly normalised.

Concerning the stigma that sanctions against migration may create in countries of destination, on the other hand, according to the Eurobarometer, the proportion of people considering migration as one of the two most important issues facing Italy has consistently increased from 2012 onwards. After an overall fall from 2005 to 2012, it spurred from 3% in 2012, to 36% in 2017.[230] While no data is available with regard to criminalisation specifically, it is likely to have contributed to the securitisation of migration, by giving irregular migration more visibility, and strengthening the link with criminal actions.

Furthermore, while contributing to more negative views of migrants in destination countries, criminalisation may also lead to other unwanted consequences, including further marginalisation and grey legal practices.[231] The Italian General Confederation of Labour (*Confederazione Generale Italiana del Lavoro*, CGIL) already argued in 2010 that article 10-bis would 'further marginalise and stigmatise migrant workers in irregular situation and increase their vulnerability to exploitation and to violations of their fundamental human rights'.[232] Moreover, across countries, criminalisation has been suspected of raising issues of racial profiling in police ID checks, and criticised for making migrants' trust in the

---

[227] Ibid.

[228] Ibid.

[229] Interviewee 12.

[230] Eurobarometer (2019).

[231] For more on this, see Chapter 6.

[232] See http://www.cgil.it/diritti-dei-migranti-lilo-invita-il-governo-a-valutare-la-revoca-del-reato-di-clandestinita/.

police and other institutional authorities low, with the result that they feel discouraged from reporting injustices or from demanding protection, in fear of being prosecuted themselves.[233]

Overall, from the available evidence, sanctions against migration, including criminalisation, do not seem to have stigmatised the act in countries of origin. On the contrary, introducing the crime of irregular entry and stay is likely to have contributed to the stigmatisation of migrants in countries of destination, with the results of not only possibly generating further irregularity, but also risking making migrants unable to seek help when needed.

Thus far, a micro-level analysis has shown that criminalisation in Italy has been affected by several issues detracting from its potential deterrent effect. Such aspects included the limited certainty of sanctions, but also the inability of information to reach and/or persuade potential migrants. Furthermore, in countries of origin, interviews suggest that migration does not seem stigmatised, even when irregular, while detention may be increasingly normalised. Still, the criminalisation of migrants in the receiving country has the serious risk of perpetuating their marginalisation, by keeping them at the margins of society and not allowing their regularisation.

Micro-level aspects are therefore important to understand the way in which criminalisation fails to significantly shape decision-making. However, these are not enough to account for the overall lack of efficacy of the norm, whose impact should be discussed in relation to external factors too. In the next section, I therefore examine the overall impact of the norm on international migration to Italy, from a macro-level perspective.

Macro-Level: International Political and Economic Determinants of Flows

Can criminalisation be considered to have had a deterrent effect on migratory flows? This question has been the driving line of the previous sections and will now be addressed from the perspective of the international political and economic context. Indeed, several interviewees mentioned the norm's scarce relevance when compared to external factors.[234] In the

---

[233] Delvino (2017, p. 41) and Delvino and Spencer (2014, p. 8).
[234] E.g., Interviewee 17.

next paragraphs, I discuss the impact of economic and political factors on migratory flows to Italy, arguing that these, rather than criminalisation, are to be held accountable for variations in yearly trends.

As seen in previous sections, sea landings to Italy dropped between 2008 and 2010,[235] and apprehensions of irregular migrants similarly fell from 2008 to 2013.[236] While the above occurred in parallel to the criminalisation of migration, the decrease in both landings and apprehensions is to be contextualised, as external factors emerge as having played a leading role.

Concerning landings, in 2009, the Treaty on Friendship, Partnership and Cooperation signed by Berlusconi and Gaddafi was ratified,[237] and likely to have had a stronger impact on unauthorised flows, by physically *preventing* them. The agreement included a provision to enable more patrolling of Libyan coasts, which Italy would co-finance,[238] as well as the transfer of three ships from Italy to Libya, and the undertaking of maritime patrols with joint crews.[239] Following the deal, Libyan coastal controls became significantly more restrictive, and 'push-backs' started being carried out: In the five days between 6 and 10 May 2009, over 471 migrants were brought back to Libya.[240] Overall, in 2009, nine push-back operations were conducted (which eventually resulted in Italy being condemned by the European Court of Human Rights).[241] By intercepting departures, Libyan authorities physically took migrants back to Libyan shores and prevented them from reaching Italian waters, with a significant containment effect. Indeed, several sources indicate the agreement as the predominant reason for the diminished sea arrivals at the

---

[235] Ismu (2014).

[236] Data on TNCs irregularly staying is from Eurostat migr_eipre.

[237] The full text is available at: http://www.repubblica.it/2008/05/sezioni/esteri/libia-italia/testo-accordo/testo-accordo.html.

[238] Chamber of Deputies, 14/04/2011.

[239] ECHR (2012), Case of Hirsi Jamaa, para. 19.

[240] Ibid., para. 13.

[241] Italy was sanctioned for exposing migrants to the risk of suffering inhuman and degrading treatment in Libya and to the danger of being returned to their respective countries of origin. The country was sentenced to pay €15,000 to each applicant. See: ECHR (2012), Case of Hirsi Jamaa.

turn of the decade.[242] In this way, criminalisation is unlikely to have been the cause of the 2009–2010 decrease in arrivals, which were, instead, contained already in Libya.

In similar fashion, the decline in apprehensions of migrants in situations of irregular stay between 2008 and 2013 may be attributed to factors other than the introduction of the crime of irregular migration, namely regularisations and economic difficulties.

Starting with regularisations, while criminalising migration in 2009, the Government also allowed for an amnesty that gave regular status to over 294,000 foreigners.[243] This, according to EMN Italy, was partly responsible for the drop in apprehensions that took place that year, since it reduced the number of irregulars on the territory.[244] A similar argument can be made with reference to 2013, building on Triandafyllidou's hypothesis that part of the contraction in irregular migration in the early 2000s was related to the de facto regularisation of Eastern Europeans following EU enlargement.[245] Indeed, between 2002 and 2006, Romanians were consistently the first nationality in terms of foreigners found irregularly present, with up to 29,825 cases in 2006.[246] Restrictions on the movement of Romanians to Italy were withdrawn in 2012,[247] which may thus explain the decrease in irregular migrants found on the territory in 2013.

On top of regularisations, the fall in apprehensions between 2008 and 2013 may also be related to the effects of the financial and economic crisis.[248] Indeed, the number of people entering Italy on working visa decreased following the economic downturn, from 360,000 in 2010 to 85,000 in 2013, showing the substantial contraction of the Italian labour market.[249] In the same period, a new tendency emerged: That of both

---

[242] See for example, Chamber of Deputies, 14/04/2011, EMN (2011, p. 2), and Colombo (2011, p. 278).

[243] Pasquinelli (2009).

[244] EMN (2012, p. 79).

[245] Triandafyllidou (2010, p. 12).

[246] Colombo (2011, p. 292).

[247] See https://www.euractiv.com/section/social-europe-jobs/news/italy-opens-up-to-romanian-bulgarian-workers/.

[248] See for example EMN (2012, p. 79). See also Chapters 5 and 6.

[249] Talani (2018, p. 237).

Italians and foreigners resident in Italy to leave the country.[250] It must be stressed that, in spite of the above, figures suggest that the number of third country nationals employed in the labour market in Italy actually increased in the years of the financial crisis, from 1,498,000 in 2007, to 2,074,000 in 2010.[251] Yet, three caveats are key. First, the increase in employment numbers was paralleled by a worsening of employment conditions for foreign workers: Unemployment rate for male foreigners almost doubled in those years, from 5.3% in 2007 to 10.4% in 2010.[252] This deterioration was substantially greater for foreigners than for citizens (whose unemployment rate passed from 5 to 7.4%), and than for foreign female workers (whose unemployment rate, although higher to start with, only increased by less than one percentage point, from 12.7 to 13.3%).[253] Second, while young foreigners have a higher activity rate than citizens, this partly reflects a higher likelihood of early school drop-out: In 2014 the rate of early school leavers was 34.4% for third country nationals, 13.5% for Italians.[254] This, in turn, grants them easier access to the job market (since they are strongly focused on vocational education), but also often translates into greater difficulties in obtaining stable long-term employment.[255] Third, according to Cingano and others, statistical issues are partly responsible for the shown increase in foreign workers: Due to the misalignment of dates of arrival, employment, and official registration, data could refer to an increase in foreigners that actually took place *before* the crisis.[256] Moreover, the authors argue, as unemployed foreigners are more likely to emigrate from Italy without deregistering, the overestimation further increases.[257] Thus, despite the apparent increase in employed foreigners, it is possible to argue that the crisis is likely to have decreased the overall attractiveness of the country for new migrants, and contributed to the fall in irregular migration.[258]

[250] Ibid.
[251] Bonifazi and Marini (2014, p. 497) and Talani (2018, p. 246).
[252] Bonifazi and Marini (2014, p. 499).
[253] Ibid.
[254] Zanfrini (2015, p. 20).
[255] Ibid.
[256] Cingano et al. (2010), as cited in Bonifazi and Marini (2014, p. 496).
[257] Ibid.
[258] See, for example, Bonifazi and Marini (2014), Talani (2018).

So far, I have argued that the drop in landings between 2009 and 2010 was not due to criminalisation, but to the agreement signed by Italy and Libya, which enabled push-backs of migrants. Likewise, the drop in apprehensions of irregularly staying migrants between 2008 and 2013 was due to regularisations and the effects of the economic crisis, more than to the threat of a monetary sanction.

Evidence of the little effect of criminalisation also emerges when looking at following years, taking account of international developments. Indeed, the decreasing trend was reversed by a peak in landings in 2011 (with over 60,000 arrivals) which, except for a low point in 2012, was destined to be on the rise in subsequent years, peaking at 181,436 in 2016.[259]

Once the agreements with Libya and Tunisia collapsed in early 2011 as a consequence of the uprisings in the north-African countries,[260] both locals and migrants who had established themselves in the two states (especially in Libya) started searching for safer places, and many took the sea route.[261] In particular, in early 2011, the rise of the flow from Tunisia intensified to 27,982 landings (as was shown in Fig. 4.8).[262] The rise in migration from Tunisia and Libya in 2011, however, should not necessarily be generalised to other North African countries: Indeed, an IOM survey finds that 44% of the Egyptians interviewed did not consider the 2011 revolution as a determinant factor in their decision to flee, which had been formed before such events and was predominantly related to economic motivations.[263] On the contrary, only 15% stated that it made them want to leave.[264]

International events proved key determinants in following years too, with the Syrian and Libyan civil wars, as well as dire conditions in several African and Middle Eastern countries, incentivising many to flee to Europe. With regards to the Syrian conflict, it is indeed possible to see

---

[259] Ismu (2014) and MoI (2017, 2018).

[260] See http://www.corriere.it/politica/11_febbraio_26/libia-berlusconi-larussa-trattato-gheddafi_7acd0620-419b-11e0-b406-2da238c0fa39.shtml and Caponio and Cappiali (2018, p. 118).

[261] Fargues and Fandrich (2012, p. 3).

[262] Parliamentary commission on the reception system (2017, pp. 54–55).

[263] IOM (2011, pp. 14–150). 75% of respondents said they intended to migrate in order to improve their economic status.

[264] Ibid. The remaining 41% said it only affected their decisions in a minor way.

a parallelism between its evolution and the fluctuations of refugee flows to Italy. Indeed, the number of Syrians who arrived in Europe through the Central Mediterranean route peaked at 39,651 in 2014.[265] The high point echoed the spiralling of casualties in their home country in 2013, and coincided with the year characterised by the most numerous deaths since the beginning of the conflict (estimated at over 70,000).[266] The decrease in the number of Syrians arriving to Italy in 2015 was initially counterbalanced by their peak in Greece, where they totalled 489,011, in that year alone.[267] Since 2016, however, Syrian arrivals to Europe (both through Italy and Greece) fell, shadowing the gradual de-escalation of the conflict (as exemplified by the still high but decreasing number of estimated yearly casualties, from over 70,000 in 2014 to roughly 20,000 in 2018),[268] as well as following the EU-Turkey statement of March 2016.[269]

While Syrian refugees were redirected to the Greek route in 2015, Eritrean flows, which had spiralled to 38,791 the previous year, kept increasing (see Fig. 4.8). Not only is Eritrea ruled by an authoritarian regime that is listed by Freedom House as the third least free country in the world (after Syria and Tibet, with a score of 2/100),[270] but it also foresees mandatory conscription for its nationals, often indefinite and worsened by physical abuses and arbitrary arrests.[271] Additionally, according to Frontex, many Eritreans had lived in Libya for some time, but then chosen to flee the country due to its insecurity and violence.[272]

As the above suggests, the Libyan civil war also played a key role in spurring migratory flows towards Italy. Indeed, before 2011, the country

---

[265] Frontex (2019, 2019).

[266] For an estimate of yearly casualties in the Syrian civil war, see EWN (2019) and SOHR (2020).

[267] Based on Frontex (2019).

[268] See EWN (2019) and SOHR (2020).

[269] Available at: https://www.consilium.europa.eu/en/press/press-releases/2016/03/18/eu-turkey-statement/.

[270] See https://freedomhouse.org/countries/freedom-world/scores?sort=asc&order=Total%20Score%20and%20Status.

[271] See https://www.hrw.org/world-report/2017/country-chapters/eritrea, https://www.refworld.org/pdfid/5c667ab84.pdf. Additionally, the country has had no national elections since 1991, and no legislature since 2002 (ibid.).

[272] Frontex ARA (2015, p. 18).

used to be the largest receiver of migrants in North Africa.[273] With the eruption of the Arab Spring, international intervention, and civil conflict, however, foreigners who were already in the country found themselves in a situation in which they could neither easily leave Libya (as most of the surrounding states implemented tough border controls), nor remain there, given the state of insecurity the country was in.[274] Foreigners were also often subject to frequent abuses by local militias and detention in camps,[275] which led many to choose to embark on the journey to Europe, even when this was not part of their initial plan, as in the case of many Eritreans. At the same time, those who would have chosen to move to Libya found themselves obliged to search for other destinations, as the unstable economy had little to offer.[276] In this context, Libya emerged as a key waypoint for those aiming to migrate towards Italy and Europe, facilitated by increasingly sophisticated smuggling networks.[277] Indeed, in recent years, the majority of migrants reaching Italy by sea left from Libyan territories (89% in 2015, 82% in 2016).[278]

When looking back at the nationalities making up the greatest numbers of arrivals to Italy following 2013 (Fig. 4.8), a final one stands out (in addition to Tunisian, Syrian and Eritrean): Nigerian. In this case too, the increase in arrivals and asylum applications in Italy (since 2012) appears strongly related to developments in the origin country itself. Specifically, the post-2012 rise parallels the beginning of the conflict with the Islamist militant group Boko Haram in North-Eastern Nigeria, and its escalation in 2014–2015. To give a sense of the evolution of the fighting, yearly casualties involving the group amounted to roughly 2000 in 2012 and over 10,000 in 2014–2015, and the UNHCR estimates that 2.4 million people were displaced by the conflict.[279] Lack of economic opportunities, as well as violence and insecurity also played a key role, with Nigeria being among the four countries most affected

---

[273] Fargues and Fandrich (2012, p. 3).

[274] Amnesty International (2015, pp. 7–8).

[275] Fargues and Fandrich (2012, p. 12).

[276] Ibid., p. 3.

[277] UNHCR (2017).

[278] See https://www.unhcr.it/risorse/carta-di-roma/fact-checking/gli-sbarchi-italia-nel-2016-dati-smentire-lallarmismo.

[279] Campbell and Harwood (2018).

by terrorism throughout 2012–2019.[280] Furthermore, the trafficking of Nigerian women and girls to Italy for sexual exploitation saw a dramatic rise since 2016, as reported by the IOM.[281] Overall, in 2012–2016, Nigerians had a rather high average asylum recognition rate in the country, averaging 44%.[282] Together, this suggests that conditions in the African state played a significant role in shaping its migratory flows.

As the above demonstrates, instability, conflicts and dire situations in countries of origin and transit played a key role in affecting flows to Italy (and, more broadly, Europe), regardless of criminalising trends.

In this context, irregular landings to Italy only dropped in the second term of 2017 and in 2018, in parallel to de-escalating tensions (such as in Syria), but also to the controversial Memorandum of Understanding[283] that Italy signed with Fayez Mustafa Serraj, President of the Libyan National Reconciliation Government.[284] The agreement explicitly aimed at reducing irregular migration, through the greater involvement of the Libyan Coast Guard,[285] and has been referred to the European Court of Human Rights for supporting human rights violations taking place in the North African country.[286]

Overall, the above shows how broader macro elements, both political and economic, provided for strong incentives to migrate, regardless of the potentiality of being subject to criminal sanctions. While criminalisation

---

[280] The Global Terrorism Index considers terrorism attacks, victims, injuries and damages. See: https://www.visionofhumanity.org/maps/global-terrorism-index/#/. On the lack of economic opportunities, see IOM (2018).

[281] IOM Italy (2017, p. 9).

[282] Based on MPI (2018). For more on Nigerian migration to Italy and Europe, see: della Valle (2018), IOM (2018).

[283] 'Memorandum d'intesa sulla cooperazione nel campo dello sviluppo, del contrasto all'immigrazione illegale, al traffico di esseri umani, al contrabbando e sul rafforzamento della sicurezza delle frontiere tra lo stato della Libia e la Repubblica Italiana'.

[284] Parliamentary commission on the reception system (2017, p. 57) and interviewee 20.

[285] EMN (2019, p. 58).

[286] See https://www.theguardian.com/world/2018/may/08/italy-deal-with-libya-pull-back-migrants-faces-legal-challenge-human-rights-violations, https://www.hrw.org/news/2020/02/12/italy-halt-abusive-migration-cooperation-libya, https://www.hrw.org/news/2019/11/13/italy-shares-responsibility-libya-abuses-against-migrants.

attempts to increase the legal and social costs of migration, the willingness to move is often determined by factors whose cost is perceived as significantly higher, than the possibility of incurring a fine.

To sum up, from the analysis so far conducted, it emerged that the initial objectives of (I) ensuring systematic immediate expulsions, and (II) reducing arrivals, were not met, in what resulted as a significant efficacy gap (Table 4.1). On one hand, the persons sanctioned with expulsion were a fraction of the overall accused, due to both practical difficulties and legal constraints. On the other hand, irregular migration to Italy continued undeterred, and in sustained numbers, in the last decade, despite the possibility of incurring a criminal sanction. This clearly emerges from the rising number of landings, peaking in 2016, and from survey results showing that, for 70% of respondents, knowing the consequences of irregular migration did not affect their choices on how to migrate.

The absent deterrent effect can be understood at the micro-level as a consequence of limited certainty of apprehension, scarce information among migrants, and possible normalisation of detention among potential and actual migrants. As I have argued, analysing the macro-level context too is key to understand fluctuations inflows as related to political and economic developments in countries of origin. As stated by a

Table 4.1 Factors contributing to the lack of effectiveness of the criminalisation of migration in Italy

Micro-level	Low certainty of sanction enforcement Medium–low degree of severity Difficulty of information to reach and/or persuade migrants Possible normalisation of detention and irregular migration in countries of origin → Lack of significance of deterrence, from a criminological perspective
Macro-level	Decrease in 2009–2013 mainly due to: - Italy-Libya partnership, - Regularisations in Italy, - Economic crisis in Italy and Europe Increase in 2013–2017 mainly due to: - Civil conflict/instability in countries of origin and transit, first of which Libya and Syria Decrease following 2017 mainly due to: - Containment in Libya through the Italian-Libyan agreement, - Gradual slow-down of conflicts, first of which in Syria → Great significance of international political and economic factors

prosecutor in Imperia, the criminalisation of migration foresees a criminal sanction, for a social phenomenon that is unstoppable[287]—or, one could say, structural.

In the previous pages, I have examined the efficacy of criminalisation in Italy, namely its ability to meet its goals. Yet, as argued in the introductory chapter, to assess the overall effectiveness of the measure, further criteria need to be examined—namely the norm's efficiency, coherence, sustainability, and utility. This is the subject of the remainder of the chapter.

Efficiency: The Cost of Criminalisation

According to a report by EMN Italy, the 2010 budget for 'Immigration, reception and rights grant' amounted to €1580 million, of which about 3% was dedicated to the management of flows.[288] Of this amount, how much is relatable to criminalisation itself? Can the norm be considered efficient, in terms of resources used? Does it create more or less cost-effectiveness in the country's broader migration-management system? In what follows, I first discuss the expenses and revenues related to the crime of irregular entry and stay, and then analyse issues related to the broader costs of the norm.

Expenses and Opportunity-Costs

When proposing criminalisation, the Italian Parliament estimated that the cost of apprehending and judging foreigners for irregular migration would amount to about €30,000,000 per year.[289] The estimate assumed that arrivals would drop by 10% compared to 2007, to 49,050. Given an estimated average unitary cost of €650 for legal aid and translation, the yearly cost was expected to amount to €31,882,500, if all new cases were processed.[290]

---

[287] Interviewee 41.

[288] EMN (2012), p. 71.

[289] Law Proposal N. 733/2008, p. 9.

[290] Ibid. The estimate was produced with regards to the norm that punished irregular entry with imprisonment, and made apprehension and direct judgement mandatory. Yet, no subsequent different estimate is available and, since the original one did not include the costs of prison, it is used here as a benchmark for the adopted norm too.

State-provided legal aid was indeed stressed by several interviewees as being a key element in the overall cost-and-benefit analysis.[291] Accepting the Government's estimates of the costs, if all the migrants who were subjectable to article 10-bis (Fig. 4.5) had been trialled, costs would have amounted to about €30 million in 2009, with fluctuations leading to a low in 2012 (€16 million), and a peak in 2014 (€85 million). Of course, however, since only limited percentages of such numbers were brought to trial, the actual expenses were much lower, although impossible to calculate given that the number of initiated trials does not reflect the total number of people indicted. Adding to the above, legal aid and translation costs need to be complemented by: (1) Expenses to accompany migrants and police officers to trial, (2) fees for the justices of the peace, and (3) costs to expel condemned foreigners.

First, transportation costs may involve both migrants and police officers who register them. A JP In Agrigento recalled that, in some occasions, migrants were brought to trial by pullman from northern Italian regions, accompanied by almost as many police officers.[292] Although it is unclear to what extent this is the norm (as other sources report trials often taking place *in absentia*),[293] to provide a benchmark, the state commissioned the transfer of 25 migrants from Liguria to Puglia by van in 2019, which cost €400 per migrant (€10,000 in total).[294] Furthermore, according to a prosecutor in Imperia, in emergency periods especially, police officers may come from other cities, with the result that not only would the state need to pay for their transportation costs, but that they would also lose working time.[295]

Second, JPs' honorarium should be included too. Based on the interviews, justices of the peace are paid extra-ordinarily based on the hearings and judgements, receiving €35 (gross) per hearing,[296] €55 per

---

[291] See Interviewees 54, 32.

[292] Interviewee 32.

[293] Interviewees 54, 55, 32, 41, 46.

[294] See https://www.linkiesta.it/it/article/2019/07/22/migranti-confine-italia-francia/42941/.

[295] Interviewee 46.

[296] Interviewees 32 and 54.

judgement[297] (according to Grignetti, however, JPs are paid €10 per judgement issued in the context of article 10-bis).[298]

Third, expulsions can be very expensive and, since they were a key goal of the norm, those occurring through criminal procedures should also be accounted for. According to a JP in Catania, costs for forced returns may reach several thousand euro,[299] and other sources similarly estimate them at €3000–5000.[300] Each migrant needs to be accompanied by two police officers and, as before, the cost is not only economic, but also of lost service time.[301] Perhaps as a consequence of the high costs involved in returns, an official of the European Commission argued that, in Italy, there is a targeted selection of return efforts, based on migrants' nationality and on whether or not they represent security problems.[302]

As evident from the above, the opportunity cost of employing public officials' time and, in particular, that of public prosecutors and police officers, should also be included in the cost analysis. With regard to police officers, not only do they register cases, but they may also be called to testify in court hearings. Public prosecutors, on the other hand, both when demanding case dismissals and when proceeding to trial, devote resources to deal with cases of irregular entry or stay. As an example, since, by the time trial takes place, concerned foreigners may have left the city where they were registered, they must be searched to be notified of the charges,[303] with a significant disbursement of both time and resources. The high volume of people who should be charged with article 10-bis have made the public prosecution office in Agrigento (responsible for Lampedusa) even create a specific office, tasked with dealing with cases of article 10-bis.

Revenues from Fines

If the above expenses represent the costs incurred by the state, the possible revenues to be derived from criminalisation stem from the money

---

[297] Interviewee 32.
[298] Grignetti (2016, p. 7).
[299] Interviewee 55.
[300] Gabanelli and Ravizza (2019) and Polchi (2017).
[301] Interviewee 55.
[302] Interviewee 15.
[303] Interviewee 32.

made through the imposition of monetary fines, the value of which can range from €5000 to €10,000. As I will demonstrate, however, the actual profits made are significantly hindered by the fact that sanctioned fines are lower than the minimum foreseen by the law, and that they are seldom repaid.

To begin with, as previously argued, there is evidence that the average sentenced fine has steadily been below the minimum of €5000, from 2009 to 2018.[304] If the average yearly fine is multiplied by the number of condemnations to monetary sanction, the total amount of fines sanctioned started from almost €280,000 in 2009, to peak at over €55 million in 2012, and then decrease to €36 million in 2015.

Yet, it is very unlikely that fines are paid. This has been the predominant viewpoint among interviewees,[305] and even the Fiorella Commission acknowledged that migrants would be unable to repay the fine.[306] Indeed, foreigners in situations of irregularity often have little financial availability, and no bank accounts or properties.[307] Sometimes, they still owe large sums to smugglers,[308] which makes it even harder for them to pay the fine. Substituting the fine with community work or house arrest is equally problematic, as the persons involved often do not have a permanent residence, and, quite evidently, cannot legally remain on Italian soil.[309]

The little likelihood that migrants pay fines is worsened by the fact that they very seldom attend trials[310]: Unless the involved migrants are in reception centres, they are impossible to find (once the crime has been contested), as they may provide fake identity details and/or run away.[311] Indeed, not having a fixed residence most of the times, their domicile is usually at their lawyers' who, if appointed by the state, generally have no actual contacts with the migrants themselves.[312] As a result, very often the

---

[304] MoJ (2018).

[305] Interviewees 15, 32, 33, 39, 46.

[306] Commissione Fiorella (2012), section 3.3.

[307] Savio (2016). Similarly, Ambrosini and Guariso (2016).

[308] Interviewee 41.

[309] Constitutional Court, sentence n. 250/2010: section 10.

[310] Interviewees 32, 41, 46, 54, 55.

[311] Interviewee 41.

[312] Interviewee 46.

crime of irregular entry and stay involves what two different magistrates called 'ghosts' trials',[313] making fines even less likely to be paid.

Counterproductive Effects: Duplicated Efforts and Longer Investigations
The above expenses, opportunity-costs and lack of profits are aggravated by the counterproductive effects the norm has on the broader migration-management system, including by overwhelming public offices and impeding smuggling investigations.

To start with, criminalisation overwhelms public offices by implying a duplication of efforts: As said previously, once a foreigner is found irregularly entering or staying, criminal and administrative proceedings begin in parallel, both with the goal of leading to the expulsion of the individual (unless the person qualifies for international protection). In this context, the duplication does not only apply to efforts, but to costs too: Expenses related to the administrative procedures should be added to those involved in criminalisation, since they apply to the same people and share the same objective. These would include not only further legal and interpretation expenses, but also the costs of holding migrants in CPRs while awaiting expulsion (which range between €30 and €35 per migrant per day).[314]

The problem is more acute, since public prosecution and JP offices are strained in their efforts and often understaffed.[315] This is especially the case in southern regions,[316] which are also the ones mostly affected by sea landings, but even in northern border-region Liguria, public offices are significantly lacking in personnel.[317] In turn, the lack of human and financial resources is particularly problematic due to the high number of individuals who should be investigated for the offence of irregular migration.[318]

---

[313] Interviewees 41 and 47. Similarly, interviewee 46.

[314] Golini et al. (2015, p. 50).

[315] Ferrero (2010), Interviewee 34, Public Prosecutor Siracusa (2017). See also Senate, 08/07/2015, p. 6.

[316] Ferrero (2010), Interviewee 34, Public Prosecutor Siracusa (2017).

[317] See https://www.ilsecoloxix.it/imperia/2019/01/17/news/uffici-dei-giudici-di-pace-a-imperia-progetto-pilota-per-impiegare-i-got-1.30289537?refresh_ce.

[318] Interviewees 34, 38, 39.

On top of duplicating efforts, diverting resources and overwhelming public offices, the application of the norm to migrants saved by SAR operations has also impeded swift investigations against migrant smugglers and traffickers, making them longer and more expensive.[319] Indeed, rescued migrants represent key sources of information for prosecutors investigating smuggling networks and are often interviewed after landing, to gather evidence.[320] In this context, in the cases in which rescued migrants are also investigated for art. 10-bis, they can only be questioned in presence of a defence lawyer, and all acts must be translated into their language.[321] As indicted persons in related processes (rather than as witnesses), they also have the possibility to refrain from answering questions,[322] therefore lengthening trials and contributing to raising their costs.[323]

As a consequence of the above, considerations relating to efficiency have played a great role in motivating the partial disapplication of the provision,[324] both among police officers and prosecutors.

To conclude, criminalisation has led to several counterproductive and inefficient outcomes, including a duplication of efforts to address irregular migration, and lengthier investigations against smugglers. Indeed, several actors lament the lack of utility of the norm to process the many cases of irregular migration, when compared to the resources needed to enforce it, and contest the merely political advantage of introducing and maintaining the provision.[325] If the above expenses, opportunity-costs and lack of profits might be justified by the state obtaining its goals, they are instead weakened by the lack of efficacy (and coherence) generated.

---

[319] Di Natale, Renato, interviewed in Ziniti (2016), Roberti, Franco, cited in Baer (2016), See also interviewee 34.

[320] Interviewees 34, 37.

[321] Di Natale, Renato, interviewed in Ziniti (2016), Roberti, Franco, cited in Baer (2016), See also interviewee 34.

[322] Savio (2016).

[323] Di Natale, Renato, interviewed in Ziniti (2016), Roberti, Franco, cited in Baer (2016), See also interviewee 34.

[324] Interviewees 41 and 47.

[325] Interviewee 34. See also section "Utility: Who Benefits from Criminalisation?", and Chapter 6, section "The Politicisation of Migration".

Coherence: Irrational Sanctions and Legal Concerns

The issue of the lack of efficiency of criminalisation is strictly linked to the norm's coherence, both within itself, and with existing law.

Starting with the former, criminalisation has often been contested for being highly (internally) incoherent, for several reasons. First, as discussed above, the inability of migrants to pay the fine is evident, given their general lack of resources and of fixed residence. More specifically, according to the Fiorella Commission, the sanction foreseen by the norm is 'irrational', since once the main penalty, the monetary fine, goes unpaid, it should be substituted by expulsion, which is however stricter than the original sanction itself.[326] Moreover, the duplication of procedures implied by criminalisation is not only inefficient, but also incoherent. As an example, charges for article 10-bis are sometimes initiated together with procedures to grant a permit to stay. According to an interviewed police official, if a migrant is pregnant, or going to apply for asylum, the police should at the same time denounce her/him and grant a permit to stay—a paradox that makes several officers avoid reporting the case.[327]

The contested incoherence of criminalisation is not limited to irrational aspects intrinsic to the law itself, but also regards possible contrasts with other pieces of legislation (external coherence). It should be stressed that, although the coherence of the norm with both the Italian Constitution and the EU Return Directive has been repeatedly questioned by local courts, it has, to date, been upheld by both the Italian Constitutional Court and the ECJ.[328] Such views have however not been endorsed by all local judicial actors, with the result that several among them opted for a less than systematic application of the norm, on the basis of its lack of external coherence. Some JPs in Siracusa, for instance, supported the unconstitutionality of the norm, on the basis that the value protected by the law is unclear, and that there is a discrimination between those who are refused entry at the border and those who (irregularly) stay on Italian soil.[329] In similar fashion, criminalisation has been deemed to be in contradiction with the Return Directive by a Catania JP since, as the fine cannot be paid by the migrants, it would need to be converted

---

[326] Commissione Fiorella (2012), section 3.3.
[327] Interviewee 14.
[328] Constitutional Court, sentence n. 250/2010 and Case C-430/11, Sagor.
[329] Interviewee 39.

into detention or another form of restriction of freedom, which would however be in contrast with the objective of EU law to privilege expulsion.[330] Issues of external coherence also emerge with regards to minors: Although article 10-bis does not discuss whether sanctions should be applied to them too, the prevalent interpretation is that this is not legally possible, in view of international norms preventing the expulsion of people younger than 18. According to the Minors' Court in Palermo, just like for the prosecution in Genoa and Milan, for instance, article 10-bis is in contrast with international conventions ratified by Italy and lacks legal offensiveness, and should therefore be disapplied in the case of minors.[331]

To sum up, criminalisation's internal coherence is to be questioned, as the nature of the sanction foreseen is intrinsically irrational and leads to unnecessary and inefficient duplication of efforts. As for its external coherence with the Return Directive and the Italian Constitution, this has been validated by both the ECJ and the Italian Constitutional Court from early on. The continuing contestation by local actors is however significant, as indicative of involved actors' attempts to limit the inefficiencies and counterproductive effects caused by the norm.

Sustainability: The Long-Term Implications

Having discussed efficacy, efficiency, and coherence, the fourth criteria to address is the sustainability of the norm, namely whether this can be maintained in the long term. According to a journalist active in Ventimiglia, it is better for the law to be written in a bad way, rather than properly, since this leads to it being disapplied.[332] While it is true that in several cases article 10-bis is not applied, a badly written norm implies severe consequences for the society.

In the case of criminalisation, these concern especially the overwhelming effect that the norm has on the offices of the public prosecution, especially in light of the high number of persons who arrive. As an example, considering that in certain periods, the port of Pozzallo (under the competence of Ragusa) could see up to three landings per

---

[330] Interviewee 54, JP Catania, Sentence 264/2018.

[331] Interviewee 40, Public prosecution at Minors' Court in Genoa (2013), Dismissal request N. 553/13, Public prosecution at Minors' Court in Milan (2010), and Dismissal request N. 2895/09.

[332] Interviewee 13.

day, each with three-hundred people, registering all migrants can become a significant endeavour.[333] At the same time, the presence of the norm has a stigmatising effect on migrants, who are increasingly associated to criminals and security threats, with the twofold effect of favouring marginalisation and escalating rhetoric. Finally, although (as will be seen in the French case), it is technically possible for criminalisation to be in place but not vastly used, this creates a potential backlash against the overall legal system, which becomes vitiated and loses credibility. This is especially more so given that, in Italy, the persecution of criminal charges is mandatory.[334]

Overall, criminalisation does not seem sustainable: If its systematic use overloads public offices (and contributes to the increased securitisation of migration), its disapplication weakens the credibility of the legal system, potentially leading to further non-compliance.

Utility: Who Benefits from Criminalisation?

To conclude the analysis of criminalisation in Italy, the last element to consider is the utility of the norm, namely whom its introduction benefitted. To do so, it is worth recalling Boswell's theoretical interpretation of 'policy gaps': For the scholar, these may in fact be policies in themselves, meant to accommodate contrasting demands, through strategies of deliberate malintegration or populist mobilisation.[335]

In this context, I argue that it is possible to interpret the decision to criminalise irregular migration in Italy as a case of deliberate malintegration, meant to convey different messages to different audiences.

On one hand, criminalisation was intended to prioritise security, by communicating a sense of safety and protection to the insecure domestic public.[336] This is shown not only by the initial proposal to include imprisonment, and eventual association of migrants with criminals that need to be sanctioned, but also by the inability to repeal the norm for fear of political repercussions, in 2014–2016. Through criminalisation, the

---

[333] Interviewee 34.

[334] Italian Constitution, art. 112.

[335] Boswell (2007).

[336] The politicisation of the norm will be further discussed in Chapter 6, section "The Politicisation of Migration".

Government gave precedence to the concerns of the domestic audience, rather than those of would-be migrants, failing to appreciate the structural elements that originated the latter's desires and capabilities to migrate.

On the other hand, criminalisation was meant to quietly allow for wealth accumulation, by avoiding to substantially affect the overall migration system (including the country's structural need for foreign work).

First, a month after passing the law on criminalisation, the Government approved an amnesty, expected to regularise over 500,000 domestic workers and carers.[337] Indeed, socio-economic interests in Italy (and the EU more broadly) vastly rely on migrants, to address what has been depicted as a 'structurally embedded' demand for alien labour,[338] often taking advantage of its low cost, scarce involvement with trade unions, and easy dismissal.[339] There is also increasing evidence that Europe will need immigration to contrast the fast ageing and low fertility rate of its population (in 2017, Italy's fertility rate was 1.32, well below the population replacement level),[340] which are already causing shortages at both the high and low ends of the labour market.[341]

Second, the 2009 Security Package was more concerned with adopting visible measures to restrict entry, than with addressing the underlying problem of the underground economy that employs (and attracts) migrants.[342] Indeed, despite the strong connections between migration and irregular work in Italy,[343] work-site inspections fell, in parallel to the adoption of the decree: By 6% in 2008, and by 25% in 2009.[344] Arguing that this would allow for a more targeted focus on 'substantial' infractions, the Government intentionally decreased the number of

---

[337] Ferrero (2010, p. 62). Although, eventually, only 294,000 applications were registered (Pasquinelli 2009).

[338] Cornelius and Takeyuki (2004, p. 9). Similarly, Overbeek (1996, p. 78).

[339] Weiner (1995, p. 199).

[340] See: https://ec.europa.eu/eurostat/databrowser/view/tps00199/default/table?lang=en. EUROSTAT TPS00199.

[341] Migration Policy Centre team, and Peter Bosch (2015, p. 3).

[342] Finotelli and Sciortino (2009, p. 127).

[343] See Talani (2018).

[344] Fasani (2010), p. 183. See also: https://ricerca.repubblica.it/repubblica/archivio/repubblica/2009/02/25/se-il-governo-premia-il-lavoro-sommerso.html.

firms to control.³⁴⁵ In the Government's justification, the financial crisis and subsequent slow-down of economic activities were among the critical aspects calling for a 'selective and qualitative invigilating action [...], meant to limit obstacles to the productive system'.³⁴⁶ In other words, the underground market was seen as playing the role of a 'social safety net'.³⁴⁷ Interestingly, according to some indicators, the underground economy actually increased, following the financial and eurozone crises.³⁴⁸

In light of the above, it is possible to interpret the choice to simultaneously criminalise migration, adopt a regularisation, and limit work-site inspections, as a decision meant not to damage the sectors of society relying on foreign work, especially in times of financial constraints. In turn, such a strategy seems to have been especially needed since the parties composing the governing coalition relied on very different electoral support, ranging from voters with anti-immigrant views, to people largely dependent on migrants, either as workers or housekeepers.³⁴⁹ Yet, without a stricter regulation of the informal market, its attractiveness significantly undermines attempts to manage migratory flows,³⁵⁰ as the underground economy has been found to be a catalyst for migration.³⁵¹

In conclusion, I suggest that the utility of criminalisation derives from the policy gap itself, used as a strategy of deliberate malintegration. While promoting tougher rhetoric on migration control, the Government avoided tackling the intrinsic relationship between the underground economy and irregular migration. As a result, people kept being attracted by work opportunities in the shadow market, and socio-economic actors prospering from it.

---

[345] Italian Ministry of Work, Health and Social Policies (2008, pp. 1–5).

[346] Ibid., pp. 4–5.

[347] See https://www.corriere.it/politica/09_marzo_07/intervista_brunetta_ammortizzatori_sociali_aldo_cazzullo_382e8f6a-0ae2-11de-a3df-00144f02aabc.shtml.

[348] Talani (2018, pp. 231–232).

[349] Zincone (2011).

[350] EMN (2012, p. 84).

[351] Talani (2018).

CONCLUSION

In this chapter, I have sought to understand the failure of the criminalisation of irregular migration in Italy to achieve its objectives, by deconstructing the norm and examining its potential discursive, implementation, and efficacy gaps. Each level highlighted specific deficiencies that eventually made the strategy less than successful. When unpacking the measure in this way, the policy gap becomes evident in all its dimensions, and shows more complex dynamics than the simple stated goal/actual outcome dichotomy.

At the discursive level, the measure underwent severe remodelling since its initial formulation, due to pressures from the civil society, practical concerns related to prisons' overpopulation, and political compromise within the governing coalition. While the final design was (in some respects) less severe than the initial one, it left many to doubt its potential for success. Indeed, in light of the analysis of the norm's utility, the incoherence of the law can be understood as a case of deliberate malintegration. This was meant to adopt a securitarian approach to address anti-immigrant concerns, while also quietly recognising the country's need for foreign (regular and irregular) labour, by granting an amnesty and reducing work-site inspections shortly thereafter. If the choice to communicate different messages to different audiences was beneficial at the political level, however, the same cannot be said with regards to the application of the norm, and its consequences for migrants.

Indeed, at the implementation level, agents' discretion played a crucial role in producing unsystematic enforcement, motivated by both practical constraints (the lack of resources and the inherent unsuitability of the norm to address migratory dynamics) and institutional ones (concerns about the compatibility of the provision with EU and national law, and about the tenuity of the offence). Courts played a notable role in favouring migrants but, interestingly, this was not so much done at the higher levels, as both the Italian Constitutional Court and the ECJ upheld the applicability of the norm, but rather at the local one, as several justices of the peace opted to dismiss or absolve cases, even when the infraction was evident. In this sense, the liberalness of the state emerged through the practical phases of implementation of the norm. The resulting uneven enforcement, however, is likely to have also partly shaped the overall effects of the norm.

As expected, at the efficacy level, criminalisation was neither able to avoid sustained migratory flows to the country in following years, nor to ensure systematic immediate expulsions. The reasons for this are to be found in the incoherent design of the measure, but also, and more fundamentally, in the greater influence on potential migrants' willingness to flee of external factors, among which dire situations in countries of origin and transit, first of which Syria and Libya. While not managing to reduce migration or increase returns, however, the norm did contribute to a number of counterproductive effects. Specifically, these included the duplication of efforts, as a consequence of the parallel administrative and criminal procedure that would start following a foreigner's apprehension, as well as the overwhelming of public offices. Investigations against smugglers and traffickers were also made longer and more difficult as a consequence of the criminalisation of migration. Certainly, this raises serious questions on the role that criminalisation can play in controlling migratory flows, and the extent to which it can prove a suitable solution.

To conclude, as the introduction of the offence was meant to address domestic concerns of security, it also emerges as paradoxical that, despite the widespread acknowledgement of its failure and of its detrimental effects, even a centre-left government was unable to repeal it, on the basis that the public would not understand. Indeed, while immediate political stakes may favour securitarian strategies such as criminalisation and deterrence, their failure is likely to lead to further concerns and demand for visible, short-term solutions, in what may result a vicious cycle of insecurity.

References

Ambrosini, Maurizio, and Alberto Guariso. 2016. Immigrazione illegale: perché è un reato inutile. *LaVoce*, February 2. http://www.lavoce.info/archives/39544/immigrazione-illegale-perche-e-un-reato-inutile/. Accessed 10 Jan 2017.

Amnesty International. 2015. *Europe's Sinking Shame: The Failure to Save Refugees and Migrants at Sea*. London: Amnesty International Ltd.

Andenaes, Johannes. 1968. Does Punishment Deter Crime? *Criminal Law Quarterly* 11: 76–93.

Baer, Giovanna. 2016. Clandestinità: Le Sette Vite di un Reato Inutile', *PaginaUno*, N. 46, February–March. http://www.rivistapaginauno.it/reato-clandestinita.php. Accessed 1 May 2018.

Bei, Francesco. 2016. Alfano: clandestinità, una norma sbagliata, ma ora deve restare. *Repubblica*, January 10. https://www.repubblica.it/politica/2016/01/10/news/angelino_alfano_riconosce_che_la_norma_anti-irregolari_del_2009_fu_un_errore_renzi_e_io_non_ideologici_usiamo_buon_sens-130935370/. Accessed 19 Sep 2019.

Bonifazi, Corrado, and Cristiano Marini. 2014. The Impact of the Economic Crisis on Foreigners in the Italian Labour Market. *Journal of Ethnic and Migration Studies* 40 (3): 493–511.

Boswell, Christina. 2007. Theorizing Migration Policy: Is There a Third Way? *International Migration Review* 41 (1): 75–100.

Camilleri, Andrea et al. 2009. Appello contro il Ritorno delle Leggi Razziali in Europa: Alla Cultura Democratica Europea e ai Giornali che la Esprimono. *Repubblica*, July 2. http://temi.repubblica.it/micromega-online/camilleri-tabucchi-maraini-fo-rame-ovadia-scaparro-amelio-appello-contro-il-%20ritorno-delle-leggi-razziali-in-europa/. Accessed 1 May 2018.

Campbell, John, and Asch Harwood. 2018. Boko Haram's Deadly Impact. *Council on Foreign Relations*, August 20. https://www.cfr.org/article/boko-harams-deadly-impact. Accessed 5 Mar 2020.

Campesi, Giuseppe. 2013. La detenzione amministrativa degli stranieri in Italia: storia, diritto, politica. *Università di Bari "Aldo Moro"*.

Canzio, Giovanni. 2016. Intervento del Primo Presidente Dott. Giovanni Canzio per la cerimonia di inaugurazione dell'anno giudiziario. *Corte Suprema di Cassazione*, January 28. http://www.cortedicassazione.it/corte-di-cassazione/it/inaugurazioni_anno_giudiziario.page. Accessed 1 May 2018.

Caponio, Tiziana, and Teresa Cappiali. 2018. Italian Migration Policies in Times of Crisis: The Policy Gap Reconsidered. *South European Society and Politics* 23 (1): 115–132.

Caputo, Angelo et al. 2009. Appello di Giuristi contro l'Introduzione dei Reati di Ingresso e Soggiorno Illegale dei Migranti. *Magistratura Democratica*, June 25. http://www.old.magistraturademocratica.it/platform/node/2216/print. Accessed 1 May 2018.

Caritas and Migrantes. 2019. *Un Nuovo Linguaggio per le Migrazioni: Sintesi*. XXVII Report 2017–2018. http://www.caritas.it/caritasitaliana/allegati/7824/Sintesi%20per%20giornalisti.pdf. Accessed 1 Jan 2020

Colombo, Asher. 2011. Gli Stranieri e la Sicurezza. In *Rapporto sulla criminalità e sicurezza in Italia 2010*, ed. Marzio Barbagli, and Asher Colombo, 269–340. Milano: Il Sole 24 ORE and Fondazione ICSA.

Commissione Fiorella per la revisione del sistema penale. 2012. *Relazione*. Released on the website of the Italian Ministry of Justice on 23 April 2013. https://www.giustizia.it/giustizia/it/mg_1_12_1.page;jsessionid=CnrqkdgEhbOXrgVen91FlTUP?facetNode_1=3_1&facetNode_2=3_1_3&fac

etNode_3=3_1_3_1&facetNode_4=4_57&contentId=SPS914197&previsiou sPage=mg_1_12. Accessed 19 Sept 2019.

Commissione Parlamentare di Inchiesta sul Sistema di Accoglienza, di Identificazione ed Espulsione, nonché sulle Condizioni di Trattenimento dei Migranti e sulle Risorse Pubbliche Impegnate. 2017. Relazione sul Sistema di Protezione e di Accoglienza dei Richiedenti Asilo (Parliamentary commission on the reception system, 2017). *Doc. XXII-bis N. 21*, Referee: on. Paolo BENI, Approved by the Commission on 20 December 2017. http://documenti.camera.it/_dati/leg17/lavori/documentiparlamentari/IndiceETesti/022bis/021/INTERO.pdf. Accessed 19 Sept 2019.

Cornelius, Wayne A., and Takeyuki Tsuda. 2004. Controlling Immigration: The Limits of Government Intervention. In *Controlling Immigration: A Global Perspective*, 2nd ed., ed. Wayne Cornelius, Takeyuki Tsuda, Philip Martin, and James Hollifield. Stanford: Stanford University Press.

Corriere del Mezzogiorno. 2014. Immigrazione, Alfano: "Problema strutturale e non più emergenziale"'. *Corriere del Mezzogiorno*, April 16. http://corrieredelmezzogiorno.corriere.it/catania/notizie/cronaca/2014/16-aprile-2014/immigrazione-alfano-problema-strutturale-non-piu-emergenziale-22382556678.shtml. Accessed 10 Jan 2017.

Decree law 113/2018. Disposizioni urgenti in materia di protezione internazionale e immigrazione, sicurezza pubblica, nonché misure per la funzionalità del Ministero dell'interno e l'organizzazione e il funzionamento dell'Agenzia nazionale per l'amministrazione e la destinazione dei beni sequestrati e confiscati alla criminalità organizzata. October 4. n. 113, *Official Journal* n. 281 of 3rd December 2018, http://documenti.camera.it/leg18/dossier/pdf/D18113B.pdf?_1561731655031. Accessed 19 Feb 2020.

Decree law 53/2019. Disposizioni urgenti in materia di ordine e sicurezza pubblica. *Official Journal* n. 138 of 14th June. http://documenti.camera.it/leg18/dossier/pdf/D19053.pdf?_1561731652312. Accessed 19 Feb 2020.

Della Valle, Gianfranco. 2018. Protezione Internazionale in Italia, il Caso Nigeria. http://cdgvr.it/wp-content/uploads/2018/07/nuova-news.pdf. Accessed 13 December 2021.

Delvino, Nicola. 2017. *The Challenge of Responding to Irregular Immigration: European, National and Local Policies Addressing the Arrival and Stay of Irregular Migrants in the European Union*. Global Exchange on Migration and Diversity Report. https://www.compas.ox.ac.uk/wp-content/uploads/AA17-Delvino-Report.pdf. Accessed 29 Mar 2019.

Delvino, Nicola, and Sarah Spencer. 2014. *Irregular Migrants in Italy: Law and Policy on Entitlements to Services*. ESRC Centre on Migration, Policy and Society (COMPAS), University of Oxford.

di Martino, Alberto, Biondi Dal Monte, Francesca, Boiano, Ilaria and Raffaelli Rosa. 2013. *La criminalizzazione dell'immigrazione irregolare: legislazione e prassi in Italia*. Pisa: Pisa University Press.

Di Robilant, Andrea. 1990. Nel '91 si potranno passare le frontiere senza documenti: Addio al passaporto. *La Stampa*, November 27. Document available at Fondation Jean Monnet pour l'Europe under code JD-0199.

EMN. 2011. Ad-Hoc Query on illegal migration in the Mediterranean Sea Basin: Responses from France, Italy, Malta and Spain (4 in Total), Requested by PL EMN NCP on 23 March 2010 (to CY, FR, GR, IT, MT and ES only), Compilation produced on 21 November 2011. https://ec.europa.eu/home-affairs/sites/homeaffairs/files/what-we-do/networks/european_migration_network/reports/docs/ad-hoc-queries/illegal-immigration/210.emn_ad-hoc_query_illegal_migration_in_the_mediterranean_sea_basin_updated_wider_dissemination.pdf. Accessed 5 Jan 2020.

EMN. 2019. *Annual Report on Migration and Asylum 2018*. European Migration Network.

EMN. (ed.). 2012. *Practical Responses to Irregular Migration: The Italian Case*. Rome: EMN Italy. https://ec.europa.eu/home-affairs/sites/homeaffairs/files/what-we-do/networks/european_migration_network/reports/docs/emn-studies/irregular-migration/it_20120105_practicalmeasurestoirregularmigration_en_version_final_en.pdf. Accessed 5 Jan 2020.

Eurobarometer. 2019. What Do You Think Are the Two Most Important Issues Facing Our Country at the Moment?. https://ec.europa.eu/commfrontoffice/publicopinion/index.cfm/Chart/getChart/chartType/lineChart/themeKy/42/groupKy/208/savFile/54. Downloaded February 2019.

European Commission. *A European Agenda on Migration*, May 13. COM(2015) 240. http://ec.europa.eu/dgs/home-affairs/what-we-do/policies/european-agenda-migration/background-information/docs/communication_on_the_european_agenda_on_migration_en.pdf. Accessed 25 Aug 2015.

European Court of Human Rights (ECHR). 2012. Case of Hirsi Jamaa and Others v. Italy, Application N. 27765/09, Judgment of the Court of 23 February 2012. https://hudoc.echr.coe.int/eng#%7B%22itemid%22:%5B%22001-109231%22%5D%7D. Accessed 27 Mar 2020.

European Parliament and Council. *Directive 2008/115/EC of 16 December 2008 on Common Standards and Procedures in Member States for Returning Illegally Staying Third-Country Nationals*, December 24. O.J. L 348/98, 2008/115/EC. http://eur-lex.europa.eu/legal-content/EN/TXT/?uri=CELEX:32008L0115. Accessed 25 Aug 2015.

European Union Court of Justice. Case C-430/11, Judgment of the Court of Justice of 6 December 2012, Sagor, Case, ECLI:EU:C:2012:777. https://eur-lex.europa.eu/legal-content/GA/SUM/?uri=CELEX:62011CJ0430. Accessed 1 May 2018.

EWN. 2019. Syria's Civil War Death Toll. *EyeWitnessNews*. https://ewn.co.za/2019/01/02/nearly-370-000-killed-in-syria-s-civil-war-since-2011. Accessed 5 Mar 2020.
Fargues, Philippe, and Christine Fandrich. 2012. *Migration After the Arab Spring*. MPC Research Report 2012/09.
Fargues, Philippe, and Sara Bonfanti. 2014. *When the Best Option Is a Leaky Boat: Why Migrants Risk Their Lives Crossing the Mediterranean and What Europe Is Doing about It*. Migration Policy Centre, EUI.
Fasani, Francesco. 2010. The Quest for la Dolce Vita? Undocumented Migration in Italy. In *Irregular Migration in Europe: Myths and Realities*, ed. Anna Triandafyllidou, 167–185. Farnham, Surrey: Ashgate Publishing Limited.
Ferrero, Giancarlo. 2010. *Contro il Reato di Immigrazione Clandestina: Un'Inutile, Immorale, Impraticabile Minaccia*, 2nd ed. Roma: Ediesse.
Finotelli, Claudia, and Giuseppe Sciortino. 2009. The Importance of Being Southern: The Making of Policies of Immigration Control in Italy. *European Journal of Migration and Law* 11: 119–138.
Fondazione Migrantes, Caritas Italiana, Comunità di Sant'Egidio, A.C.L.I., Fondazione Centro Astalli, Comunità Papa Giovanni XXIII. 2009. Solo Una Legge Giusta Può Dare Più Sicurezza. http://www.santegidio.org/downloads/SoloUnaLeggeGiustaPuoDareSicurezza.pdf. Accessed 1 May 2018.
Frontex. 2015. *Annual Risk Analysis 2015* (ARA 2015). Warsaw: Frontex.
Frontex. 2017, 2019. *Detections of Illegal Border-Crossings Statistics*. https://frontex.europa.eu/along-eu-borders/migratory-map/ (downloaded on 7 October 2017 and 6 August 2019).
Gabanelli, Milena, and Simona Ravizza. 2019. Migranti irregolari: quanti ne ha rimpatriati Salvini in 8 mesi di governo? Corriere della Sera. March 3. https://www.corriere.it/dataroom-milena-gabanelli/migranti-irregolari-quando-ne-ha-rimpatriati-salvini-8-mesi-governo/accb467e-3c37-11e9-8da9-1361971309b1-va.shtml. Accessed 7 Jan 2020.
Golini, Antonio et al. 2015. *Rapporto sull'accoglienza di migranti e rifugiati in Italia. Aspetti, procedure, problem*. Italian Ministry of the Interior, Rome, October. https://www.asylumineurope.org/sites/default/files/resources/ministry_of_interior_report_on_reception_of_migrants_and_refugees_in_italy_october_2015.pdf. Accessed 10 Aug 2019.
Grignetti, Francesco. 2016. I giudici di pace ostaggio del reato di clandestinità. *La Stampa*, January 11, p. 7.
Grillo, Beppe, and Gianroberto Casaleggio. 2013. Reato di clandestinità. *Il Blog delle Stelle*, October 10. www.ilblogdellestelle.it/2013/10/reato_di_clandestinita_4.html. Accessed 26 Apr 2018.
Group Salaires et Questions Sociales. 1950, IT. Réponse au questionnaire relatif aux Mouvements de Main d'œuvre (Migration): Italie, Conversation sur le

Plan Shuman. September 28. Document available at Fondation Jean Monnet pour l'Europe under code AMG 7/5/25.

IOM. 2011. *Egypt After January 25. Survey of Youth Migration Intentions.* Cairo. http://www.migration4development.org/sites/default/files/iom_2011_egypt_after_january_25_survey_of_youth_migration_intentions.pdf. Accessed 31 May 2019.

IOM. 2018. Enabling a Better Understanding of Migration Flows (and its Root Causes) from Nigeria towards Europe. http://www.dtm.iom.int/reports/enabling-better-understanding-migration-flows-and-its-root-causes-nigeria-towards-europe. Accessed 13 December 2021.

IOM Italy. 2017. La Tratta di Esseri Umani attraverso la Rotta del Mediterraneo Centrale: Dati, Storie e Informazioni Raccolte dall'Organizzazione Internazionale per le Migrazioni. http://www.italy.iom.int/sites/default/files/news-documents/RAPPORTO_OIM_Vittime_di_tratta_0.pdf. Accessed 9 October 2017.

IOM and MoI. Date unknown. Campagna Informativa Aware Migrants 2015–2019. http://www.libertaciviliimmigrazione.dlci.interno.gov.it/sites/default/files/allegati/campagna_informativa_aware_migrants.pdf. Accessed 4 Aug 2019.

Ismu. 2014. 'Sbarchi e Richieste di Asilo: Serie Storica Anni 1997–2014' (Ismu elaboration upon MoI data). http://www.ismu.org/wp-content/uploads/2015/02/Sbarchi-e-richieste-asilo-1997-2014.xls. Aaccessed 1 May 2018.

ISTAT. 2018. Data related to article 10-bis TUI, 2009–2016 (obtained in May 2018).

Italian Constitution. [1947] 2012. https://www.senato.it/documenti/repository/istituzione/costituzione.pdf. Accessed 20 Mar 2020.

Italian Constitutional Court. 2010. Sentenza 250/ 2010. *Gazzetta Ufficiale*, No. 28, July 14. https://www.cortecostituzionale.it/actionSchedaPronuncia.do?anno=2010&numero=250. Accessed 24 May 2018.

Italian Minister of Work, Health and Social Policies. 2008. Direttiva del Ministro (Maurizio Sacconi), Servizi Ispettivi e Attività di Vigilanza, September 18. Annex II to Circular n. 111 of 17th December. https://www.inps.it/circolariZip/Circolare%20numero%20111%20del%202017-12-2008_Allegato%20n%202.pdf. Accessed 5 Mar 2020.

Italian Ministry of the Interior (MoI). 2007. *Rapporto sulla criminalità in Italia, analisi, prevenzione e contrasto.* https://www.poliziadistato.it/statics/45/rapporto_criminalita_appendice_2006.pdf. Accessed 10 Jan 2019.

Italian Ministry of the Interior (MoI). 2015. Centri per l'immigrazione. *Interno.gov.it.* Updated on July 28. http://www.interno.gov.it/it/temi/immigrazione-e-asilo/sistema-accoglienza-sul-territorio/centri-limmigrazione. Accessed 20 June 2016.

Italian Ministry of the Interior (MoI). 2016, 2017, 2018, 2019. Cruscotto Giornaliero (update on landings) as of 31/12/2016, 31/12/2017, 31/12/2018, 30/6/2019. http://www.libertaciviliimmigrazione.dlci.interno.gov.it/it/doc umentazione/statistica/cruscotto-statistico-giornaliero. Accessed 2017–2019.
Italian Ministry of Justice (MoJ). (2018). Data related to article 10-bis TUI, 2009-2018 (obtained in May 2018).
Italian Ministry of Work, Health and Social Policies. 2008. Documento di Programmazione dell'Attività di Vigilanza per l'Anno 2009, Direzione generale per l'attività ispettiva. http://www.dplmodena.it/Documento% 20di%20programmazione%202009.pdf. Accessed 5 Mar 2020.
Italian Parliament. 2008. Audizione del Ministro dell'interno, Roberto Maroni, in merito alle misure avviate per migliorare l'efficacia della normativa in materia di immigrazione. October 15. (Chamber of Deputies, 15/10/2008) http://www.camera.it/_dati/leg16/lavori/stenbic/ 30/2008/1015/s000r.htm. Accessed 27 Apr 2018.
Italian Parliament. 2011. Urgent Interpellation n. 2/01053. Presented by Gabriella Carlucci, meeting n.464. April 14. (Chamber of Deputies, 14/04/2011). http://banchedati.camera.it/sindacatoispettivo_16/showXh tml.asp?highLight=0&idAtto=37940&stile=6. Accessed 5 Jan 2017.
Italian Parliament. 2014. Resoconto Stenografico dell'Assemblea n. 166, meeting n. 166. Wednesday, January 15. (Chamber of Deputies, 15/01/2014). http://www.senato.it/service/PDF/PDFServer/BGT/00735336.pdf. Accessed 5 Jan 2018.
Italian Parliament. 2016. Resoconto Stenografico dell'Assemblea, Seduta n. 547, Thursday, January 14 (Chamber of Deputies, 14/01/2016) http://www. camera.it/leg17/410?idSeduta=0547&tipo=stenografico#sed0547.stenograf ico.tit00070.sub00060. Accessed 23 June 2016.
Italian Parliament, Law 39/1990 (Law Martelli). Conversione in legge, con modificazioni, del decreto-legge 30 dicembre 1989, n. 416, recante norme urgenti in materia di asilo politico, di ingresso e soggiorno dei cittadini extracomunitari e di regolarizzazione dei cittadini extracomunitari ed apolidi gia' presenti nel territorio dello Stato. Disposizioni in materia di asilo. Gazzetta Ufficiale, N. 49. February 28.
Italian Parliament, Law 40/1998. Disciplina dell'immigrazione e norme sulla condizione dello straniero. *Gazzetta Ufficiale*, No. 59. Ordinary Supplement No. 40. March 12. http://www.camera.it/parlam/leggi/98040l.htm. Accessed 10 Jan 2017.
Italian Parliament, Law 67/2014. Deleghe al Governo in materia di pene detentive non carcerarie e di riforma del sistema sanzionatorio. Disposizioni in materia di sospensione del procedimento con messa alla prova e nei confronti degli irreperibili. *Gazzetta Ufficiale*, N. 100. May 28. http://www.gazzettau fficiale.it/eli/id/2014/05/02/14G00070/sg. Accessed 10 Jan 2017.

Italian Parliament, Law 94/2009. Disposizioni in Materia di Sicurezza Pubblica (Dispositions on Public Security). *Gazzetta Ufficiale*, No. 170, Ordinary Supplement No. 128. July 24. Law 94/2009. http://www.asgi.it/wp-content/uploads/public/legge.15.luglio.2009.n.94.pdf. Accessed 16 June 2016.

Italian Parliament, Law No. 173/2020. Conversione in legge, con modificazioni, del decreto-legge 21 ottobre 2020, n. 130, recante disposizioni urgenti in materia di immigrazione, protezione internazionale e complementare, modifiche agli articoli 131-bis, 391-bis, 391-ter e 588 del codice penale, nonché misure in materia di divieto di accesso agli esercizi pubblici ed ai locali di pubblico trattenimento, di contrasto all'utilizzo distorto del web e di disciplina del Garante nazionale dei diritti delle persone private della libertà personale. *Gazzetta Ufficiale*, No. 314. December 18.

Italian Parliament, Law 189/2002. Modifica alla normativa in materia di immigrazione e di asilo. *Gazzetta Ufficiale*, No. 199. Ordinary Supplement. July 30. http://www.parlamento.it/parlam/leggi/02189l.htm. Accessed 10 Jan 2017.

Italian Parliament, Law Proposal N. 733/2008. Disposizioni in materia di sicurezza pubblica. Communicated to the President on 3rd June 2008. http://www.senato.it/service/PDF/PDFServer/BGT/00302495.pdf. Accessed 10 Jan 2017.

Italian Parliament, Law Proposal N. 5808. Modifiche al testo unico delle disposizioni concernenti la disciplina dell'immigrazione e norme sulla condizione dello straniero, emanato con decreto legislativo 25 luglio 1998, n.286. Presented on 15 March 1999. http://briguglio.asgi.it/immigrazione-e-asilo/1999/maggio/pdl-fini.html. Accessed 1 May 2018.

Italian Parliament, Legislative Decree 286/1998. Testo Unico Delle Disposizioni Concernenti La Disciplina Dell'immigrazione E Norme Sulla Condizione Dello Straniero. *Gazzetta Ufficiale*, N. 191. August 18, 1989. Ordinary Supplement n. 139. http://www.camera.it/parlam/leggi/deleghe/98286dl.htm. Accessed 1 May 2018.

Italian Senate. 2015. Resoconto Stenografico del Senato n. 14, Indagine Conoscitiva sui Temi dell'Immigrazione, meeting n. 295. July, Wednesday 8 (Senate, 08/07/2015).

Italian Supreme Court of Cassation, 1st Section, Pen., Sentence of 17/07/2013, n. 35588.

Italian Supreme Court of Cassation, 1st Section, Pen., Sentence of 16/11/2016, n. 53691.

Jacobson, Hyla. 2016. *The Criminalization of Irregular Migration in Italy*, Mediterranean Migration Mosaic. May 13.

JP Agrigento. 2016. Sentence 195/2016, 04/11/2016.

JP Agrigento. 2016. Settlement 12/11/2016, related to process N. 20/2012.

JP Agrigento. 2017. Sentence 208/2017, 28/11/2017.

JP Catania. 2018. Sentence 264/2018, 27/09/2018.
JP Monza. 2012. Sentence 2560/2012, 22/10/2012. https://www.penale contemporaneo.it/d/2099-art-10-bis-e-direttiva-rimpatri-dopo-la-sentenza-sagor-della-corte-di-giustizia. Accessed 1 May 2018.
JP Rome. 2011. Unknown sentence number, 16/06/2011. https://www.penalecontemporaneo.it/d/877-giudice-di-pace-di-roma-1662011-giud-chiassai-reato-di-clandestinita. Accessed 1 May 2018.
JP Turin. 2011. Sentence 314/2011, 22/02/2011. https://www.penalecontemporaneo.it/d/557-giudice-di-pace-di-torino-22-febbraio-2011sent-est-polotti-di-zumaglia-reato-di-clandestinita. Accessed 1 May 2018.
Live Sicilia. 2013. Lampedusa: Bilancio di 311 vittime. October 9. *Live Sicilia*. https://livesicilia.it/2013/10/09/lampedusa-289corpi-nicolini-barroso-letta_381916/. Accessed 5 Jan 2017.
Migration Policy Centre team, and Peter Bosch. 2015. *Towards a Pro-Active European Labour Migration Policy: Concrete Measures for a Comprehensive Package*. Migration Policy Centre, EUI.
MPI. 2018. Asylum Recognition Rates in the EU/EFTA by Country, 2008–2017. Migration Policy Institute. https://www.migrationpolicy.org/programs/data-hub/charts/asylum-recognition-rates-euefta-country-2008-2017. Accessed 5 Jan 2020.
Nicolai, Chiara. 2013. *Estratto*: I Dati Più Significativi Emersi Dall'indagine Svolta Presso Gli Uffici Del Giudice Di Pace Di Milano, Diritto Penale Contemporaneo. https://www.penalecontemporaneo.it/upload/1390468345Chiara%20Nicolai_Tesi%20art10bis_Estratto.pdf. Accessed 1 May 2018.
Overbeek, Henk. 1996. L'Europe en quête d'une politique de migration: Les contraintes de la mondialisation et de la restructuration des marchés du travail. *Études Internationales* 27 (1): 53–80.
Pansa, Alessandro. 2016. Interview with Repubblica: "Reato di clandestinità, Pansa: 'Abolirlo? Meglio riformarlo". *Repubblica*. January 10. http://www.repubblica.it/politica/2016/01/10/news/clandestinita_resta_reato_pansa_cosi_com_e_intasa_procure_-130954064/. Accessed 1 May 2018.
Paoli, Simone. 2018. La legge Turco-Napolitano: Un lasciapassare per l'Europa. *Meridiana* 91: 121–149. Retrieved from http://www.jstor.org/stable/9002202., Accessed 5 Mar 2019.
Pasquinelli, Sergio. 2009. Perché la Sanatoria Ha Fatto Flop. La voce, October 9. http://www.lavoce.info/archives/25930/perche-la-sanatoria-ha-fatto-flop/. Accessed 1 May 2018.
Polchi, Vladimiro. 2017. In 74 per scortare 29 migranti: così funzionano le espulsioni. Repubblica, January 18. https://www.repubblica.it/cronaca/2017/01/18/news/in_74_per_scortare_29_migranti_cosi_funzionano_le_espulsioni-156271202/. Accessed 7 Jan 2020.

Provera, Mark. 2015. *The Criminalisation of Irregular Migration in the European Union*. CEPS Paper in Liberty and Security in Europe, No. 80/February 2015.
Public Prosecution at Minors' Court in Genoa. 2013. Dismissal request N. 553/13 R.G.N.R. mod. 21, 08/07/2013.
Public Prosecution at Minors' Court in Milan. 2010. Dismissal request N. 2895/09 R.G.N.R, 10/02/2010.
Public Prosecutor Genoa. 2011. Acquittal request N. 3294 RG PM 21bis, 22/07/2011.
Public Prosecutor Genoa. 2014. Dismissal request N. 4755/14 mod 21bis, 2014.
Public Prosecutor Siracusa. 2017. Relazione Annuale sull'Andamento della Giustizia nel Distretto di Catania (Periodo 1 Luglio 2016–30 Giugno 2017). October 15.
Public Prosecutor Turin. 2009. Questione di legittimità costituzionale dell'art. 10 bis D.L.vo n.286198, come introdotto dall. art.1 co.16 L.15.7.2009 n.94 in relazione agli artt.2,3 co.1, e25 co.2 Cost. September 15. http://briguglio.asgi.it/immigrazione-e-asilo/2009/settembre/pg-to-cost-art10bis.pdf. Accessed 1 May 2018.
Report Committee on Migration, Refugees and Displaced Persons. 2014. *The Large-Scale Arrival of Mixed Migratory Flows on Italian Shores*. Council of Europe Parliamentary Assembly, Doc. 13531. June 9. Rapporteur: Mr Christopher Chope, United Kingdom, European Democrat Group. http://www.assembly.coe.int/nw/xml/XRef/X2H-Xref-ViewPDF.asp?FileID=20941&lang=en. Accessed 25 Aug 2015.
Repubblica. 2016. Clandestinità, depenalizzazione del reato: è scontro. Alfano: "Meglio non abrogare". *Repubblica*, January 8. https://www.repubblica.it/politica/2016/01/08/news/clandestinita_pronto_il_provvedimento_del_governo_maroni_sara_invasione_-130824651/. Accessed 5 Jan 2019.
Richardson, Roslyn. 2010. Sending a Message? Refugees and Australia's Deterrence Campaign. *Media International Australia* 135: 7–18.
Roadmap Italiana. 2015. (Italian Roadmap), Italian Ministry of the Interior. September 28. http://www.statewatch.org/news/2015/nov/italian-Roadmap.pdf. Accessed 5 Apr 2017.
Savio, Guido. 2016. Reato di Clandestinità. Inutile e Dannoso, Ecco Perché Va Cancellato. *Stranieri in Italia*, January 11. https://stranieriinitalia.it/attualita/reato-di-clandestinita-inutile-e-dannoso-ecco-perche-va-cancellato/. Accessed 1 May 2018.
Simone. 2019. *Schemi & Schede di Diritto Processuale Penale*. Metodo schematico Simone, Gruppo Editoriale Simone. https://www.simone.it/catalogo/v7_2.htm. Accessed 10 Sept 2019.

Syrian Observatory of Human Rights (SOHR). 2020. *Nearly 585,000 people have been killed since the beginning of the Syrian Revolution*. January 4. https://www.syriahr.com/en/152189/. Accessed 13 December 2021.

Talani, Leila Simona. 2018. *The Political Economy of Italy in the Euro. Between Credibility and Competitiveness*. London: Palgrave Macmillan.

Talani, Leila Simona, Matilde Rosina, Orsola Torrisi, Giulia Monteleone, and Rita Deliperi. Forthcoming. *Onward Migration and the Case of the Italo-Bangladeshi Community's Relocation to the United Kingdom*. London: COMITES.

Thielemann, Eiko. 2011. *How Effective Are Migration and Non Migration Policies That Affect Forced Migration?* LSE Migration Study Unit Working Paper No. 2011/14.

Triandafyllidou, Anna, ed. 2010. *Irregular Migration in Europe: Myths and Realities*. Farnham, Surrey: Ashgate Publishing Limited.

UNHCR. 2017. Mixed Migration Trends in Libya: Changing Dynamics and Protection Challenges. https://www.unhcr.org/595a02b44.pdf. Accessed 5 Mar 2020.

UNHCR. 2018. Malta Asylum Trends Real Time. http://www.unhcr.org.mt/charts/. Accessed 1 May 2018.

UNHCR. 2019. Operational portal: Mediterranean Situation. https://data2.unhcr.org/en/situations/mediterranean/location/5205. Accessed 4 Aug 2019.

Vesci, Pietro. 2008. L'immigrazione nei programmi elettorali per le elezioni politiche del 2008: UDC, Lega Nord, La Destra e La Sinistra Arcobaleno. *Neodemos.info*, April 2. http://www.neodemos.info/pillole/limmigrazione-nei-programmi-elettorali-per-le-elezioni-politiche-del-2008udc-lega-nord-la-destra-e-la-sinistra-arcobaleno/?print=pdf. Accessed 1 May 2018.

Weiner, Myron. 1995. *The Global Migration Crisis: Challenge to States and to Human Rights*. New York: Harper Collins College Publishers.

Zanfrini, Laura. 2015. The Labour Market. In *The Twenty-First Italian Report on Migrations 2015*, ed. Vincenzo Cesareo. Milano: McGraw-Hill Education and Ismu.

Zincone, Giovanna. 2011. The case of Italy. In *Migration Policymaking in Europe: The Dynamics of Actors and Contexts in Past and Present*, ed. Giovanna Zincone, Rinus Penninx, and Maren Borkert, 247–290. Amsterdam University Press.

Ziniti, Alessandra. 2016. Il procuratore di Agrigento, Renato Di Natale: 'Con il reato di clandestinità abbiamo gli uffici ingolfati'. *Repubblica – Palermo*. May 29. http://palermo.repubblica.it/cronaca/2016/05/29/news/il_procuratore_di_agrigento_renato_di_natale_con_il_reato_di_clandestinita_abbiamo_gli_uffici_ingolfati_-140854592/. Accessed 1 May 2018.

CHAPTER 5

France: Instrumentalisation, Courts, and Marginalisation

INTRODUCTION

France has quite a long history of immigration, compared to other European countries. Indeed, industrialisation and limited population growth spurred the need for foreign labour already in mid-nineteenth century, which continued throughout the following century.[1] In the twentieth century, the country saw sustained inflows of migrants, both for labour purposes and family reunification, and in significant part from its former colonies.[2] While, until the 1980s, these trends were paralleled by a general support for an open migration regime, an emphasis on deterring 'unwanted migrants' emerged in the two last decades of the twentieth century, following the 1970s oil crisis and the rise of the far right, with the result that border controls became increasingly relevant in the country's political landscape.[3]

In this context, the criminalisation of irregular migration emerged as a significantly severe measure, among the most severe in the EU, which foresees both a prison term and a fine, together with possible expulsion.

---

[1] Hollifield (2014, pp. 158ff.).
[2] Hollifield (2014) and Wihtol de Wenden (2010).
[3] Hollifield (2014, p. 164) and Wihtol de Wenden (2010, p. 115).

What were the consequences of criminalising irregular entry and stay? Did this imply greater deterrence of irregular migration?

In this chapter, I address these and related questions, delving into the effectiveness of criminalisation as a tool of deterrence in France. To do so, I first present the evolution of the French immigration legislation as related to criminalisation, and then investigate whether it is possible to identify any gaps in how the measure was designed and implemented, and in its overall effectiveness. In doing so, I rely on a range of sources including interviews, questionnaires, legislative texts, court rulings, and official data, to depict a comprehensive picture of how the norm was used, and with what consequences, in the country.

The Evolution of Criminalisation

1938 to 2004: Introducing and Strengthening Criminal Sanctions for Irregular Migration

Aiming to address the need for foreign workers caused by industrialisation, by low population growth and, later, by the loss of manpower during World War I, France kept an open-door policy towards foreigners in the early twentieth century, becoming what someone called 'Europe's refugee haven'.[4] In the early 1930s, however, the economic recession that hit the country, coupled with a sudden increase in refugee flows from Germany, spurred the rise of xenophobic concerns among the population, often picked up by the far-right press too.[5] Anti-foreigners sentiments increased among the working class too, and Catholic and left-wing trade unions ended up demanding lower levels of migration too.[6]

Against this backdrop, in April 1938, Eduard Daladier was nominated as prime minister. Less than a month after the appointment, on May 2nd, the new centre-right government issued a decree-law restricting new migrants' rights and granting more power to border guards, arguing that France had reached its 'saturation point' and criticising other countries' restrictive migration policies.[7]

---

[4] Maga (1982, pp. 424–425).

[5] Perry (2004, pp. 339, 342) and Maga (1982).

[6] Perry (2004, p. 343).

[7] Maga (1982, pp. 435–437).

Importantly, the decree-law also turned a consistent number of actions into crimes, including irregular entry, irregular stay, the facilitation of irregular migration, re-entry after expulsion, and the use of false documents.[8] In particular, the new crime of irregular entry and stay became punishable by a monetary sanction of between 100 and 1000 francs, as well as by a prison term ranging between one month and one year.[9] On top of that, foreigners would be expelled at the expiration of the penalty.[10]

Criticised by courts and MPs, Daladier backtracked slightly, in the hope of preventing the issue from escalating, and agreed to endorse suspended sentences for migrants who did not apply for documents within the established time limits (at the discretion however of the relevant judges).[11] In spite of the above concession, Daladier continued with his turn towards more restrictionism, creating 'special centres' to host irregular migrants, and subjecting new refugees to mandatory military service and work, aiming to deter them from moving to France.[12]

Officially, the stated goal of the 1938 decree-law was to balance France's openness to foreigners coming to study, visit, work or demand asylum, while adopting a 'just and necessary rigour' against those entering or staying in the country in unauthorised fashion.[13] According to Riadh Ben Khalifa, both the law and the following circulars were aimed at providing clear procedures to address migration, making their implementation homogenous and 'facing the [potential] empathy of [border guards]' towards foreigners.[14]

As the above suggests, migration was increasingly seen as a security issue: Not only could it have promoted violent anti-foreigner protests,[15] but migrants and refugees themselves may have been infiltrates who would have weakened the French state from within. In Daladier's words, the continued acceptance of large numbers of migrants and refugees

---

[8] Decree of 2 May 1938, articles 2, 4, 9, 12.
[9] Decree of 2 May 1938, art. 2.
[10] Ibid.
[11] Maga (1982, pp. 438–439).
[12] Ibid.
[13] Decree of 2 May 1938.
[14] Ben Khalifa (2012, p. 15).
[15] Maga (1982, p. 428).

would invite a 'Trojan Horse of spies and subversives'.[16] Additionally, he argued, through its open-door policy, France was attracting migrants and refugees, who were neither any longer needed, nor could be economically supported by the state.[17]

Overall, in the view of Ben Khalifa, the toughening of legislation in the late 1930s may be explained by the high numbers of migrants, the unwillingness of other countries to relax immigration conditions, and the rise of the nationalist far right.[18] It was justified at the time by the impending war and increasing number of foreigners arrested and suspected of being infiltrators.[19]

Subsequent legislation either confirmed or reinforced the penalties against irregular migration. In particular, the Ordonnance of 1945 (on which much of the French immigration legislation has been based until the establishment of the *Code de l'entrée et du séjour des étrangers et du droit d'asile*, Ceseda, in 2004), reaffirmed the crime and increased the possible amount of the fine to a range of between 180 and 3600 francs.[20]

While the 1930s and 1940s saw slightly decreasing immigration flows, the trend reversed in the 1950s, which became increasingly characterised by labour-related settlement (rather than temporary migration),[21] and continued throughout the second half of the century. Following the change of trend, a shift of focus also took place in policy priorities. The centre-right started placing more emphasis on expulsions, rather than criminal sanctions, through the 1980 law Bonnet, which allowed foreigners to be expelled if they could not prove having entered regularly, or if they had stayed in France for more than three months without having obtained a permit to do so.[22]

Deviating from such intention and wishing to limit reliance on direct expulsions,[23] in 1981 the new centre-left government headed by Pierre Mauroy under Socialist President François Mitterand proposed changes

[16] As cited in Maga (1982, p. 429).

[17] Maga (1982, p. 434).

[18] Ben Khalifa (2012, p. 25).

[19] Ibid., p. 12.

[20] Ordonnance n° 45-2658, art. 19.

[21] Hollifield (2014, p. 160).

[22] Law 80-9, art. 6.

[23] Projet de loi n. 366 (1981, p. 3).

to the immigration law, aiming to soften some of the strict measures introduced by the law Bonnet.[24]

While restraining the use of expulsions, however, the Government also strengthened the sanctions for irregular entry and stay. In particular, it increased the fine to a maximum of 8000 francs, maintained the possibility for the competent court to condemn the person to expulsion after the expiration of the prison term, and introduced a ban from re-entering the French territory (*interdiction du territoire français*, ITF) for up to a year, in case of repeated offences.[25] According to Schain, the tightening of migration rules in 1981 confirms the government's willingness to restrict entries, and represents a first step towards the substantial conflation of the border-control policies supported by the left and right, which eventually materialised by the late 1980s.[26]

Indeed, the willingness to maintain and extend criminalisation persisted, across the political spectrum, in following years: The 1986 Pasqua law increased the fine and extended the entry ban to up to three years,[27] and law n. 93-1420 set the penalty at one year imprisonment and 25,000 francs.[28] The latter constituted the basis for article L.621-1 of Ceseda,[29] which, in 2004, defined a year imprisonment and a €3750 fine as penalties for foreigners irregularly entering or staying in the country. It also allowed authorities to issue an expulsion order at the end of the prison term, and to ban the foreigner from the French territory for three years.[30]

Overall, the goal of the norm criminalising migration includes a strong element of deterrence. To begin with, when the norm was adopted, Daladier placed significant emphasis on reducing the attractiveness that France had for foreigners,[31] making it possible to see criminalisation as one of the measures meant to achieve such goal. Moreover, the French

---

[24] Schain (2012, p. 53).
[25] Law 81-973, art. 4.
[26] Schain (2012, p. 53).
[27] Law 86-1025, arts. 4 and 20.
[28] Law 93-1420, art. 11.
[29] Etude d'Impact (2012).
[30] Ceseda, art. L.621-1, initial version.
[31] See Maga (1982, p. 434).

penal code states that the sanctioning and rehabilitating functions of criminal penalties are aimed, *inter alia*, at preventing new infractions,[32] hence including an element of deterrence. More recently, in 2008, the country explicitly indicated the articles criminalising foreigners' irregular entry or stay (L.621-1 and L.621-2 of the Ceseda) as fulfilling the Schengen Code requirement to have in place deterrent measures against unauthorised movements across Schengen states.[33]

Post-2004: The Influence of the ECJ

Following the introduction of Ceseda in 2004, the legislation on the criminalisation of irregular migration was not meant to be changed soon. Indeed, it did not undergo significant changes for several years. It was only due to the intervention of the European Court of Justice (ECJ) that the norm was amended again, in the 2010s. The influence of the ECJ can be seen in three main respects, which I address in turn.

First, on 6 December 2011, the ECJ Achughbabian sentence declared the French practice of imprisoning migrants on the sole grounds of irregular presence to be incompatible with the Return Directive (according to which the return of irregular migrants is to be prioritised).[34] Following the Court's reasoning, imposing a prison term for irregular stay would contradict the goals of the Directive, by impeding and delaying return procedures.[35] It should be noted that the ECJ's verdict was not based on the penal and restraining nature of the French sanction.[36] The Court explicitly stated that the Return Directive 'does not preclude the law of a Member State from classifying an illegal stay as an offence and laying down penal sanctions to deter and prevent such an infringement'.[37] Instead, it rested on considerations of efficiency, meant to avoid the postponement of returns. On 13 December 2011, the French Government acknowledged in a circular the incompatibility of the norm with EU law, and instructed public prosecutors not to criminally pursue instances of

---

[32] Penal Code, art. 130-1.
[33] Etude d'Impact (2012).
[34] ECJ, Case C-329/11.
[35] Ibid., para. 39.
[36] Ibid., para. 32.
[37] Ibid., para. 28.

irregular stay when it represented the only infraction. A year later, on 31 December 2012, the Parliament eventually repealed the crime of irregular stay.[38]

The Achughbabian sentence affected a second element related to the criminalisation of migration in France: The *garde à vue* (GAV), a penal regime that allows for a time-limited custody of an individual (currently set at maximum forty-eight hours), whenever it is reasonable to think he/she committed a crime.[39] The GAV's compatibility with the EU Return Directive in regards to irregular stay has become the object of intense debate in recent years due to the fact that, since June 2011, the offence that justifies the GAV must be punishable with imprisonment.[40] As the ECJ ruled in December 2011 that irregular stay should not be sanctioned with imprisonment, the GAV would become inapplicable to irregular stay. The Government initially rejected such view,[41] only to have its interpretation dismissed by the Supreme Court (*Cour de Cassation*) in July 2012.[42] As a consequence, the GAV was replaced in 2012 by the *retenue*, an administrative practice meant to allow verification of a person's documents and permits, while retaining him or her in police or gendarmerie offices for up to 16 hours.[43] We will see in later pages whether and to what extent this made a difference.

The ECJ Affum ruling contributed to a third significant change of the criminal provisions related to irregular migration. On 7 June 2016, the Court argued that subjecting to imprisonment foreigners who have entered irregularly by trespassing an internal Schengen frontier, would slow down the implementation of returns, and therefore prove in contrast with the EU Return Directive.[44] As a consequence,[45] in September 2018, the Parliament abrogated paragraph 2 of L. 621-2 of Ceseda, making

---

[38] Law 2012-1560, arts. 2 and 8.
[39] Code of Criminal Procedure, arts. 62 and 63.
[40] Ibid.
[41] Circular 11-04-C39, p. 2.
[42] Cour de Cassation, 2012: C100965.
[43] Law 2012-1560, art. 2.
[44] ECJ, Case C-47/15.
[45] Projet de Loi n. 714 (2018), exposé des motifs.

irregular entry no longer a crime, *if* the foreigner is coming directly from the territory of another Schengen state.[46]

As the above demonstrates, the French system of criminalisation has been greatly affected by the jurisdiction of courts, and in particular that of the European Court of Justice, through the Achughbabian and Affum rulings. In the next section, I examine the significance of such changes.

A Univocal Trend Towards Decriminalisation?

The two above-mentioned ECJ rulings led to the partial abrogation of the offence of irregular migration in France. Yet, decriminalisation should not be seen as a decisive and univocal trend.

Indeed, decriminalisation was pursued only when strictly necessary. As an example, because the return directive allows Member States not to apply its content to third country nationals who are subject to a refusal of entry, or are apprehended at the *external* border of a member state,[47] the crime of irregular entry and the connected penalties were maintained in 2012.[48]

Additionally, the partial repeal of the crimes of irregular entry or stay was often paralleled by an increase in the criminalisation of other acts. This has occurred (or threatened to occur) in at least three cases.

First, while accepting to abrogate the crime of irregular stay in 2012, the Parliament simultaneously introduced the so-called *délit de mantien*, meant to sanction persons who remain on French territory contravening an expulsion order.[49] The foreseen punishment involves a €3750 fine and one-year imprisonment—like irregular entry and stay originally.[50]

Second, although the crime of irregular entry from a Schengen member state was abolished in 2018, sanctions for other infractions were simultaneously reinforced. As examples, the infraction related to the fraudulent use of ID or travel documents was expanded both in geographical and application terms,[51] the collection of fingerprints and

---

[46] Law 778-2018, art. 35, para. 1, point 3; Circular INTV1824378J.
[47] Directive 2008/115/EC, art. 2, para. 2 lett. a.
[48] Etude d'Impact (2012).
[49] Law 2012-1560, art. 9 (amending art L. 624-1 of CESEDA).
[50] Ibid.
[51] Projet de Loi n. 714 (2018).

photographs was made systematic in all cases in which *retenue* takes place to verify foreigners' right to stay, and the interdiction from the French territory for up to three years was added as a sanction to the refusal to be fingerprinted.[52]

Third, in the draft law of 30 January 2018, the French Ministry of the Interior accompanied to the repeal of the crime of irregular entry from a Schengen state, the proposal to introduce a crime of irregular border crossing.[53] The foreseen sanctions (here too, one-year imprisonment and €3750 fine) would have been applicable to foreigners crossing an external Schengen frontier, but also an internal one in case of temporary controls.[54] Indeed, in view of the border checks re-established by France on 13 November 2015, the provision would have made it possible to apply the new criminal offence (and the related GAV) to foreigners attempting the journey from Italy to France.[55] In February 2018, however, the *Conseil d'État* opposed the introduction of the new infraction[56] and, indeed, in the actual *projet de loi* of 21 February 2018, the provision was modified, with no reference to the *délit de franchissement non authorisé*.

As a final point, while the introduction of the *retenue* in 2012 signalled a shift from a criminal to an administrative regime, it has itself been criticised, as it has been perceived as a mere replication of the GAV, in administrative language.[57] Among others, Danièle Lochak argued that, while the goal is administrative, the way in which the new measure is framed strongly resembles criminal procedures, as not only is the *retenue* under the responsibility of the public prosecution, but people in such regime also have very similar guarantees to those granted to people held in GAV.[58] In fact, as I discuss in section 'Efficiency: Prison, Expulsions,

---

[52] Law 2018-778, art. 35.

[53] The crime would have regarded foreigners entering the country without legally passing through a border crossing point, and been named *délit de franchissement non authorisé des frontières de l'espace Schengen*. See Avant-projet de loi 2018, art. 16, para. 3.

[54] Avant-projet de loi 2018, art. 16, para. 3.

[55] Gisti (2018, p. 7).

[56] Conseil d'état avis 394206 (2018), Title II, 59.

[57] Interviewees 49 and 52; Lochak (2016, p. 4) and Henriot (2013, p. 12).

[58] Lochak (2016, p. 4). In 2020, the CESEDA was further modified by Ordonnance n. 2020-1733 of 16 December 2020. The crimes of migration are now listed in Book VIII, Title II. Such changes are outside the scope of this research. However, relevant crimes

and Fines', such guarantees were sometimes diminished, as in the case of legal assistance.

Overall, the 2012 and 2018 legislative changes led to the partial repeal of the crime of irregular migration. At the same time, however, they introduced the *délit de mantien*, and tightened the sanctions related to the use of false documents and refusal of being fingerprinted, in what looks like a continuous path towards criminalisation.

Having seen how the norm surrounding criminalisation evolved through time, it is possible to now ask how criminalisation has been presented in the Government's rhetoric, and implemented in practice, looking for possible policy gaps. In the next two sections, I investigate these elements, paving the way for the analysis of the measure's effectiveness in section 'Effectiveness: Beyond Goals and Outcomes'.

Rhetoric and Instrumentalisation

Having been in place for over 80 years, the introduction of measures to criminalise irregular migration predates the focus of this study. From the analysis of the previous pages, however, as well as of the situation in more recent years, the topic does emerge as extremely controversial in France, beyond specific legal[59] and humanitarian[60] networks. In particular, in the twenty-first century, the main focus of the public debate and government efforts seems to have revolved around the criminalisation of facilitators of irregular migration[61] and the effectiveness of returns,[62] rather than the crime of unauthorised migration itself. Even in the context of the legal quarrel with the ECJ over decriminalisation in the early 2010s, the matter does not appear to have been greatly politicised,[63] with

---

listed include irregular entry, refusal of being fingerprinted, hindrance to the execution of a refusal of entry, *délit de mantien*. See: https://www.legifrance.gouv.fr/codes/id/LEGITEXT000006070158/.

[59] See Gisti (2012), Henriot (2012), and Lochak (2016).

[60] Interviewee 56.

[61] Dieu (2015, p. 24).

[62] Interviewee 49.

[63] As an example, when carrying out a research on Google News on 'Sarkozy and Achughbabian' between 2010 and 2014, only one article emerges. Moreover, based on the Manifesto Project's Data Dashboard, from 1945 to 2019, only twice did a French party mention '*séjour irrégulier*' (irregular stay) in its electoral manifestos. Specifically, this

more emphasis being placed by then-president Nicolas Sarkozy on identity matters and the expulsion of Roma.[64] Overall, as Luca D'Ambrosio notices, criminalisation of migration in France has never been a 'taboo'.[65]

Despite the norm not being greatly contentious, its sanctions, as written on paper, have been among the harshest in Europe,[66] including both a fine and prison term. As a result, the distance between politicians' rhetoric and the norm on paper does not appear to have been greatly significant in the French case, although more evidence on the former in the 1930s would be needed, to draw a definite conclusion. In this context, what is relevant to note is the *instrumental* use that the security apparatus made of the measure.

Instrumentalisation is a recurring theme in the French literature on criminalisation. Indeed, according to Danièle Lochak, the criminal law is being instrumentalised in the sphere of migration control, in order to exploit its power to investigate and intimidate, while the sanction itself is not the actual goal of the legislator.[67]

The mechanism through which such instrumentalisation occurs is strongly related to the use of the GAV. In short, although irregular migrants are frequently placed in criminal custody (for unauthorised entry or stay), they are then only rarely tried. More specifically, this is because criminal custody is used by police officers to withhold foreigners while checking their credentials and permission to stay,[68] and to thus prevent them from absconding. Once the irregular status is confirmed, the Public Prosecution does not however proceed to raise criminal charges against the migrants, but rather transfers them to a detention centre, from where they will be expelled.[69] In this way, the criminal law (and in particular the GAV) is used to increase the efficacy of the administrative law (and in

---

was done in 2012 by the Left Front and by Europe Ecology—The Greens, both in favour of repealing the infraction. No other political group mentioned the matter, just like no reference was made by any party to either '*entrée irrégulier*' or '*décriminalisation*', since the mid-1940s (based on Krause et al. 2019: Manifesto Project Data Dashboard, available at https://visuals.manifesto-project.wzb.eu/mpdb-shiny/cmp_dashboard_corpus/).

[64] Schain (2012, pp. 116–117).

[65] D'Ambrosio (2010, pp. 5–6).

[66] For an overview of EU member states' sanctions, see FRA (2014).

[67] Lochak (2016, p. 3). Similarly, Henriot (2013), Interviewee 45.

[68] Lochak (2016, p. 3).

[69] Ibid.

particular of expulsions). In other words, the former is instrumentalised, to support the goals of the latter.

Thus, while the discursive gap does not seem to have been predominant in the French case, the literature suggests that the criminalisation of migration is often employed to increase the effectiveness of other measures and, in particular, of returns. While the trend has been discussed by several academics, a systematic analysis of nationwide data to assess its extent is still missing, to the knowledge of the author.[70] To what degree is such argument supported by empirical evidence? To address such issue, and proceed towards the identification of potential implementation gaps, the next part of the chapter discusses the execution of the norm criminalising irregular entry and stay in France, posing the following questions: How does the criminalisation of entry and stay take place, in practice? How extensive has its use been, over time?

Implementation

This section relies on official data, interviews, and previous studies to shed light into the way in which the criminalisation of migration has been applied throughout France, since the early 2000s.[71] Specifically, the first subsection addresses procedural aspects, outlining how the process leading to the sanctioning of foreigners for irregular entry or stay unfolds. Based on such understanding, the second subsection then examines data regarding the practical implementation of the norm, to identify the main trends.

Criminalisation in Practice

The first step in the process leading to the condemnation of migrants for irregular entry or stay involves the apprehension of foreigners. On one hand, procedures to sanction irregular stay could either start with a police identity check, or following the arrest of a person for another crime,

---

[70] While Scherr (2011) analyses data on the use of GAV, he neither includes the study of condemnations, nor discusses the issue of instrumentalisation. On the contrary, the Étude d'Impact (2012) only refers to year 2009.

[71] Due to limited data following the 2018 legislative changes, the analysis mainly focuses on trends until that year.

such as drug smuggling or theft.[72] In the case of irregular entry, on the other hand, apprehension may occur when foreigners either cross a border outside of border crossing points, or attempt to enter with false or no documents (since 2012, irregular entry needs to be caught in flagrance).[73]

In case of non-conformity with the requirements set by the law, the concerned foreigners are questioned, and the cases registered. Migrants are then indicted and held in *garde à vue* in the police offices for up to 48 hours (since 2012, this is no longer possible for irregular stay). This is the very first step of the criminal process, during which the police investigate migrants' background and situation.[74]

While in GAV, migrants have a number of rights, including meeting a lawyer, requesting an interpreter, being visited by a doctor, and contacting their family and consulate.[75] In the meantime, police officers carry out a number of procedures to assess the regularity of the persons' entry or stay, including fingerprinting and interrogating them, searching national and international databases, and investigating their situation with regards to possible asylum requests.[76]

Following the GAV, if the person is found to be irregular, the public prosecutor can either raise charges and initiate the criminal process, or avoid doing so and allow the administrative procedures to unfold (Fig. 5.1).[77] The latter option has traditionally been the most widely used, as reported by an official working for the Border Police in Nice, and leads to an order of expulsion (*obligation de quitter le territoire*, OQTF) being issued by the prefect.[78] If expulsion is not immediately possible, the person is held in an administrative detention centre (*centre de rétention administrative*, CRA).[79] According to Patrick Henriot, former magistrate and national secretary of the French Magistrates Union (*Syndicat de la Magistrature*), the administration already prepares the documents for administrative expulsion while foreigners are still in GAV, so that they

---

[72] Interviewee 56.

[73] Law 2012-1560, art. 8.

[74] See Henriot (2013, p. 2) and Scherr (2011).

[75] Étude d'impact (2018, p. 150), Circular INTK1300159C, Sec II.2, pp. 4–5.

[76] Karamanli (2012, p. 23) and Étude d'impact (2018, p. 150).

[77] Interviewees 44 and 45.

[78] Interviewee 44.

[79] Interviewee 44.

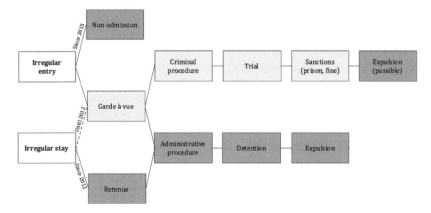

Fig. 5.1 Irregular entry and stay in France—criminal and administrative pathways (*Source* Author's elaboration. Criminal procedures in light grey; administrative ones in dark grey)

can be placed in detention right at the end of custody.[80] Likewise, there is evidence of calls between prefects and prosecutors, in which the latter would indicate to proceed administratively, and make the GAV end only once the persons are already in detention.[81] It must be stressed that in France, differently from in Italy, prosecutors do not always have the legal obligation of pursuing offences.[82]

On the contrary, if the criminal process is pursued (which seems rather uncommon), a trial begins, with migrants summoned to court with immediate comparison.[83] In case of condemnation, they are sentenced to imprisonment, fined and, at the end of the prison term, possibly expelled. According to the responsible for prison-related issues at La Cimade, condemnations for irregular entry can be very swift: A person could be apprehended on Thursday, undergo trial on Saturday, and be brought to prison right after.[84] Once the prison term expires, migrants leave but,

---

[80] Henriot (2013, p. 2).

[81] Interviewee 52.

[82] Code of Criminal Procedure, art. 79. In particular, they have the obligation to pursue *crimes*, but not *délits*.

[83] Interviewee 48.

[84] Interviewee 56.

if they were irregular before entering, they are immediately expelled or placed in a detention centre waiting for expulsion.[85]

The above is how the criminal procedure for irregular entry or stay would have normally unfolded, until the early 2010s. Yet, a few changes occurred, which affected the proceedings.

As far as irregular stay is concerned, both the criminal pursuit of the infraction and the GAV preceding it were abrogated in 2012 (as seen in section 'Post-2004: The Influence of the ECJ'). Consequently, since 2013, foreigners found in contravention of rules on stay are placed in *retenue* for up to 16 hours, following which the administrative procedures leading to expulsion begin. The criminal process cannot be started, unless other infractions are registered.

Regarding irregular entry, since 2015, the reintroduction of border controls in France has enabled the police to directly subject foreigners to push-backs (*non-admissions*) to the country from which they are coming. According to the Departmental Director of the Alpes-Maritimes *Police aux Frontières* (PAF), this practice has indeed replaced the use of GAV along the French-Italian border, and will remain the predominant one until borders are opened again.[86]

Criminalisation in Numbers

Having seen the process through which criminalisation takes place, it is now possible to ask to what extent the norm has been used. In this section, I discuss the actual implementation of the norm, based on figures released by the French Government.

Registered Infractions

Starting with the infractions against the general conditions of entry and stay that were registered by the French police and gendarmerie, these have fluctuated since 2000 (Fig. 5.2).[87] Indeed, registered cases of entry and stay substantially increased in the first decade of the twenty-first century, and especially in the first seven years, when the number of annual cases

---

[85] Interviewee 56.

[86] Interviewee 43.

[87] Unless otherwise specified, data regarding registered infractions of irregular entry and stay derives from: Departmental figures registered by police and gendarmerie (2018).

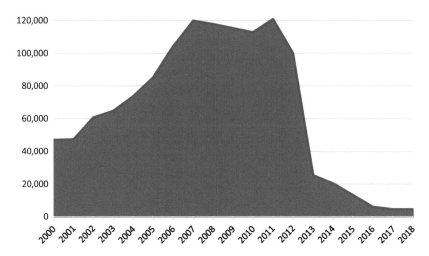

Fig. 5.2 Infractions to general conditions of entry and stay registered by French police and gendarmerie, 2000–2018 (*Source* Author's elaboration based on Departmental figures registered by police and gendarmerie [2018])

almost doubled, from 46,886 in 2000 to 117,603 in 2007. The figures then stabilised over the following five years (2007–2011), but dropped significantly after 2011: In 2013, the number of crimes registered plummeted by 74.8%, down to 25,079, and then further fell to 5861 in 2016, and to 4331 in the first 10 months of 2018.

Most apprehensions took place in provinces close to the border, and in particular in four areas: On the south-eastern French-Italian border (characterised by migrants crossing from Ventimiglia to Menton), on the south-western French-Spanish border (a crossing point for migrants making use of the Western Mediterranean route from Morocco to the EU through Spain), in Paris (a key attraction point for migrants seeking opportunities), and Calais (a crucial aggregation point for migrants aiming to attempt the journey to the United Kingdom). Indeed, in the period 2000–2018, the four departments of Paris, Alpes-Maritimes, Pyrénées-Orientales and Pas-de-Calais registered a total of 300,047 infractions to the general conditions of entry and stay, i.e. a third of the entire continental France (which registered 918,597 cases in those 19 years).[88]

Overall, registered infractions increased between 2000 and 2007, then plateaued until 2011, and finally sank since 2013 (in parallel to the repeal of the crime of irregular stay in December 2012). How many of the registered cases were brought to trial? How many involved criminal custody? This is the subject of the next sub-section.

Indictments and GAV

Data on the use of the GAV for migration-related offences is scarce. Figures from INHESI (the French National Institute for Higher Studies in Security and Justice) and the Parliament show that in 2010, GAV was employed in 61,184 instances for infractions against general conditions of entry and stay, for a total of 85,137 people.[89] In 2011, the number of cases was 59,629 (most of which were based on art. L.621-1 of Ceseda).[90]

Yet, official data on the number of cases of GAV related to irregular entry or stay that are comparable across a substantial number of years are not publicly available, to the knowledge of the author. Therefore, in this part of the analysis, I focus on the number of people against whom charges have been raised, using it as a proxy for the number of GAV related to irregular entry or stay.[91]

In 2005–2013, the number of people charged with irregular entry or stay saw a pattern similar to the one described above for registered crimes (Figs. 5.3 and 5.7).[92] The number gradually increased after 2005, peaking at 111,692 in 2008. After that year, it decreased significantly, especially from 2011 onwards. More specifically, 2009 and 2010 saw an average 12% decrease compared to the respective previous years, whereas, after

---

[88] Departmental figures registered by police and gendarmerie (2018).

[89] Rapport 2011 de l'INHESI/ONDRP, cited in Lacaze (2012, p. 199).

[90] Karamanli (2012, p. 23). Similarly, Rastello (2012).

[91] Indicted persons are placed in GAV for the above-mentioned questioning and controls. Moreover, the number of people put in GAV, and that of people being indicted, were both 85,137 in 2010, implying a substantial overlap of the two indicators. See Lacaze (2012, p. 199) and Crimes to foreigners' police (2013).

[92] Unless otherwise specified, data regarding indictments for irregular entry and stay derives from: Scherr (2011, p. 151), for years 2005–2007, and Crimes to foreigners' police (2011 and 2013), for years 2008–2013.

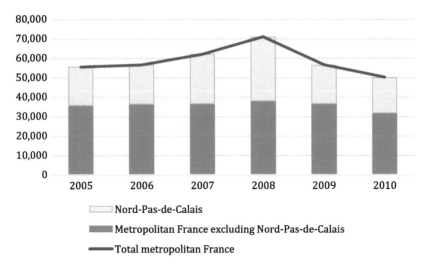

Fig. 5.3 Persons indicted by the Police aux Frontières for infractions of entry and stay, in metropolitan France and Calais, 2005–2010 (*Source* Author's elaboration based on data from Crimes to foreigners' police [2011 and 2013] and Scherr [2011])

a short window of stable figures in 2010–2011, 2012 saw a 20% drop compared to 2011. More steeply, 2013 witnessed a 74% decrease, which led to 18,198 people being indicted (i.e. 16.29% of the 2008 levels).[93] As noted by the French Government, the decrease in indictments post-2012 was a direct consequence of the jurisprudence of the ECJ and the French Supreme Court on criminalisation and the use of GAV.[94]

Zooming in on regional trends, Scherr attributes the 2008 peak in indictments for irregular entry/stay to the single region of Nord-Pas-de-Calais (Fig. 5.3).[95] Indeed, the region appears as the key driving force behind the changes at the national level that took place in both 2008 and 2009, being responsible for 88.5% of the total increase in the former year,

---

[93] Following Law 2012–1560, index 69 only concerns the crimes of irregular entry, not of stay (Crimes to foreigners' police 2011). All percentages in the section were calculated by the author, based on the above-mentioned databases.

[94] Crimes to foreigners' police (2013).

[95] Scherr (2011, p. 149).

and for 80% of the decrease in the latter.⁹⁶ On the contrary, looking at indictments for the rest of metropolitan France, figures were in fact rather stable from 2005 to 2009, with numbers fluctuating between 35,551 and 37,431.⁹⁷

Aiming to explain the above trend, Scherr hypothesises the 2008 increase to be related to the growing number of migrants aggregating in Calais to cross the border to the United Kingdom, due to worsening international conflicts (although he does not specify which ones).⁹⁸ With regard to the 2008–2009 decrease, then, he notices a few potential reasons.⁹⁹ First, not only did police officers in Calais likely start looking for more efficient solutions (e.g., by prioritising administrative expulsions), but the dismantling of the 'Jungle of Calais' in September 2009 may also have contributed to falling indictments. Second, the stable pattern in the rest of metropolitan France following 2007, he suggests, was likely a result of the police aiming to meet the targets reached in that year. Third, he stresses the relevance of the international context, including the decreased attractiveness of Europe following the global financial crisis (for more on this, see section 'Macro-level: International Political and Economic Determinants of Flows').

Following the repeal of GAV for irregular stay, the use of *retenue* as an alternative method to check foreigners' regularity of stay increased. Starting from 29,947 cases in 2013, these rose to 43,765 in 2016, and to 33,711 in the first six months of 2017.¹⁰⁰ Whether or not the GAV is still used to question migrants at the border is contested.¹⁰¹ Two trends however emerge from the interviews and secondary literature. First, already in 2012, Maugendre noted an increase in the number of people being held in GAV for *double délits* (double crimes), meaning irregular stay *and* another one (such as insult against public officials), which he interprets as a stratagem employed by the police to still make

---

[96] In 2008, indictments at the national level increased by 8136, of which 7198 took place in Nord-Pas-de-Calais. In 2009, national figures registered a decrease of 15,583, of which 12,967 in Nord-Pas-de-Calais (ibid.).

[97] Ibid., p. 154.

[98] Ibid., p. 152.

[99] Ibid., pp. 152–153.

[100] Etude d'impact (2018, p. 148).

[101] See interviewee 45, as opposed to interviewees 44 and 52.

use of the GAV, while technically they could not.[102] He called this the *garde à vue 'de confort'*,[103] and Lochak similarly argued that foreigners are often held for meaningless infractions (such as for smoking in public places), used as pretexts to interrogate them on their residence status.[104] Second, an interviewee reported that, although requisition orders would not be possible (since irregular stay is no longer a crime and irregular entry needs to be caught in flagrance), in Calais the police still make use of special powers to interrogate migrants, under the instruction of the public prosecutor. The above highlights the significant resistance of the police to abandon the GAV as a tool of migration management, and points to the different expedients used to avoid doing so.

Overall, from the above, an inverted U-shape trend emerges as characterising the number of cases of irregular entry and stay indicted by French officials, which peaked in 2008. How many of these resulted in actual trials? And how many in condemnations?

Condemnations
Condemnations for irregular stay drastically fell, from the 1990s to today (Fig. 5.4). According to data released by the French Ministry of Justice, these saw a downward trend starting in the mid-1990s, when they steeply decreased from 11,845 in 1994 to 4979 in 1998.[105] The following years then experienced a fluctuating trend, with roughly 4000- to-6000 condemnations per year in 1999–2008. Finally, in 2009, the yearly number dropped to 2275, and then further to 1124 in 2016.[106]

Looking at the type of sanctions given by courts, the majority of condemnations consistently resulted in prison terms from 2012 to 2016, with only fewer fines (Fig. 5.5). Specifically, while, in 2012–2016, the average yearly condemnations to imprisonment for irregular entry or stay

---

[102] Rastello (2012).

[103] Ibid.

[104] Lochak (2016, p. 3).

[105] Condamnations selon la nature de l'infraction de 2009 à 2016 (Condemnations by infraction, 2016). Although irregular stay was decriminalised in December 2012, the data report levels different from zero for the years following 2012. It is possible that, despite the name ('*Condamnations pour séjour irrégulier des étrangers*'), the dataset includes not only condemnations for irregular stay, but for other crimes too, and in particular irregular entry.

[106] Condemnations by infraction (2016).

5 FRANCE: INSTRUMENTALISATION, COURTS, AND MARGINALISATION   193

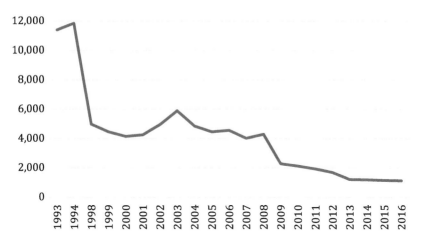

Fig. 5.4 Condemnations for irregular stay in France, 1993–2016 (*Source* Author's elaboration based on data from Condemnations by infraction [2016])

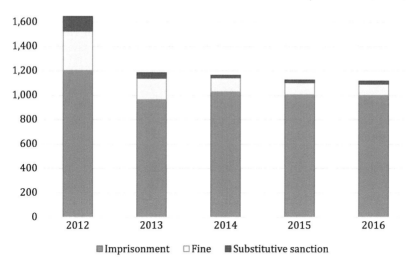

Fig. 5.5 Condemnations for irregular entry or stay in France, by type of sanction, 2012–2016 (*Source* Author's elaboration based on Tables of condemnations [2012–2016: Code 26101])

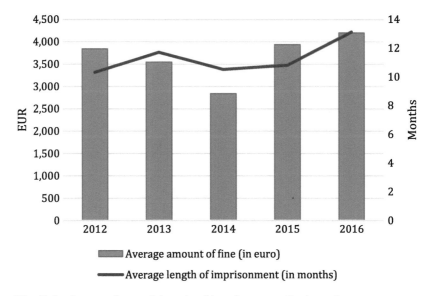

Fig. 5.6 Average fine and length of imprisonment for irregular entry or stay, 2012–2016 (*Source* Author's elaboration based on Tables of condemnations [2012–2016: Code 26101])

were 1038, those sanctioning a fine were 163, and the ones leading to expulsion 46.[107] As a result, although the Ceseda foresees that fines and imprisonment should be sanctioned together,[108] this was not always the case.

As far as the severity of sanctions is concerned, it appears that, despite a decrease in the use of the norm in 2012–2016, sanctions slightly worsened (Fig. 5.6). On one hand, the average length of imprisonment (one year according to the law) was 10 months in 2012, 13 in 2016. On the other hand, average fines (€3750 according to the law) fluctuated between 3833 in 2012, 2830 in 2014, and 4188 in 2016. There is evidence suggesting that sanctions were less severe before 2012, with

---

[107] Tables of condemnations (2012–2016). Unless otherwise specified, data regarding the type of condemnations for irregular migration derive from: Tables of condemnations (2012–2016'), for years 2012–2016, and Annuaire Statistique de la Justice (2006–2012), for years 2006–2010.

[108] Ceseda, art. 621-1, initial version.

average fines never higher than €1200 in 2000–2010, and average imprisonments of about 6 months in 2000–2009. Yet, pre-2012 data refers not only to irregular entry and stay, but to other migration-related crimes too, and is thus not perfectly comparable to post-2012 figures.[109]

The limited number of sanctioned expulsions may be explained by the general preference of local actors (primarily prosecutors, encouraged by the Ministry of Justice directly)[110] to proceed through the administrative pathway leading to removals, rather than achieving them through the often-lengthier criminal system. Indeed, in 2016, the average length of criminal processes for irregular entry or stay was over one year (16 months),[111] whereas administrative detention could last up to 45 days.

Putting Data into Perspective

In the above paragraphs, I have highlighted how the number of registered crimes, of indicted people, and of final condemnations changed throughout time. By comparing the number of condemnations with that of registered crimes, it becomes apparent that only an extremely small percentage of the latter actually resulted in a sentence of guilt (Fig. 5.7).

Taking 2009 as a reference year, a study of impact by the French Government noted the following[112]:

- 103,817 people were accused of an infraction related to irregular entry or stay;
- 80,063 were placed in GAV for the same charges (77% of the total);
- 5306 were condemned on the basis of article L. 621-1 of CESEDA (irregular entry or stay) (6.6% of people in GAV);

---

[109] See Annuaire Statistiques de la Justice (2006–2012). Specifically, the Annuaire statistiques (2011–2012: 198) refer to code 7 on 'irregular entry or stay', as also including unauthorised entry after an entry-ban, the *délit de solidarité*, and the failure to respect house arrest for people who were issued an expulsion order. The Tableaux des condamnations, however, use code 26101, related specifically to the 'Irregular entry or stay of a foreigner' as processed under the Police des étrangers – nomads.

[110] See Circular JUSD0630020C, section II, 1.1.1.

[111] Tables of condamnations (2016).

[112] Data from Etude d'impact (2012, pp. 33–34). Percentages calculated by the author.

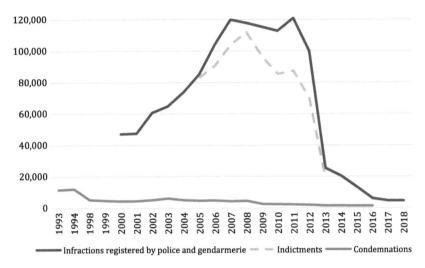

Fig. 5.7 Overview of data concerning irregular entry and stay in France, 1993–2018 (*Source* Author's elaboration based on Departmental figures registered by police and gendarmerie [2018], Crimes to foreigners' police [2011 and 2013: Index 69 of état 4001], and Condemnations by infraction [2016])

- 597 were condemned, with art. L621-1 of CESEDA representing the only infraction (11% of people condemned for art. L621-1 in general);
- 197 were condemned to imprisonment, with art. L621-1 of CESEDA representing the only infraction (33% of people condemned for art. L621-1 only);
- Of these, 50 were cases of second offence (*recidive*) (25% of people imprisoned for art. L621-1 only).

Consequently, 147 people were imprisoned on the sole basis of committing an infraction of irregular stay for the first time, representing 0.14% of the 103,817 people who were initially implicated for such offence. On the contrary, most of the people condemned were so for both irregular migration and other offences. Interestingly, the study also reports that the average fine for art. 621-1 of Ceseda in 2009 was €368,[113] much less than the foreseen €3750 that the Code foresees.

Considering the broader 2000–2016 period, the percentage of people condemned for irregular entry or stay, over those registered for it, had a U-shaped trend. Starting from an average of 8–9% in the early 2000s, it dropped to 1.60% by 2011, to then reach 9% in 2015 and 19% in 2016.[114] Of note, the increased proportion of the last years took place at a time when both the number of crimes registered and of actual condemnations were decreasing, reaching unprecedented low levels in 2016. As a result, the rise in the percentage, since 2012, was caused by the faster drop in registered crimes than in condemnations, but does not imply an increase in the latter.

To sum up, the above data show that, while the GAV was used in significant proportions in the first decade of the twenty-first century, the full penal response was actually 'in retreat'.[115] In other words, the thesis that the criminal law and GAV were instrumentalised to increase returns finds ample support in the analysis of official figures.

Indeed, the preference for a selective use of criminal proceedings to sanction migration-related offences was explicitly declared already in a 2006 circular by the Ministry of Justice. This recommended public prosecutors to abandon the charges against simple irregular entry or stay infractions, and only pursue them in the case of second (or multiple) offences, or when additional crimes were concerned.[116] In 2011, the Government re-confirmed such guidelines, inviting prosecutors to only pursue infractions to general rules of entry and stay when these were apprehended in connection with other crimes.[117]

Such data is also consistent with the information reported by an interviewed officer of the *Police aux Frontières* in Nice. According to him, prosecutors generally pursue irregular entry or stay through the administrative pathway, unless the person under investigation is caught with false

---

[113] Etude d'impact (2012, p. 34).

[114] Author's elaboration based on Departmental figures registered by police and gendarmerie (2018), and Condemnations by infraction (2016).

[115] Étude d'Impact (2012, pp. 33–34).

[116] Circular JUSD0630020C, section II, 1.1.1.

[117] Circular 11-04-C39.

documents, or has committed other crimes.[118] Interestingly, according to the Director of the Alpes-Maritimes PAF, criminally pursuing irregular migration would involve a paradox, since the person would be forced to remain on French territory, rather than be expelled.[119]

I will return to such paradox in section 'Coherence: The Contrast with the Return Directive'. In the meantime, further questions need to be addressed: What have been the effects of criminalisation, both in terms of deterrence and more broadly? How can they be related to the patterns of implementation described above? In the next section, I address such matters, to understand whether it is possible to identify a deterrent effect stemming from the criminalisation of migration, and assess the latter's overall effectiveness.

Effectiveness: Beyond Goals and Outcomes

To delve into the outcomes of the criminalisation of migration, and its effectiveness in terms of deterrence, in this section I analyse the norm by referring to the five evaluation criteria discussed in the introduction of the book—namely efficacy, efficiency, coherence, sustainability, and utility. As a first step, I provide some background regarding the main trends of irregular migration to France: What have been the fluctuations regarding unauthorised movements to the country, in the twenty-first century?

Irregular Migration in France: 2008–2017

Before discussing patterns of irregular migration to France, a caveat is necessary. In Italy, the numbers involving irregular migration are rather easy to identify in recent years, due to the relative visibility of sea arrivals and the detailed reporting of both Frontex (since 2008)[120] and the Italian Ministry of the Interior (especially since 2013).[121] The geography of France, however, makes it harder to obtain definite numbers,

---

[118] Interviewee 44.

[119] Interviewee 45.

[120] Through the database on Detections of illegal border-crossings statistics, available at https://frontex.europa.eu/along-eu-borders/migratory-map/.

[121] Through the Cruscotto giornaliero, available at http://www.interno.gov.it/it/sala-stampa/dati-e-statistiche/sbarchi-e-accoglienza-dei-migranti-tutti-i-dati.

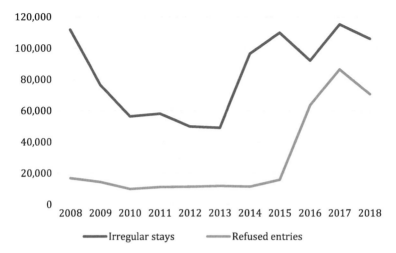

Fig. 5.8 Third country nationals found irregularly present in France, and refused entry at French external borders, 2008–2018 (*Source* Author's elaboration based on Eurostat migr_eipre and migr_eirfs)

to the extent that the Government does not release official figures[122] on irregular migration, due to their controversial nature. Eurostat provides post-2008 data on migrants who are apprehended in irregular situations and refused entry at the border (Fig. 5.8).[123] However, the lack of border controls (until 2015) at the frontier with other Schengen countries makes them likely underestimated.

Indeed, while in 2008 (with no border controls), three quarters of the people refused entry in France were so at airports,[124] in 2017 (with border controls), 87% of them were refused entry at a land border (mostly at the French-Italian one).[125] In absolute numbers, refusals at air and sea borders were rather constant in 2008–2017 (9444 on average at air

---

[122] Wihtol de Wenden (2010, p. 119).

[123] Respectively: Eurostat migr_eipre and migr_eirfs. Unless otherwise specified, figures on apprehended irregular migrants and third country nationals refused entry comes from these two databases.

[124] EMN (2011, p. 16).

[125] La Nouvelle République (2018).

borders, 675 at sea borders),[126] whereas those at land borders spiralled from 3135 in 2008, to 75,610 in 2017. While the steep rise may be due to an actual increase in attempted crossings, it is also likely to be the result of more controls.

The limited number of years that both had border controls in place and that fall within the scope of this study (2015–2018), makes the analysis of the post-2015 period of little help. Therefore, data on apprehensions at the French border are not used in the analysis of the effectiveness of criminalisation, which relies instead on the numbers of migrants found irregularly present inside the country. While, clearly, not even these can be expected to be perfectly representative, they offer greater consistency through time.

France traditionally had a rather high population of irregular migrants, and was consistently among the top EU member states by number of foreigners apprehended in irregular situation, between 2008 and 2017.[127] While maintaining such position, the country registered a drastic decrease in apprehended migrants around the turn of the decade, from 111,690 in 2008 to 48,965 in 2013. Following that year, however, the numbers increased again in 2014 and 2015, almost reaching 2008 levels in 2015 (with 109,720 cases) and overcoming them in 2017, with 115,085 people found irregularly present.

Looking at the countries of origin of people detected as irregular in the country, Afghanistan, Eritrea, and Iraq were the three main ones in 2008–2009 (Fig. 5.9). Indeed, it is changes in migration from those three countries that were largely responsible for the overall decrease in apprehensions in the late 2000s: Between 2008 and 2009, Eritrean and Iraqi numbers diminished by over 10,000 each; between 2009 and 2010, Afghani figures dropped by over 13,000.[128] Overall, these figures account for over half of the total reduction in apprehensions for both 2009 and 2010. 2011 then saw an increase in Tunisians, and, in 2018, most apprehended migrants were from Algeria and Iraq.

To what extent can changes in these trends be attributed to criminalisation? In the remainder of the chapter, I employ the five evaluation

---

[126] Based on Eurostat migr_eirfs.

[127] Based on Eurostat migr_eipre.

[128] Based on Eurostat migr_eipre.

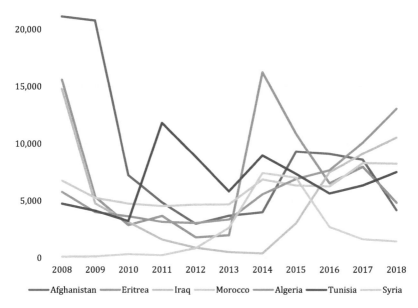

Fig. 5.9 Third country nationals found to be irregularly present in France by nationality (rounded, selected nationalities only), 2008–2018 (*Source* Author's elaboration based on migr_eipre)

criteria to assess the effectiveness of criminalisation. The discussion represents the last of the three gaps highlighted by Czaika and de Haas (the efficacy gap), and is enriched by the consideration of both criminological and IPE arguments.

As argued earlier, the goal of the criminalisation of migration in France was twofold. On one hand, the measure was employed to facilitate and increase returns, through the instrumental use of criminal custody. On the other hand, it aimed to decrease irregular flows to the country, through deterrence. Have such goals been achieved? Assessing the efficacy of the norm, I look at both objectives, starting with that of enhancing repatriations.

Efficacy (I): Criminalisation and Expulsions

Did criminalisation contribute to increasing the number and proportion of returns carried out by France? To anticipate this section's argument, it does not seem to have done so significantly.

To begin with, as seen above (section 'Criminalisation in Numbers'), the number of people sanctioned to expulsion following a trial for irregular entry or stay was limited. In 2012–2016, on average 46 cases of irregular entry or stay were sanctioned with expulsion each year, signalling a limited use of criminalisation to pursue returns.[129]

It is true that GAV may have played a role by making administrative detention and expulsion orders easier. Indeed, if we look at the rate of effective returns to third countries,[130] this increased from 15.5% in 2008 to 19.5% in 2012 (in parallel to the extensive use of GAV), to then fall after that year (in parallel to the repeal of GAV for irregular stay).[131]

However, a few caveats are worth mentioning. First, even the intense use of criminal custody did not manage to significantly reduce the gap between the French rate of effective returns to third countries, and that of the EU (see Figs. 5.10, and 6.5). Indeed, the French effective return rate is surprising low: Despite the increase from 2008 to 2012, the proportion was never higher than 19.5%. Instead, it was consistently about half the overall EU average, in all years between 2008 and 2017.[132] Second, *retenue* grants police officers powers that are similar to those they had through the GAV, which makes it difficult to argue that the repeal of the latter in 2012 was behind the reduced effectiveness of returns following that year. Third, examining the above in absolute terms, it appears that returns to third countries in 2008–2012 were in fact not increasing consistently, as they were in 2012 no higher than they had been in 2008.[133]

Overall, the extensive use of criminal custody in 2006–2012, although paralleled by a slight increase in the rate of effective returns, did not

---

[129] Tables of condemnations (2012–6).

[130] The 'effective return rate' refers to the proportion of issued expulsion orders that are actually carried out. Specifically, I focus here on repatriations to third countries, rather than to other EU member states.

[131] Data on returns is based on Eurostat migr_eirtn and migr_eiord.

[132] Ibid.

[133] Eurostat migr_eirtn.

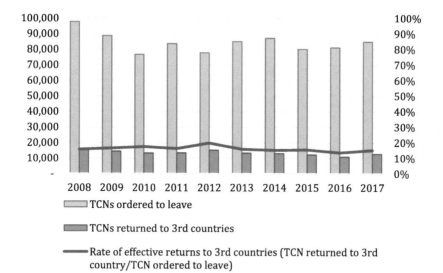

Fig. 5.10 Third country nationals (TCNs) ordered to leave and returned to third countries by France, 2008–2017 (*Source* Author's elaboration based on Eurostat data)

manage to close the gap between French and EU effective return rates. The full criminal response did not contribute to increasing returns either, as numbers of expulsions following criminal trials were, on average, rather low.

Efficacy (II): Criminalisation and Irregular Migration

Having discussed whether the criminalisation of migration in France may be seen as having contributed to enhancing returns, I now shift to the second goal of the norm: Deterring potential (irregular) migrants. Has the measure achieved such goal?

Focusing on the decrease in foreigners found in irregular situations from 2008 to 2013, some sources have associated such trend to the deterrence measures put in place by Sarkozy as Minister of the Interior first (2002–2004 and 2005–2007), President then (2007–2012), of which the extensive use of GAV may be seen as an example. For instance, according to Frontex, the French (and Greek) 'stricter law-enforcement measures' are among the likely explanations for the 2008–2009 decrease of detections of Afghan and Iraqi nationals on the eastern Mediterranean route.[134] Indeed, during Sarkozy's mandates, harsh rhetoric was employed, and stricter legislation supported: Expulsion of undocumented immigrants was made easier, return targets were set, and detention was extended from 12 to 32 days.[135]

At the same time, however, the above looks problematic, and several arguments counter the view of criminalisation having had a significant deterrent effect. In the next pages, I first outline the key points presented in official documents and reported in the interviews, as well as problems related to data, to then analyse the issue from a double perspective: On one hand, I apply the micro-level criminological criteria defined in Chapter 3. On the other hand, I look at macro-level international political and economic determinants of migratory flows.

Data Issues and Stakeholders' View

The first problematic aspect with perceiving criminalisation as having reduced migratory flows since 2008 concerns figures. Not only has the number of condemnations for irregular entry or stay been rather low in the analysed period, but further issues surface too.

To start with, Frontex itself highlighted that lower apprehensions did not necessarily signal lower presences. Instead, the 2009 decrease was likely due to the dispersal of migrants following 'the dismantlement of the camps […] in northern France', and 'the end of the border controls with Switzerland where many migrants staying illegally in France used to be detected on exit'.[136] 'Thus', the agency continues, 'the decreasing detections *do not necessarily reflect a actual decrease in the number of illegal migrants*'.[137]

Furthermore, the 2008–2010 decrease was largely due to shrinking numbers of registered Eritrean, Iraqi and Afghan citizens, all of whom

---

[134] Frontex ARA (2010, p. 16).

[135] See Marthaler (2008), Schain (2012, pp. 40–41, 117), and Henley (2003).

[136] Frontex ARA (2010, p. 30).

[137] Ibid., emphasis added.

had high asylum recognition rates.[138] Because, according to the Refugee Convention, refugees are exonerated from sanctions for irregular entry or stay,[139] criminalisation was not applicable to great part of the Iraqi, Eritrean, and Afghan migrants found irregular. Therefore, most of the decrease in the arrival to France of people with such nationalities (which together accounted for 58% of the total decrease in 2009, and 88% in 2010),[140] cannot be explained by the stricter enforcement of migratory laws, since these were not applicable to such population.

In line with the above, most of the people I interviewed reported not thinking of criminalisation as having a deterrent effect.[141] Different lawyers argued that criminalisation cannot prevent people from desiring a better life,[142] and that its effects are very limited when compared to the situations that make people leave their homes.[143] In particular, after travelling across the Sahara desert, being held in Libyan detention centres, and crossing the Mediterranean sea, criminal sanctions are hardly dissuasive.[144] Even the Director of the Alpes-Maritimes *Police aux Frontières*, while arguing that criminalisation in general does have a deterrent effect, recognised that the effect of the specific criminalisation of irregular entry and stay in France has been limited in practice.[145]

Institutional bodies seem to acknowledge the limited efficacy of criminalisation to provide a significant deterrent effect too. These include the Senate Commission on Conditions in French Prisons,[146] as well as the Ministry of the Interior itself. Indeed, when proposing the decriminalisation of irregular entry from a Schengen state in 2018, the latter's *étude d'impact* foresaw no effect of such choice (since the French services

---

[138] In 2008 and 2009, Eritrean asylum seekers had a recognition rate of 69% in France, and Iraqi applicants of 82%. Less staggeringly but still significantly, Afghani applicants had a recognition rate of 30% in 2008 and 37.5% in 2009. See MPI (2018).

[139] Refugee Convention (1951, art. 31).

[140] Based on migr_eipre.

[141] E.g., Interviewee 52.

[142] Interviewee 45.

[143] Interviewee 48.

[144] Interviewee 50.

[145] Interviewee 43.

[146] French Senate (2000, Rapport n. 449, p. 35).

were already meant to disapply the norm as in contrast with EU law).[147] No mention of a decreased efficacy, or more limited instruments for the administration, was made.

Overall, as seen so far, data, interviews, and institutional documents do not appear to support the claim that the criminalisation of migration had a deterrent effect on migratory flows. Why has this been the case? In the next sections, I discuss both criminological and IPE aspects to provide an answer to such question.

Micro-Level: Criminological Criteria

A first potential interpretation of criminalisation's lack of efficacy in reducing migration may be derived by looking at the micro-level. Specifically, how did criminalisation fare, compared to the criminological factors identified in Chapter 3?

Starting with the legal costs, it is noticeable that the severity of the norm has been great in the French case, with the foreseen sanction involving both a fine, a prison sentence, and the possible interdiction from the territory. Certainty of incurring the sanction, however, has not always been privileged.

As far as severity is concerned, criminological studies have highlighted that it can be counterproductive if excessive, as magistrates may decide not to condemn if they perceive sanctions to be unnecessary.[148] Interestingly, this seems to have been the case in France since the very introduction of the crime of irregular entry or stay. As Henri Batiffol argued already in 1946, the excessive severity of the 1938 decree led many public prosecutors to refrain from criminally pursuing such crimes, defeating the very purpose of the law.[149] This still resonates today, and may indeed have affected the certainty of incurring punishment.

More precisely, concerning the certainty of the sanctions, a distinction must be made: GAV has been used rather systematically, especially in 2006–2012, as it was the traditional way through which orders to leave the French territory were issued.[150] Yet, it has very seldom led to the full application of the criminal process. Indeed, as seen previously, the

---

[147] Etude d'Impact (2018, p. 161).

[148] Nagin (2013, p. 20).

[149] Batiffol (1948, p. 101). Similarly, Loyer (2016, p. 15).

[150] Karamanli (2012, pp. 23, 25).

number of condemnations for irregular entry or stay—already decreasing since 1994 and never higher than 2300 per year since 2009—has consistently been rather low, and would be hard to hypothesise any meaningful impact on migrants' willingness to flee.

In fact, even with regard to the GAV, following the Achughbabian sentence, the procedure was characterised by a period of uncertainty. The Government initially provided a restrictive interpretation of the ECJ jurisprudence[151] which was however not shared by all.[152] Police and gendarmerie services in different areas thus applied different standards in the process of charging foreigners with irregular stay,[153] and different jurisdictions issued contradictory rulings,[154] turning the Achughbabian sentence into what Henriot called a 'relative' concept.[155]

As a result, the requirement of certainty of punishment seems to have been satisfied only in part, and specifically with regard to the application of criminal custody in 2006–2012. After that year, however, uncertainty increased. Moreover, throughout the last two decades, condemnations were not consistently pursued.

Looking at migrants' own perspective, criminalisation does not appear to have played a significant role in affecting their migration decisions. To begin with, it is questionable whether they knew of the measure itself. According to a staff member of the Association d'Aide aux Jeunes Travailleurs (AAJT)[156] in Marseille, who has experience with both asylum seekers and unaccompanied minors, migrants generally have very general knowledge of policies.[157] In particular, he reports that while some migration is organised (through smuggling networks), not all of it is, and it is not uncommon to find people without any precise strategy about what country to migrate to.[158]

Indeed, none of the respondents to the questionnaires I conducted acknowledged knowing the possibility of being imprisoned for irregular

---

[151] Circular 11-04-C39.
[152] Henriot (2013), Lacaze (2012, p. 202), Ferré (2012, pp. 96–97).
[153] Crimes to foreigners' police (2011 and 2013).
[154] Henriot (estimated 2012, p. 2) and Rastello (2012).
[155] Henriot (estimated 2012, p. 2).
[156] http://aajt.fr/AAJT/index.php?p=accueil.
[157] Interviewee 50.
[158] Interviewee 50.

entry or stay. Two respondents reported having no knowledge at all of any of the sanctions in place. Out of the three who reported having at least some knowledge, all were aware of repatriations, but only two of detention. Interestingly, only one knew about such measures before leaving.[159]

Importantly, the respondent who indicated knowing about the sanctions in place before emigrating, said that such knowledge did not make him change his decision to flee.[160] This is consistent with the result of interviews conducted by Guiraudon among Moroccan migrants in France, for whom the decision to leave their country for France was not affected by the latter's migration policies, but rather by economic or political reasons, and usually made possible by the help of relatives.[161]

While the above is anecdotal evidence, and I do not aim to say it can be representative, it is significant to highlight that the gap between (a) governments' policies aimed at deterring inflows, and (b) the effective knowledge held by some of the migrants themselves, may be substantial. In this context, the frequent changes in legislation (see section 'Coherence: The Contrast with the Return Directive') are likely to further negatively affect the degree of knowledge of policies and sanctions, requiring detailed understanding and frequent updates.

If the legal costs are only relatively consistent, and information on criminalisation scarce, the social costs associated to the norm in receiving countries seem higher. Indeed, both the GAV[162] and the full criminal procedure[163] are meant to stigmatise foreigners, linking them with criminals and people who deserve being punished.

The French Government itself recognised the stigmatising effect of the norm criminalising migration, acknowledging that the related imprisonment has the characteristic of associating foreigners with a 'suspicion of potential delinquency'.[164] The Eurobarometer indeed suggests that negative views of migrants are quite widespread in France, where more than

---

[159] Surveys were carried out in September and October 2018 in Paris.

[160] Survey FR02.

[161] Guiraudon (2008, p. 376).

[162] Interviewee 48.

[163] Interviewee 52.

[164] Etude d'Impact (2012, p. 33).

half of the surveyed population (57.11% in 2014, 54.85% in 2018) associate migration of people from outside the EU to either 'fairly negative' or 'very negative' feelings.[165]

Is this considered a concern by migrants themselves? In analysing the above-mentioned interviews with Moroccan immigrants in France in the late 2000s, Guiraudon reports that all of them noted a significant worsening of everyday discrimination, rhetoric on migration, and policies.[166] Importantly, however, none of the respondents, or of their families, preferred going back to Morocco, and in fact all declared being happy to be in France.[167]

This might find an explanation in the fact that, as argued in Chapter 3, the sense of stigmatisation may be limited by the perceived lack of legitimacy of the receiving country's migratory system.[168] Indeed, this seems to have been the case in France, as the responsible for prisons at La Cimade reports that foreigners in prison, more than being deterred by the country's criminal sanctions, are disgusted by them.[169] Likewise, Guiraudon noted that the migrants she interviewed perceived the way in which the French consulates treated Moroccans as being humiliating, racist, and unequal.[170]

Still, the toughening of migration policies may itself be the cause of further marginalisation of migrants, and therefore of further engagement with irregular activities. According to Claire Saas, criminalisation 'paralyses' the individual, making his/her journey much longer, and reintegration into society extremely difficult: In the absence of valid documents and permits, it is very hard for foreigners to obtain a regular job, even at the end of the criminal procedure, which as a consequence pushes them closer to other irregular networks and activities.[171] This is made

---

[165] See http://ec.europa.eu/commfrontoffice/publicopinion/index.cfm/Chart/getChart/themeKy/59/groupKy/279.

[166] Guiraudon (2008, p. 377).

[167] Ibid., pp. 377–378. Although Guiraudon's study only relies on three respondents, her findings suggest a lack of strong significance of stigmatisation in receiving countries as perceived by migrants themselves, and is relevant in that it includes different regional realities, including northern, central and south-eastern France.

[168] See Ryo (2015) in Chapter 3.

[169] Interviewee 56.

[170] Guiraudon (2008, p. 379).

[171] Saas (2012, p. 41).

worse by the fact that the possibility to sentence foreigners to expulsion at the expiration of the prison term provides few incentives for the state to invest in supporting their reintegration into the French market and society,[172] despite the latter being one of the objectives of criminal sanctions.[173]

Passing through the penal procedures also means that the person will be registered as having a criminal record in a number of national and European databases on both criminal records and migratory information (such as VIS, Eurodac, SIS I and II), which will therefore affect their subsequent life, including in seeking a residence permit and a job.[174] This, of course, is the case even when such penal procedures are only used as an attribute of the expulsion apparatus.[175]

The above takes place in a context in which migrants are already more likely to be unemployed, than French citizens. Indeed, third country nationals' unemployment rate increased from 18.1% in 2008 to 25.8% in 2014 in France.[176] The figure has slightly diminished in more recent years, although it is still considerably higher than the EU average, amounting to 23.7% in 2017 (against a 16.5% EU average).[177] In this context, the gap between the unemployment rate affecting third country nationals and French citizens is also significant, as the former was 15 percentage points higher than the latter from 2011 to 2017.[178]

In Saas's view, it would be better to term the marginalising and related effects of the criminalisation of migration as 'opposite' consequences, rather than 'undesired' ones, since they are well-known.[179] Indeed, this seems to have already been the case for the 1938 decree-law which first introduced the crime of irregular entry and stay: As many of the migrants already present in the country could not regularise their status, they were forced to live in irregularity.[180] The Government itself partly recognised

---

[172] On this point, see Maugendre (2012, p. 55).
[173] Penal Code, arts. 130–131.
[174] Saas (2012, p. 41).
[175] Ibid.
[176] Based on Eurostat lfsa_urgan (2019).
[177] Ibid.
[178] Ibid.
[179] Saas (2012, p. 41).
[180] Ben Khalifa (2012, p. 26).

such possibility, acknowledging that people who were supposed to leave the country may not able to obtain a visa to do so, and foreseeing alternative solutions for such instances (although still based on surveillance).[181] Yet, according to Ben Khalifa, as a consequence of such institutional setting, France became a true '*fabrique de clandestins*' (factory of irregular migrants), in the years 1938–1940.[182]

Finally, the existence of the *délit de solidarité*[183] for people supporting foreigners in irregular situations also contributes to creating a sense of isolation, making both individuals and associations refrain from providing humanitarian support, in fear of being penalised themselves. According to D'Ambrosio, the goal of such norm was both to prevent foreigners from being helped, and to send a signal to the local population that immigration laws must be respected, or consequences would ensue.[184] Indeed, the head of the French branch of Refugees Welcome, an NGO that helps refugees integrate in receiving communities, revealed feeling discouraged to provide assistance to migrants in situations of irregularity, lest the organisation be prevented from continuing to operate.[185] This can have significant consequences for migrants themselves, pushing them further underground.[186]

Overall, the criminalisation of migration is likely to increase the stigma associated by citizens to migrants. While the latter might in some cases shield themselves from such factor by emphasising the lack of legitimacy of the French migration system, consequences are still likely to be significant. Specifically, criminalisation may promote foreigners' marginalisation, including by affecting their criminal record.

As shown in this section, the ineffectiveness of criminalisation to bring about deterrence may be explained at the micro-level by the limited certainty of sanctions, but also by the questionable ability of information to persuade potential migrants not to emigrate. In this context, although

---

[181] Decree of 2 May 1938, p. 4967.

[182] Ben Khalifa (2012, p. 26).

[183] Humanitarian assistance to migrants in need was declared not to be criminally pursuable in July 2018 by the Conseil Constitutionnel. See: https://www.vie-publique.fr/focus/decrypter-actualite/delit-solidarite.html.

[184] D'Ambrosio (2010, p. 6).

[185] Interviewee 57.

[186] Interviewee 48.

foreigners' stigmatisation in the receiving country might be filtered by their perception of the migration system as illegitimate, it nonetheless can have significant consequences.

Macro-level: International Political and Economic Determinants of Flows

Beyond the micro-level aspects mentioned above, the lack of success of the criminalisation of entry and stay in deterring migration may also be understood with reference to macro-level international political and economic factors. As I will argue, at this level of analysis, interpretations that attribute the fluctuations in irregular migration to France to factors *other than* criminalisation seem to hold better explanatory power. In particular, in the next pages I focus on the likely external determinants of the flows in three periods: 2008–2013, 2013–2017, and post 2017.

To start with, a comparison with trends at the EU level is needed. Irregular migration has seen a decrease in the period 2008–2013 not just in France, but in the whole of the EU: Combining the number of third country nationals found irregularly present with that of TCNs refused entry at the external borders, a net decrease can be seen at the EU level, as these figures combined dropped from 1,218,230 in 2008 to 739,800 in 2013.[187] Similarly to the Italian case, diminished irregular migration in the EU has been associated to two factors: EU enlargement, and the global financial crisis.

Looking at the first hypothesis, according to Triandafyllidou, the decrease in irregular migration to Europe following 2000 was due to the EU's 2004 and 2007 enlargements. Indeed, in the 1990s and early 2000s, Central-Eastern European migration to EU15 member states was significant, with part of it being irregular. Once such states joined the EU, however, their citizens' status became automatically regularised, which led to a reduction in the irregular foreign population in other EU countries.[188] In France, restrictions to free movement for the countries that joined in 2004 were only removed in 2008,[189] which could help explain the decrease in foreigners found irregularly staying, following that year (although more data would be needed to test the hypothesis).

---

[187] Eurostat Migr_eipre and migr_eirfs.
[188] Triandafyllidou (2010, p. 12).
[189] http://news.bbc.co.uk/1/hi/world/europe/3513889.stm#france.

More evidence seems to be available to test the second hypothesis, i.e. that following 2008, the global financial crisis reduced the attractiveness of several European countries, where less opportunities were available. According to Fix and others, irregular migrants are highly 'sensitive to changes in economic and labour market conditions'.[190] Indeed, studies in the United States have found significant correlation between economic cycles, in particular the wage-gap between the United States and Mexico, and irregular migration.[191] Likewise, ILO and OECD reports have argued that foreigners' unemployment, or the gap between their and citizens' unemployment rates, may influence migratory patterns.[192] For Castles and Miller, multiple factors drive the possible decrease in irregular migration following economic crises.[193] Not only are the sectors in which migrants typically work (agriculture, construction, catering...) highly impacted by economic fluctuations, but migrant networks also pass on such information. Further, often not being able to access welfare support (if irregular), migrants have little incentive to migrate without a job, and receiving states tend to increase border controls.[194]

Indeed, through statistical analysis on data regarding the period between January 2008 and May 2009, Frontex finds the unemployment rate (at the EU level) to have contributed to reducing the number of foreigners found irregular in the EU.[195] In their words: 'for every thousand people more unemployed, two people fewer are detected for staying illegally in the EU'.[196] In this context, not only did the financial crisis greatly increase third country nationals' unemployment levels throughout the EU (which surged from 14.5% in 2008, to 22.3% in 2013), but, among EU15 countries, France also stood out for experiencing a great and increasing gap between foreigners' and citizens' unemployment levels.[197]

---

[190] Fix et al. (2009, p. 19).
[191] Papademetriou et al. (2009, p. 4).
[192] Herm and Poulain (2012, pp. 146–147).
[193] Castles and Miller (2011, p. 6).
[194] Ibid.
[195] Frontex (2009, p. 15).
[196] Ibid.
[197] Reyneri (2009).

It is important to highlight how such trends are linked to specific sectors and gender, however: Post-crisis unemployment has affected male migrants more than female ones.[198] Indeed, in the EU, the former tend to be mainly occupied in cyclical occupations (construction, hospitality, low-tech manufacturing) with generally temporary contracts (or no contract at all).[199] Female migrants, on the other hand, tend to dominate the non-cyclical domestic work sector (health, care, and housekeeping), which is less related to economic fluctuations.[200] At the same time, women from the Middle East, North, and sub-Saharan Africa migrate predominantly for family reunification purposes, rather than for labour opportunities.[201] Indeed, the great majority of the decline in apprehensions between 2008 and 2013 regarded male migrants, with female ones remaining rather constant, only marginally decreasing.[202]

In the context of decreasing migration to France in 2008–2013, the year 2011 represents a slight exception, with marginally higher numbers of foreigners found irregularly staying. The increase can be explained by referencing the inflows from Tunisia following the Arab Spring. Indeed, especially in the early months of the year, many in the country took advantage of the lack of order that followed the revolution to cross the Mediterranean, reach Italy, and travel further north to France.[203] Although France temporarily reintroduced border controls in April 2011,[204] figures show an increase of 8575 apprehensions of Tunisians between 2010 and 2011.[205]

Focusing now on the increase in apprehensions starting in 2013, it should first of all be reminded that 2013 and the following years (2015 especially) were the ones in which the EU as a whole saw increasing levels of irregular migration through the Mediterranean Sea, due predominantly to the Syrian and Libyan civil wars.

---

[198] Ibid.

[199] Ibid.

[200] Ibid., Frontex (2009, p. 25).

[201] Blangiardo (2012).

[202] Eurostat migr_eipre.

[203] Fargues and Fandrich (2012, p. 4).

[204] See https://www.ft.com/content/8d963a4c-7422-11e0-b788-00144feabdc0.

[205] Based on Eurostat migr_eipre.

According to the French interior ministry, while France has not been directly hit by the increase in migratory flows to southern European countries, the indirect effect has been notable, with an increase in asylum requests, more pressure on key border points (notably the frontier with Italy), and big agglomeration of migrants.[206] As an example, in 2017, France received 121,200 asylum requests, 50% more than in 2014.[207]

Importantly however, the increase in such flows has been strongly related to worsening conditions in the Middle East, in particular in Syria. As a matter of fact, the number of Syrians found irregularly present in France increased substantially after 2010, from 845 in 2012, to 7415 in 2014. The apprehension of Eritreans also rose in 2014, when it reached 16,240 (compared to 2630 the previous year).[208]

Moreover, as for Italy, the worsening of the Libyan situation also played a key role in incentivising migrants to attempt the journey across the Mediterranean Sea. Looking at the nationality of the people demanding asylum, a study by the interior ministry stresses the increase especially in people from Sub-Saharan Africa (and the Balkans): In 2017, Albanians represented 11% of the requests, followed predominantly by sub-Saharan African countries such as Sudan, Guinea, Ivory Coast, and the DRC.[209] Among the top 10 countries by asylum applications in France, five have vast parts of the populations speaking French (Haiti, Guinea, Ivory Coast, DRC, Algeria), which seems to highlight the relevance of cultural and linguistic ties in explaining migrants' chosen destination country.[210]

Finally, looking at the last couple of years, the criminalisation of irregular entry does not seem to have played a decisive role either. Indeed, the number of people refused entry at the border was quite substantial (and increasing) in 2016–2017.[211] In 2017, over 87% of entry refusals took place at a land border,[212] and more than half at the south eastern border

---

[206] Etude d'Impact (2018, p. 7).
[207] Etude d'Impact (2018, p. 7).
[208] Eurostat migr_eipre.
[209] Etude d'Impact (2018, pp. 9–11).
[210] See, for example, Thielemann (2003 and 2011).
[211] Etude d'Impact (2018, p. 10).
[212] La Cimade (2018).

with Italy.[213] These trends clearly indicate persisting secondary movements within the EU, and there is evidence of migrants making up to twenty attempts to cross the French-Italian border,[214] which very much contradicts the idea of them being dissuaded.

Overall, irregular migration to France saw a decrease in 2008–2012, followed by an increase since 2013. As I have argued, these changes are unlikely to be a result of a deterrent effect stemming from criminalisation. Indeed, a substantial number of actors, including public officials, argued that the norm should not be seen as having a deterrent effect, due not only to its limited application, but also to its scarce relevance when compared to the reasons that incentivise migrants to flee. Indeed, the fluctuations in migratory patterns seem to be better explained by international political and economic developments, first of which the effects of the global financial crisis in Europe, and the worsening situation in the Libyan and Syrian civil conflicts.[215]

Having discussed the efficacy of criminalisation in meeting its goals of reduced migration and increased returns, in the next section I consider the costs and benefits involved in the application of the norm in France.

Efficiency: Prison, Expulsions, and Fines

The structure dedicated to the management of migration in France has seen a notable expansion over the last decades, passing from having 320 people to patrol the borders in 1938,[216] to 10,825 in 2017.[217] Likewise, the budget dedicated to the management of irregular migration is estimated to have increased to over €4.4bn in 2010.[218] In this context, is criminalisation a cost-effective strategy for migration control? Looking at the different steps involved in the norm, it seems relevant to analyse the cost of the GAV/*retenue*, the cost of prison, the cost of repatriations, and the revenues from fines.

---

[213] La Nouvelle République (2018).
[214] La Cimade (2018, p. 39).
[215] Similarly, see Scherr (2011, pp. 152–153).
[216] Ben Khalifa (2012, p. 17).
[217] French Senate avis n. 114 (2014, p. 32).
[218] Gourevitch (2012, p. 63).

To begin with, the move from *garde à vue* to *retenue* seems to have brought a reduction of costs, although not necessarily in a positive way for foreigners themselves. While the provisions relating to meals, interpreters, medical assistance, and rights to contact relatives and the embassy were rather similar, those regarding legal support did change, in a restrictive way. Indeed, lawyers, whose support indicted people can request, represented a significant cost of GAV for the state. Taking 2011 as a reference year, the ministry of the interior estimated that, out of 59,629 people placed in GAV for irregular entry or stay, 34% asked for the support of a lawyer, which effectively occurred in 85% of the requested cases (17,233).[219] Most frequently, legal aid covered not only an initial 30 minute meeting, but further discussions too.[220] The total expenses for the state to support the legal assistance of people in GAV for irregular entry and stay in 2011 was thus €4,042,720.[221] With the shift to the administrative *retenue* since early 2013, the above-mentioned legal costs were expected to diminish, since the foreigners involved would no longer have the right to a lawyer for the whole duration of custody, but only to the initial thirty-minute meeting.[222] Being the cost for such activity €61, the new expenses were foreseen to be €1,051,213 (17,233 cases * €61), involving a net decrease of costs of €2,991,507 (−74%).[223] While the decrease of costs is notable, the outcome is not necessarily positive as, if the *retenue* is just a GAV in administrative terms, as argued by Lochak,[224] then the reduction of legal protection for foreigners is to be questioned.

A second cost related to criminalisation is that of the prison term to be served by the foreigner condemned. According to the French Senate, in 2013, the cost of one day in prison (for any offence), averaged €99.49.[225]

---

[219] Étude d'Impact (2012).

[220] Ibid.

[221] Ibid. The overall cost results from the sum of: Initial interviews (5170 cases * 61€ = 315,370€), support throughout the first 24 hours of GAV (11,340 cases * 300€ = 3,402,000€), and support in cases of extended GAV (723 cases * 450€ = 32,535€).

[222] Étude d'Impact (2012).

[223] Ibid.

[224] Lochak (2016, p. 4).

[225] French Senate, avis n. 114 (2014, p. 42).

This means that in 2013, for the 287 cases of condemnations to imprisonment where irregular entry was the only infraction,[226] which lasted on average ten months, the overall cost was of €8,566,089 (287 cases * 300 days * €99.49). The cost is likely to have been higher in previous years, when condemnations were more frequent (Fig. 5.4).

Third, expulsions are very expensive too, especially when forced.[227] Estimates on the actual costs vary widely, from €10,000 to €27,000. If for Gourevitch, forced returns cost on average €10,000 in 2009 and 2011,[228] according to the Ministry of Immigration, the unitary cost in 2009 was approximately €12,000.[229] Both the *Cour de Comptes* and the Senate's Budgetary Commission are more pessimistic, estimating the cost at €13,220 and €20,970, respectively.[230] Finally, La Cimade goes beyond all such estimates, maintaining that the cost of an expulsion in 2008 was instead closer to €27,000, as any estimate should include costs such as migrants' detention, transportation, and other.[231] In 2012–2016, on average 46 expulsions were sanctioned each year, in consequence of infractions related to irregular entry or stay,[232] which would make the overall expenditure for such cases range from €460,000 (46 cases * €10,000) to €1,242,000 (46 cases * €27,000).

Overall, the costs of criminalisation are hard to quantify. Yet, the above shows they would be significant, if the norm was fully enforced. To the factors highlighted, the cost of police officers, judges, and administrative personnel dealing with cases of criminalisation should be added, which would result in a further increase in the overall amount of expenses related to criminalisation.

Furthermore, the high cost of prison is to be combined with the fact that after serving the prison term, the person could be expelled. Although from the data this does not seem to often be the case, it is indeed a possibility. This would add to the argument of the scarce efficiency of criminalisation: Quite evidently, the high costs of maintaining a person in

---

[226] Tables of Condemnations (2012–2016).
[227] According to Gourevitch (2012, pp. 54–63), voluntary returns cost less to the state.
[228] Gourevitch (2012, pp. 54–63).
[229] Harzoune (2012).
[230] Ibid.
[231] Ibid.
[232] Tables of condemnations (2012–2016).

prison would find no apparent goal, if the person was to be expelled at the end of the prison term, since repatriation would impede any chances of social reintegration of the person condemned.[233] This would seem to only have a goal of punishing the foreigner (in Henriot's words, 'making them pay')[234] and deterring further migration, which is however problematic for (at least) two reasons. First, the absence of plans for the reintegration of the convicted person in the French society creates insecurity both for the person herself, and for the society more broadly. Indeed, there may be a risk of prison leading to further engagement with irregular activities[235] and, potentially, radicalisation.[236] Second, if, as argued above, deterrence is not working, the justification of the measure is nullified. The above makes the expenses related to imprisonment even less efficient and defensible.

Although the above costs are to be balanced against the potential revenues from criminalisation (the fines to be paid by foreigners), the latter's significance is questionable. In 2012–2016, on average 163 people were condemned for irregular entry or stay to a fine, for an average amount of €3663,[237] which represents a total of €597,069 in revenues per year, on average. However, it is highly doubtful that such sums were paid. On one hand, the affected migrants often have only very limited availability of funds. On the other hand, due to the status of irregularity, they do not generally own assets that could be confiscated as collaterals (such as houses or other) in France.

Overall, therefore, although unsystematic enforcement reduces the costs of criminalisation, these would be substantial if the norm was applied extensively. Furthermore, the potential revenues from fines are likely to mostly go unpaid.

[233] Maugendre (2012, p. 55).
[234] Henriot (2012, p. 62).
[235] Saas (2012, p. 41).
[236] Neumann and Rogers (2007).
[237] See Tables of condemnations (2012–2016).

Coherence: The Contrast with the Return Directive

Coherence is a key parameter for the analysis of criminalisation in France, as both its internal and external dimensions have proved quite problematic in the country, due to several reasons.

First, the frequent changes in French immigration law make maintaining the internal coherence of a norm quite hard. Indeed, the French migration-management system is extremely complex. As of 2018, it included 9 different categories of returns, and transposed the 4 types of residence permit foreseen at the EU level into 17 different ones.[238] Changes to existing legislation are also very frequent: Since 1980, the system was modified by 16 major laws and, since 2005, the Ceseda was amended on average every two years.[239] The DEMIG Policy database (concerning migration policy changes in forty-four countries), reports that France was the country with the highest number of changes between 1945 and 2014, with over 250.[240] Even the *Conseil d'état* expressed its hope for a drastic simplification of migration legislation.[241] Indeed, the consequences of such complexity and frequent changes are notable, particularly for the coherence of norms, their implementation, and evaluation.[242]

Second, and crucially, criminalisation had a problematic relationship with EU legislation and particularly the Return Directive (external coherence). This indeed led to the repeal of the crime of irregular stay first, and that of irregular entry from a Schengen state then, following ECJ rulings (see section 'Post-2004: The Influence of the ECJ'). Interestingly, Henriot notes a paradox: while the French Government, judges, and administration tried to resist the changes to criminalisation brought about by the ECJ jurisprudence, the goal of the latter was to prioritise expulsions, which, in fact, was the very objective of the French Government too.[243] This points to the inherent internal contradiction of a norm

---

[238] Conseil d'état avis 394206 (2018), sec. 8.
[239] Ibid., sec. 7.
[240] de Haas et al. (2015, p. 19).
[241] Conseil d'état avis 394206 (2018), sec. 4–10.
[242] Ibid., sec. 7.
[243] Henriot (2013, p. 13).

that, wanting to expel foreigners, forces them to remain in prison in its territory,[244] paying for the expenses and risking further marginalisation.

Despite the strong resistance to change, once this became inevitable, the French Government used coherence to justify it. In 2012, decriminalisation of irregular stay was presented as increasing the coherence of the French immigration system with what was already a widespread practice of not pursuing the related crimes through penal means. In the words of the Government: '[T]here's true coherence in connecting the criminal legislation with the societal evolutions so that the law does not consider a foreigner as committing a crime for the only purpose of him/her staying irregularly in France'.[245] Likewise, when irregular entry from another Schengen state was decriminalised in 2018, the Government presented the choice as increasing the coherence of the CESEDA with European law, following the Affum sentence.[246]

Overall, internal incoherence of criminalisation of irregular entry and stay (due to numerous changes and contrasting goals), and external incoherence with the Return Directive (due to the lack of prioritisation of expulsions), proved key detractors from the effectiveness of the norm, to the extent of leading to the repeal of substantial parts of it.

Sustainability: Long-Term Effects of Criminalisation

The French criminalisation of irregular entry and stay seems to lack in sustainability too, due to several reasons. First, building on the argument just made concerning internal coherence, the many changes of the migration management legislation make it hard to evaluate the long-term effects of policies.

Furthermore, while instrumentalisation meant that the norm was present but not vastly applied, the trend was criticised by the criminal chamber of the Supreme Court. For the latter, instrumentalisation did not represent a good practice, since GAV should have the goal of leading to the related criminal procedure.[247] This not only implies that, in the long term, it should not be possible to maintain an instrumental use of

---

[244] Similarly, Interviewee 43.
[245] Étude d'impact (2012, p. 34). Author's translation.
[246] Étude d'Impact (2018, p. 161).
[247] Henriot (2013, p. 8).

the criminal law, but also reasserts the key intervention of courts, in the French case.

Finally, the burden that a full-scale criminalisation would pose on the criminal and prison system is substantial. This would be so in financial terms (as seen in section 'Efficiency: Prison, Expulsions, and Fines'), but also in terms of human resources needed. For the Senate Commission on Conditions in French Prisons, criminalisation has a 'perverse' effect, overcrowding prisons, and degrading detention conditions.[248] Interestingly, when the Ministry of Justice recommended public prosecutors to abandon the charges against cases of simple irregular entry or stay, the stated goal was to 'avoid an overload of jurisdictions',[249] demonstrating the unsustainability of the full application of the norm.

Utility: Law-Enforcement and Political Concerns

Despite the limited salience of criminalisation in France, the adoption and continuation of the norm seem to have served political and bureaucratic goals.

To better understand this, we need to go back to the late nineteenth century, when the country had significant labour shortages, and low population growth.[250] This is helpful to understand the emergence, in the early 1920s, of two leading interest groups aiming to affect migration policy: Employers, who were interested in obtaining access to foreign labour, and pronatalists, who aimed to increase the French population by increasing fertility rate.[251] Interestingly, migration was vastly privately organised at the time: Employers would directly hire persons from abroad, asking for their regularisation only once in the country (to the extent that, in the 1960s, about 90% of new immigrants were regularised after entry).[252] While pronatalists are no longer at the forefront of migration policy today, employers are still very active, with substantial numbers favouring more liberalisation, to provide the 'structurally

---

[248] French Senate (2000), Rapport n. 449, p. 191.
[249] Circular JUSD0630020C, section II, 1.1.1.
[250] Hollifield (2014, pp. 158–159).
[251] Ibid.
[252] Ibid., p. 163 and Schain (2012, p. 47).

necessary' foreign labour to meet the demand for flexible employment.[253] Several trade unions, on the other hand, have traditionally opposed irregular migrants, perceiving them as competitors and a threat to wage levels.[254] This seems to have been the case both at the time of the introduction of the criminalisation of migration in 1938, and for a long time afterwards, although it changed somewhat in the late 2000s.[255] Thus, if, at the time of the introduction of criminalisation, employers tended to favour migration, an increasing share of the population was growing wary of it (see section 'The Evolution of Criminalisation' too).

Fast-forward to the 2010s, the attempts to maintain criminalisation seem to have had one clear goal: Exploiting the strength of the criminal system, to increase expulsions. Although this is largely an administrative and bureaucratic matter, I argue it has a political side too.

To begin with, as already demonstrated, criminalisation was used in an instrumental way, to increase returns. As reported by the ministry of the interior, when the number of people imprisoned for irregular stay in 2009 (197) is compared to those expelled and ordered to leave in the same year (29,288 and 94,693, respectively), it becomes evident that the true goal was making foreigners leave the country, rather than sanctioning them.[256] Indeed, the early 2010s witnessed widespread discomfort among local actors at the idea of decriminalising migration, to the extent that several courts are thought to have issued more expulsion orders in the first part of 2012, anticipating that the Supreme Court would no longer allow the use of GAV for irregular stay.[257]

Returns had a strong political dimension. By the time Nicolas Sarkozy became president, he had made expulsion targets a key political goal.[258] Indeed, already as Minister of the Interior in 2003, he had set expulsion targets for local authorities, indicating that they should at least double in 2004, and stating that '[t]he credibility of any public policy on immigration depend[ed] on effective execution of repatriation decisions'.[259] As

---

[253] Wihtol de Wenden (2010, p. 121).
[254] Ibid.
[255] Perry (2004, p. 343) and Wihtol de Wenden (2010, p. 121).
[256] Étude d'Impact (2012, pp. 34–35).
[257] France TV Info (2013).
[258] Marthaler (2008).
[259] Henley (2003).

president, Sarkozy then decided to publicly display the number of expulsions carried out, to strengthen support in view of the 2012 elections.[260] According to Henriot, gaining public support through returns was likely among the reasons for which the administration was so adamant to retain the use of GAV and the crime of irregular stay.[261]

Thus, the above may be interpreted as supporting the idea of the criminal law being instrumentalised to obtain more returns, and of these, in turn, being exploited by politicians to win the public's support.

The political side of criminalisation, however, also becomes apparent when focusing on the crime of irregular *entry* specifically. Recalling that (1) the GAV could be substituted by the *retenue*, with very similar purposes and effects, (2) evidence of criminal pursuit not being largely used was known by the legislators,[262] and (3) decriminalisation of irregular stay was said to increase coherence with societal practices,[263] why would the Government maintain criminal measures against irregular entry in 2012?

It seems that the reason for maintaining irregular entry as a criminal offence relates to the greater symbolic weight of the criminal law, as well as to the political difficulty of justifying decriminalisation. In 2000, for instance, centre-right Minister of Justice Clément argued in favour of maintaining the crime, stressing not only the importance of GAV to carry out expulsions, but also that decriminalisation 'would risk being interpreted as the willingness of the government to fight less vehemently irregular migration, to somehow lower the guard'.[264]

Thus, the instrumentalisation of the GAV, the political willingness to appear firm on migration (both through criminalisation and returns), and the symbolic power of the criminal law, look as the principal causes for maintaining irregular entry as a crime. In turn, criminalisation may be seen as benefitting two main groups: Law-enforcement actors (who could more easily meet their targets), and political actors (especially those emphasising expulsions).

---

[260] Henriot (2013, p. 3).
[261] Ibid.
[262] Étude d'Impact (2012, pp. 33–34).
[263] Ibid.
[264] French Senate (2005), Rapport n. 300, p. 222. Author's translation.

Criminalisation and the Purpose of Punishment

As seen in Chapter 3, the major philosophical purposes of criminal punishment have traditionally been identified with incapacitation, retribution, rehabilitation, and deterrence.[265] What is their relevance, in the context of criminalisation of irregular entry and stay in France?

Starting with incapacitation, in the case of the French criminalisation of migration, this goal would result from the imprisonment of the concerned persons, or their expulsion, since these are the only measures that would physically prevent irregular migrants from perpetuating their infraction, namely their situation of irregularity. As shown in section 'Criminalisation in Numbers', however, yearly figures on imprisonment for irregular entry and stay have traditionally been rather low, never overcoming 1200 since 2012. Likewise, as I have argued above, the GAV has not been able to drastically increase returns to third countries (section 'Efficacy (I): Criminalisation and Expulsions'). Consequently, the goal of incapacitation does not seem to have been successfully supported by the criminalisation of migration.

With regard to retribution, this implies punishing irregular migrants in order to make them atone for their guilts.[266] Several authors have stressed the non-proportionality of the sanctions related to breaches to migration rules in France[267] and, according to Henriot, it is the stigmatisation of migrants as dangerous and criminal individuals that is able to justify the punitive goal of criminalisation.[268] Indeed, criminalisation itself (and in particular imprisonment) may contribute to migrants' stigmatisation, in a vicious cycle that makes such measures seem justified and hard to repeal. It is however to be remembered that irregular entry and stay are rarely pursued criminally in practice (section 'Criminalisation in Numbers'), which does not seem to support the idea that the foreseen sanctions are intended to punish migrants.[269]

Moving to rehabilitation, this would imply the social and economic integration of migrants into hosting communities at the expiration of the prison term, and after the fine has been paid. Yet, as Maugendre argues,

---

[265] Banks and Pagliarella (2004, pp. 105–120).
[266] Henriot (2012, p. 62).
[267] See, for example, Saas (2012).
[268] Henriot (2012, p. 62).
[269] Similarly, Lochak (2016, p. 5).

the fact that migrants in France can be repatriated at the expiration of their prison term, directly counters the 'possibilities of amendment and social reinsertion of the sentenced person',[270] thus substantially nullifying the rehabilitative potential of the sanction. As shown throughout the chapter, this contradiction is in fact one of the elements that make criminalisation an inefficient and incoherent policy, since its goals are directly in contrast with one another: While the state forces migrants to remain in its territory through imprisonment, it hopes to expel them; while it disburses money to keep them in prison, it does not let them stay once prison ends; while it risks marginalisation through imprisonment, it does not foresee reintegration. In other words, no personal- or societal-amelioration goal is met, through the criminalisation of migrants in France.

This leaves us with the goal of deterrence, the hope of preventing further irregular migration through the threat of criminal sanctions. According to Lochak, the police's 'punctual operations' targeted against specific areas and nationalities are meant not so much to criminally sanction the persons arrested, but rather to interrogate the highest number of foreigners in irregular situation with the hope of intimidating both them and potential future offenders.[271] In other words, the real objective of the legislation on the criminalisation of migration is deterrence, and the stigmatisation of the concerned foreigners it carries with it.[272] As I have argued throughout this chapter, however, deterrence is not achieved, due to both micro- and macro-level factors (see section 'Efficacy (II): Criminalisation and Irregular Migration').

Overall, the above shows the lack of significance not only of deterrence, but also of any of the other three goals traditionally associated with criminal punishment, including incapacitation, retribution, and rehabilitation. Arguably, this points to the actual purpose of criminalisation and its continuation: The political and symbolic value of employing the criminal law to sanction migration-related offences, in a de facto instrumentalisation of the measure for electoral and bureaucratic purposes.

---

[270] Maugendre (2012, p. 55, author's translation).
[271] Lochak (2016, p. 3).
[272] Lochak (2016, p. 5).

Conclusion

In this chapter, I have analysed the evolution and effects of the criminalisation of migration in France. Starting from the analysis of the development of the norm since the late 1930s, I have then investigated its implementation, and finally assessed its overall effectiveness and potentially problematic aspects. In doing so, I have drawn from interviews with stakeholders, written questionnaires with migrants, and official databases.

With regard to the development of the norm, the criminalisation of migration does not seem to have been affected by significant discursive gaps, although a lack of data does not allow definite claims. Interviews and relevant literature suggest that the criminalisation of entry and stay has tended not to be greatly politicised in France. At the same time, the country is among the few in Europe foreseeing a mandatorily double punishment including both imprisonment and a fine, thus resulting with severe sanctions in place. Together, these two aspects do not seem to suggest the presence of notable discursive gaps.

What is interesting to note, however, is the significant *instrumentalisation* of the criminalisation of migration, to increase expulsions. By relying on the criminal custody that the norm allows, law-enforcement agents may withhold foreigners while checking their documents, and then swiftly transfer them to a detention centre to be repatriated.

Based on figures released by the French Government, this chapter has confirmed the hypothesis of criminalisation being used instrumentally. While, in 2006–2012, foreigners indicted for irregular entry or stay were about or over 70,000 per year (with a peak of over 110,000 in 2011),[273] annual condemnations for the same offence were never more than 4600,[274] thus signalling a great divergence between the number of people held in custody, and those actually sentenced. Following 2012, both indictments and registered offences dropped, in parallel to changes in the law that led to the repeal of the crime of irregular stay (but not of that of entry).

As a result, implementation gaps have been substantial, in the case of the criminalisation of migration in France. In this context, the European Court of Justice played a key role, in inducing the partial repeal of the norm in 2012.

---

[273] Crimes to foreigners' police (2013).
[274] Condemnations by infraction (2016).

Shifting now to efficacy gaps, these also emerge as relevant: Criminalisation only marginally increased returns to third countries, and does not seem a predominant factor in explaining fluctuations in migration to the country. Specifically, the lack of deterrence may be explained by a number of elements. From a micro-level perspective, not only has the certainty of punishment been generally low, but the great severity of the sanctions may also have contributed to judges' frequent decision not to fully enforce the measure. Further, surveyed migrants were scarcely aware of criminalisation. From a macro-level viewpoint, on the other hand, fluctuations in the number of migrants apprehended in France since 2008 appear to be explained more convincingly by international political and economic factors. Among these, the financial and Eurozone crises, as well as the conflicts in Syria and Libya, seem to hold more explanatory power in accounting for the 2008–2013 drop in migration to France and subsequent increase, respectively.

Beyond the limited efficacy of criminalisation in reducing irregular migration, the measure resulted problematic also with regard to the evaluation parameters of efficiency, coherence, sustainability, and utility. Several of such aspects are related to the fact that, while aiming to expel foreigners, criminalisation foresees a sanction to keep them in prison, on French territory. The contradiction not only makes the costs of prison redundant, but also drastically deducts from the internal and external coherence of the norm, to the extent that it was deemed incompatible with EU law by the ECJ. Quite paradoxically, even though the French administration vehemently fought to maintain irregular stay as a criminal offence in 2012, its objective was in fact aligned to that of the EU, as both aimed to prioritise expulsions. This was however not acknowledged by the French Government, which only reluctantly accepted to repeal the norm. Finally, the sustainability of the norm was negatively impacted by the high costs that the full-scale use of criminalisation would imply.

If, as argued so far, criminalisation has neither been successful at meeting its targets, nor coherent, why then has it been maintained, when possible? In other words, who benefits from it? As this chapter has argued, the instrumentalisation of the measure has profited not only the security apparatus, which under the Sarkozy presidency aimed to reach pre-defined return targets, but also the political actors who made expulsions a key electoral issue (first of whom Sarkozy himself). Despite the limited direct politicisation, the political and symbolic aspects of the norm have therefore been extremely relevant. This is further shown by the fact that, having the possibility of substituting the criminal GAV with the administrative *retenue* with substantially little differences, the Government still decided not to do so in 2012.

In conclusion, the French case well represents the paradox of deterrence, that is to say how securitisation and criminalisation are politically easy to pursue, but become extremely difficult to undo, even in the face of evident ineffectiveness. While not being effective, however, the criminalisation of migration is likely to lead to negative consequences, including a higher stigma associated to migrants, and their potentially greater marginalisation and engagement with underground activities, as will be discussed in more detail in the next chapter.

References

Annuaire Statistique de la Justice. 2006. Secrétariat Général, Direction de l'Administration générale et de l'Équipement, Sous-Direction De La Statistique, Des Études Et De La Documentation, Paris. http://www.justice.gouv.fr/art_pix/1_annuairestat2006.pdf. Accessed 13 Nov 2018.

Annuaire Statistique de la Justice. 2007. Secrétariat Général, Direction de l'Administration générale et de l'Équipement, Sous-Direction De La Statistique, Des Études Et De La Documentation, Paris. http://www.justice.gouv.fr/art_pix/1_annuaire2007.pdf. Accessed 13 Nov 2018.

Annuaire Statistique de la Justice. 2011–2012. Secrétariat Général, Direction de l'Administration générale et de l'Équipement, Sous-Direction De La Statistique, Des Études Et De La Documentation, Paris. http://www.justice.gouv.fr/art_pix/stat_annuaire_2011-2012.pdf. Accessed 13 Nov 2018.

Banks, Cyndi, and Irene Pagliarella. 2004. *Criminal justice ethics: Theory and practice*. Thousand Oaks: Sage Publications.

Batiffol, Henri. 1948. L'ordonnance du 2 novembre 1945: Communication de M. Batttifol. In *Travaux du Comité français de droit international privé*, 7e année, 1945–1946, 93–132. https://www.persee.fr/doc/tcfdi_1158-3428_1948_num_7_1945_1735. Accessed 12 Apr 2019.

Ben Khalifa, Riadh. 2012. La Fabrique des Clandestins en France, 1938–1940. *Centre d'information et d'études sur les migrations internationales*, 2012/1 N° 139: 11–26. https://www.cairn.info/revue-migrations-societe-2012-1-page-11.htm. Accessed 19 Sept 2019.

Blangiardo, Gian Carlo. 2012. *Gender and Migration in Southern and Eastern Mediterranean and SubSaharan African Countries*. Research Reports—Gender and Migration Series Demographic and Economic Module, CARIM-RR 2012/01. http://cadmus.eui.eu/bitstream/handle/

1814/20834/CARIM_RR_2012_01.pdf?sequence=1&isallowed=y. Accessed 21 May 2019.
Castles, Stephen, and Mark J. Miller. 2011. Migration and the Global Economic Crisis: One Year On. http://www.age-of-migration.com/uk/financialcrisis/updates/migration_crisis_april2010.pdf. Accessed 20 Sept 2019.
Condemnations by Infraction. 2016. French Ministry of Justice. (*Condamnations selon la nature de l'infraction de 2009 à 2016*). http://www.justice.gouv.fr/art_pix/stat_condamnations_infractions_2016.ods. Accessed 13 Nov 2018.
Conseil d'état. 2018. Avis sur un projet de loi pour une immigration maitrisée et un droit d'asile effectif, N° 394206 (Conseil d'état avis 394206), NOR: INTX1801788L, 15 February 2018. http://www.conseil-etat.fr/Decisions-Avis-Publications/Avis/Selection-des-avis-faisant-l-objet-d-une-communication-particuliere/Projet-de-loi-pour-une-immigration-maitrisee-et-un-droit-d-asile-effectif. Accessed 19 Feb 2019.
Cour de Cassation. 2012. Arrêt n° 965 du 5 juillet 2012 (11-30.530), Première Chambre Civile - ECLI: FR: CCASS: 2012: C100965. https://www.courdecassation.fr/jurisprudence_2/premiere_chambre_civile_568/965_5_23804.html. Accessed 27 Feb 2019.
Crimes to Foreigners' Police. 2011 and 2013. (*Délits à la police des étrangers - Évolution de l'action des services - index 69 2008–11, and 2010–13*), https://www.data.gouv.fr/fr/datasets/15064-delits-a-la-police-des-etrangers-evolution-de-laction-des-services-index-69/, https://www.data.gouv.fr/fr/datasets/2013-40-delits-a-la-police-des-etrangers-evolution-de-laction-des-services-index-69/. Accessed 12 Nov 2018.
D'Ambrosio, Luca. 2010. Quand l'immigration est un délit. *La vie des idées*, November 30. https://laviedesidees.fr/Quand-l-immigration-est-un-delit.html. Accessed 12 Apr 2019.
de Haas, Hein, Katharina Natter, and Simona Vezzoli. 2015. Conceptualizing and Measuring Migration Policy Change. *Comparative Migration Studies* 3 (1): 15.
Departmental Figures Registered by Police and Gendarmerie. 2018. (*Chiffres départementaux mensuels relatifs aux crimes et délits enregistrés par les services de police et de gendarmerie depuis janvier 1996*). https://www.data.gouv.fr/fr/datasets/chiffres-departementaux-mensuels-relatifs-aux-crimes-et-delits-enregistres-par-les-services-de-police-et-de-gendarmerie-depuis-janvier-1996/. Accessed 12 Nov 2018.
Dieu, François. 2015. La lutte contre l'immigration irrégulière. Quelques reprises sur l'expérience française. *Rivista di Criminologia, Vittimologia e Sicurezza* IX (1): 18–27.
EMN. 2011. *Rapport statistique annuel sur les migrations et la protection internationale 2008*. Elaboré par le Point de contact français du Réseau européen des migrations. https://ec.europa.eu/home-affairs/sites/homeaf

fairs/files/what-we-do/networks/european_migration_network/reports/docs/migration-statistics/asylum-migration/2008/09b._france_annual_protection_statistics_2008_final_version22aug2011_fr.pdf. Accessed 5 Jan 2020.

European Parliament and Council. 2008. *Directive 2008/115/EC of 16 December 2008 on Common Standards and Procedures in Member States for Returning Illegally Staying Third-Country Nationals*, December 24. O.J. L 348/98, 2008/115/EC. http://eur-lex.europa.eu/legal-content/EN/TXT/?uri=CELEX:32008L0115. Accessed 25 Aug 2015.

European Union Court of Justice, Case C-329/11, Alexandre Achughbabian v Préfet du Val-de-Marne, Judgment of the Court (Grand Chamber) of 6 December 2011, European Court Reports 2011-00000. http://eur-lex.europa.eu/legal-content/EN/TXT/?uri=CELEX%3A62011CJ0329. Accessed 29 Mar 2017.

European Union Court of Justice, Case C-47/15, Sélina Affum v Préfet du Pas-de-Calais and Procureur général de la Cour d'appel de Douai, Judgment of the Court (Grand Chamber) of 7 June 2016. https://eur-lex.europa.eu/legal-content/en/TXT/PDF/?uri=uriserv%3AOJ.C_.2016.296.01.0011.01.ENG. Accessed 29 Apr 2019.

Eurostat. 2018. Statistics on Enforcement of Immigration Legislation (including migr_eipre, migr_eirfs, migr_eirtn, migr_eiord). http://appsso.eurostat.ec.europa.eu/nui/show.do?dataset=migr_eipre&lang=en. Accessed Dec 2018.

Eurostat. 2019. lfsa_urgan. http://appsso.eurostat.ec.europa.eu/nui/show.do?dataset=lfsa_urgan&lang=en. Accessed Feb 2019.

Fargues, Philippe, and Christine Fandrich. 2012. *Migration after the Arab Spring*, MPC Research Report 2012/09.

Ferré, Nathalie. 2012. Les usages du droit pénal contre les étrangers. In *Immigration: un régime pénal d'exception*, ed. Gisti, 92–104. Paris: Gisti. https://www.gisti.org/publication_som.php?id_article=2781#1lutte. Accessed 23 Feb 2017.

Fix, Michael, Demetrios G. Papademetriou, Jeanne Batalova, Aaron Terrazas, Serena Yi-Ying Lin, and Michelle Mittelstadt. 2009. *Migration and the Global Recession*. Washington, DC: Migration Policy Institute.

FRA. 2014. *Criminalisation of Migrants in an Irregular Situation and of Persons Engaging with Them*. Vienna: FRA.

France TV Info. 2013. En 2012, les expulsions de sans-papiers ont atteint un record. *France TV Info*, January 22. https://www.francetvinfo.fr/france/en-2012-les-reconduites-a-la-frontiere-de-sans-papiers-ont-atteint-un-record_209117.html. Accessed 18 Feb 2019.

French Ministry of Justice. 2006. Circular JUSD0630020C. February 21.

French Ministry of Justice. 2011. Circular 11-04-C39. December 13. https://www.gisti.org/IMG/pdf/crim_no11-04-c39.pdf. Accessed 10 Jan 2019.

French Ministry of the Interior. 2012. *Etude d'Impact: Projet de loi relatif à la retenue pour vérification du droit au séjour et modifiant le délit d'aide au séjour irrégulier pour en exclure les actions humanitaires et désintéressées.* September 21 (Etude d'Impact 2012). http://www.senat.fr/leg/etudes-imp act/pjl11-789-ei/pjl11-789-ei.html. Accessed 10 Jan 2017.

French Ministry of the Interior. 2013. Circular INTK1300159C, January 18.

French Ministry of the Interior, Avant-projet de loi 'pour une immigration maîtrisée et un droit d'asile effectif' (Avant-projet de loi 2018), NOR: INTX1901788L/Rose-1, first version known, made public by the Gisti on 30 January 2018. https://www.gisti.org/IMG/pdf/pjl2018_pjl-norint x1901788l-rose-1.pdf. Accessed 10 Jan 2019.

French Ministry of the Interior, *Etude d'Impact : Projet de loi pour une immigration maîtrisée et un droit d'asile effectif*, 20 February 2018 (Etude d'Impact 2018). https://www.gisti.org/IMG/pdf/pjl2018_etude-impact_20180221.pdf. Accessed 10 Jan 2019.

French Parliament, Code de l'Entrée et du Séjour des Étrangers et du Droit d'Asile, as of 1 January 2017 (Ceseda). https://www.legifrance.gouv.fr/aff ichCode.do?cidTexte=LEGITEXT000006070158. Accessed 22 Jan 2016.

French Parliament, Code of Criminal Procedure, as of 17 February 2017. https://www.legifrance.gouv.fr/affichCode.do?cidTexte=LEGITEXT0 00006071154. Accessed 23 Feb 2017.

French Parliament, Décret sur la police des étrangers. *Journal Officiel* of May 1st, 2nd, 3rd 1938 (Decree of 2 May 1938), 4967–4969. https://gallica. bnf.fr/ark:/12148/bpt6k20313224/f23.image. Accessed 10 Jan 2019.

French Parliament, Law 80-9 of 10 January 1980. Relative à la Prévention de l'Immigration Clandestine et Portant Modification de l'Ordonnance n° 45-2658 du 2 Novembre 1945 Relative aux Conditions d'Entrée et de Séjour en France des étrangers et pour la Création de l'Office National d'Immigration. *Journal Officiel de la Republique Française*, 11 January 1980, 71–72.

French Parliament, Law 81-973 of 29 October 1981. Relative Aux Conditions D'entree Et De Sejour Des Etrangers En France. *JORF*, 30 October 1981.

French Parliament, Law 86-1025 of 9 September 1986, Relative aux Conditions d'Entrée ed de Séjour des étrangers en France, JORF, 12 September 1986, 11035.

French Parliament, Law 93-1420 of 31 December 1993, Portant modification de diverses dispositions pour la mise en oeuvre de l'accord sur l'Espace économique européen et du traité sur l'Union européenne. https://www. legifrance.gouv.fr/affichTexteArticle.do;jsessionid=552096381D2C122299 B868848237B5D6.tpdila10v_2?idArticle=LEGIARTI000006336362&cid Texte=JORFTEXT000000699737&categorieLien=id&dateTexte=19980511. Accessed 10 Jan 2017.

French Parliament, Law 778-2018 of 10 September 2018. Pour une immigration maîtrisée, un droit d'asile effectif et une integration réussie', JORF n° 0209 of 11 September 2018. https://www.legifrance.gouv.fr/affichTexte. do?cidTexte=JORFTEXT000037381808&categorieLien=id. Accessed 29 Feb 2020.

French Parliament, Law 2012-1560 of 31 December 2012. Loi relative à la retenue pour vérification du droit au séjour et modifiant le délit d'aide au séjour irrégulier pour en exclure les actions humanitaires et désintéressées. https://www.legifrance.gouv.fr/affichTexte.do;jsessionid=1368A8 F4515099719D3F22684307AB15.tpdila10v_2?cidTexte=JORFTEXT0000 26871211&dateTexte=20140525. Accessed 10 Jan 2017.

French Parliament, LOI n° 2018-778 du 10 septembre 2018 pour une immigration maîtrisée, un droit d'asile effectif et une intégration réussie (Law 2018-778), JORF, 11 September 2018. https://www.legifrance.gouv.fr/aff ichTexte.do?cidTexte=JORFTEXT000037381808&categorieLien=id.

French Parliament, Ordonnance n° 45-2658 of 2 November 1945. Relative Aux Conditions D'entrée Et De Séjour Des Étrangers En France. https://www. legifrance.gouv.fr/affichTexteArticle.do;jsessionid=552096381D2C122299 B868848237B5D6.tpdila10v_2?idArticle=LEGIARTI000006336357&cid Texte=JORFTEXT000000699737&categorieLien=id&dateTexte=19811029. Accessed 10 Jan 2017.

French Parliament, Penal Code, as of 1 January 2017. https://www.legifrance. gouv.fr/affichCode.do?cidTexte=LEGITEXT000006070719. Accessed 10 Jan 2017.

French Parliament, Projet de Loi N° 714, Pour une Immigration Maîtrisée et un Droit d'Asile Effectif, Enregistré à la Présidence de l'Assemblée nationale le 21 février 2018, PRÉSENTÉ au nom de M. Édouard PHILIPPE, Premier ministre, par M. Gérard COLLOMB, ministre d'État, ministre de l'intérieur. http://www.assemblee-nationale.fr/15/projets/pl0714.asp. Accessed 10 Jan 2019.

French Senate. 1981. Projet de Loi Relatif aux Conditions d'Entrée et de Séjour des Etrangers en France, N. 366, 10 September 1981. https://www.senat.fr/ leg/1980-1981/i1980_1981_0366.pdf. Accessed 10 Jan 2019.

French Senate. 2000. Rapport de la commission d'enquête (1) sur les conditions de détention dans les établissements pénitentiaires en France, créée en vertu d'une résolution adoptée par le Sénat le 10 février 2000, Tome 1 (Rapport n. 449). *Journal Officiel*, June 29. https://www.senat.fr/rap/l99-449/l99-449. html. Accessed 14 Mar 2019.

French Senate. 2005. Rapport de la commission d'enquête (1) sur l'immigration clandestine, créée en vertu d'une résolution adoptée par le Sénat le 27 octobre 2005, Tome II: Annexes (Rapport n. 300). *Journal Official*, 7 April

2006. https://www.senat.fr/rap/r05-300-2/r05-300-20.html. Accessed 14 Mar 2019.

French Senate. 2014. SESSION ORDINAIRE DE 2014–2015 Enregistré à la Présidence du Sénat le 20 novembre 2014, AVIS PRÉSENTÉ *au nom de la commission des lois constitutionnelles, de législation, du suffrage universel, du Règlement et d'administration générale (1) sur le projet de* loi de finances *pour* 2015, *ADOPTÉ PAR L'ASSEMBLÉE NATIONALE* (Avis n. 114). http://www.senat.fr/rap/a14-114-8/a14-114-8_mono.html#toc144.

Frontex. 2009. *The Impact of the Global Economic Crisis on Illegal Migration to the EU*. Warsaw, Poland: Frontex.

Frontex. 2010. *Annual Risk Analysis 2010* (ARA 2010). Warsaw: Frontex. https://frontex.europa.eu/assets/Publications/Risk_Analysis/Annual_Risk_Analysis_2010.pdf. Accessed 30 Apr 2019.

Gisti, ed. 2012. *Immigration: un régime pénal d'exception*. Paris: Gisti. https://www.gisti.org/publication_som.php?id_article=2781#1lutte. Accessed 23 Feb 2017.

Gisti. 2018. Commentaire de l'Avant Projet de Loi 'Pour Une Immigration Maîtrisée Et Un Droit D'asile Effectif', made public on 30 January 2018. https://www.gisti.org/IMG/pdf/plj20180130_analyse-2-e_loignement.pdf. Accessed 10 Jan 2019.

Gourevitch, Jean Paul. 2012. L'immigration en France: dépenses, recettes, investissements, rentabilité. *Contribuables Associés*, Monographie n°27, December 2012, p. 63.

Guiraudon, Virginie. 2008. Moroccan Immigration in France: Do Migration Policies Matter? *Journal of Immigrant & Refugee Studies* 6 (3): 366–381.

Harzoune, Mustapha. 2012. Combien coûte une expulsion? http://www.histoire-immigration.fr/questions-contemporaines/politique-et-immigration/combien-coute-une-expulsion. Accessed 10 Jan 2019.

Henley, Jon. 2003. France Sets Targets for Expelling Migrants. *The Guardian*, October 28. https://www.theguardian.com/world/2003/oct/28/france.jonhenley. Accessed 13 Apr 2019.

Henriot, Patrick. 2012. Les formes multiples de l'enfermement, une nouvelle forme de 'punitivité. In *Immigration: un régime pénal d'exception*, ed. Gisti, 60–71. Paris: Gisti. https://www.gisti.org/publication_som.php?id_article=2781#1lutte. Accessed 23 Feb 2017.

Henriot, Patrick. 2013. Dépénalisation du séjour irrégulier des étrangers: l'opiniâtre résistance des autorités françaises. *La Revue des droits de l'homme* [online], 3 | 2013, 1–14.

Henriot, Patrick. Estimated 2012. Garde à vue et séjour irrégulier : Les enseignements de l'arrêt «Achughbabian» sont limpides. http://www.syndicat-magistrature.org/IMG/pdf/Achughbabian.pdf. Accessed 14 Mar 2019.

Herm, Anne, and Michel Poulain. 2012. Economic Crisis and International Migration. What the EU Data Reveal? *Revue européenne des migrations internationales* [Online] 28 (4): 145–169.

Hollifield, James F. 2014. Immigration and the Republican Tradition in France. In *Controlling Immigration: A Global Perspective*, ed. James F. Hollifield, Philip L. Martin, and Pia M. Orrenius, 3rd edition, 157–187. Stanford: Stanford University Press.

Karamanli, Marietta. 2012. Communication de Mme M. KARAMANLI relative aux jurisprudences européenne et française en matière de garde à vue des étrangers mis en cause pour entrée ou séjour irrégulier. In Commission des Affaires Européennes (2012), Rapport d'information n° 129 sur des actes de l'Union européenne sur des textes soumis à l'Assemblée nationale en application de l'article 88-4 de la Constitution du 6 juin au 13 juillet 2012, 17–26. http://www2.assemblee-nationale.fr/documents/notice/14/europe/rap-info/i0129/(index)/rapports-information-ce. Accessed 5 Jan 2017.

La Cimade. 2018. Statistiques 2017: Ce Que Disent Les Chiffres De L'expulsion Des Personnes Exilées. La Cimade, August 25. https://www.lacimade.org/statistiques-ce-que-disent-les-chiffres-de-lexpulsion-des-personnes-exilees/. Accessed 5 Jan 2019.

La Nouvelle République. 2018. De plus en plus de migrants refoulés aux frontières en France. La Nouvelle République, June 28. https://www.lanouvellerepublique.fr/a-la-une/de-plus-en-plus-de-migrants-refoules-aux-frontieres-en-france. Accessed 5 Jan 2019.

Lacaze, Marion. 2012. La privation de liberté 'initiale' des étrangers en situation irrégulière. *Politeia* 22: 197–228.

Lochak, Danièle. 2016. 'Pénalisation', in 'L'étranger et le Droit Pénal', *AJ Pénal*, Janvier 2016, 10–12. https://hal-univ-paris10.archives-ouvertes.fr/hal-01674295/document. Accessed 5 Jan 2019.

Loyer, Elie-Benjamin. 2016. The Repression of Immigration Infractions in the French Criminal Courts (1880–1938). In *Eleventh European Social Science History Conference, International Institute of Social History*, March 2016, Valence, Spain. halshs-01296294.

Maga, Timothy. 1982. Closing the Door: The French Government and Refugee Policy, 1933–1939. *French Historical Studies* 12 (3): 424–442.

Marthaler, Sally. 2008. Nicolas Sarkozy and the Politics of French Immigration Policy. *Journal of European Public Policy* 15 (3): 382–397.

Maugendre, Stéphane. 2012. Interdiction du territoire: histoire d'une exception. In *Immigration: un régime pénal d'exception*, ed. Gisti, 43–56. Paris: Gisti. https://www.gisti.org/publication_som.php?id_article=2781#llutte. Accessed 23 Feb 2017.

MPI. 2018. *Asylum Recognition Rates in the EU/EFTA by Country, 2008–2017.* Washington, DC: Migration Policy Institute. https://www.migrationpolicy.org/programs/data-hub/charts/asylum-recognition-rates-euefta-country-2008-2017. Accessed 5 Jan 2020.

Nagin, Daniel S. 2013. Deterrence in the Twenty-first Century: A Review of the Evidence, Carnegie Mellon University Research Showcase @ CMU, Working Paper, Heinz College Research.

Neumann, Peter, and Brooke Rogers. 2007. *Recruitment and Mobilisation for the Islamist Militant Movement in Europe.* European Commission.

Papademetriou, Demetrios G., Madeleine Sumption, and Will Somerville. 2009. *Migration and the Economic Downturn: What to Expect in the European Union.* Washington, DC: Migration Policy Institute.

Perry, Matt. 2004. 'Sans Distinction de Nationalité?' The French Communist Party, Immigrants and Unemployment in the 1930s. *European History Quarterly* 34 (3): 337–369.

Rastello, Celine. 2012. Garde à vue des étrangers: Qu'est-ce Qui Change? *L'OBS*, July 6. https://www.nouvelobs.com/societe/20120705.OBS6298/garde-a-vue-des-etrangers-qu-est-ce-qui-change.html. Accessed 5 Jan 2019.

Reyneri, Emilio. 2009. Immigration and the Economic Crisis in Western Europe. In *Presented at the VI Conference on Migrations in Spain, A Coruna*, September 17–19.

Ryo, Emily. 2015. Less Enforcement, More Compliance: Rethinking Unauthorized Migration. *UCLA Law Review* 62: 622–670.

Saas, Claire. 2012. L'immigré, cible d'un droit pénal de l'ennemi. In *Immigration: un régime pénal d'exception*, ed. Gisti, 32–42. Paris: Gisti. https://www.gisti.org/publication_som.php?id_article=2781#1lutte. Accessed 23 Feb 2017.

Schain, Martin A. 2012. *The Politics of Immigration in France, Britain, and the United States*, 2nd ed. New York: Palgrave Macmillan.

Scherr, Mickaël. 2011. L'évolution des phénomènes d'immigration illégale à travers les statistiques sur les personnes mises en cause par la police et la gendarmerie nationale. Dossier INHESJ, 144–158. https://inhesj.fr/sites/default/files/ondrp_files/contributions-exterieures/m_scherr.pdf. Accessed 12 Dec 2018.

Tables of Condemnations. 2012–2016. Ministry of Justice. (*Tableaux des Condamnations en 2012, 2013, 2014, 2015, 2016*). http://www.justice.gouv.fr/statistiques-10054/donnees-statistiques-10302/les-condamnations-27130.html. Accessed 13 Nov 2018.

Thielemann, Eiko. 2003. Does Policy Matter? On Governments' Attempts to Control Unwanted Migration. IIIS Discussion Paper No. 09. https://www.tcd.ie/triss/assets/PDFs/iiis/iiisdp09.pdf. Accessed 13 Dec 2021.

Thielemann, Eiko. 2011. *How Effective Are Migration and Non Migration Policies that Affect Forced Migration?* LSE Migration Study Unit Working Paper No. 2011/14.

Triandafyllidou, Anna, ed. 2010. *Irregular Migration in Europe: Myths and Realities.* Farnham, Surrey: Ashgate Publishing Limited.

UN General Assembly. 1951. *Convention Relating to the Status of Refugees* (Refugee Convention), 28 July, United Nations, Treaty Series, Vol. 189, p. 137. Available at: http://www.refworld.org/docid/3be01b964.html. Accessed 20 Aug 2015.

Wihtol de Wenden, Catherine. 2010. Irregular Migration in France. In *Irregular Migration in Europe: Myths and Realities*, ed. Anna Triandafyllidou, 115–124. Farnham, Surrey: Ashgate Publishing Limited.

CHAPTER 6

The Effects and Counter-Effects of Criminalisation: On Skinny Balloons and Vicious Cycles

INTRODUCTION

> Everyone has the right to leave any country, including his own, and to return to his country.
> Universal Declaration of Human Rights (1948), art. 13

As stated by the UNDHR, the right to (e)migrate is a fundamental right, to be universally protected. Much more contentious is, however, the right to (im)migrate into a country. Indeed, in what has been called 'Fortress Europe', the criminalisation of migration is embedded in a context of securitisation and deterrence.[1]

At the EU level, emphasis on deterring migration dates back to the 1970s. Indeed, it is possible to see reference to such goal already in the 1976 Council Resolution on an Action Programme for Migrant Workers and Members of their Families, which called for 'a common approach to deterrent measures', to be facilitated by the Commission.[2] This was

---

[1] See, for instance, Huysmans (2000).
[2] Council Resolution (1976, pp. 13, 21).

soon associated to asylum seekers too: According to the Commission, in 1991, member states were prioritising 'deterrent actions', to make asylum seekers' situation less attractive, including by limiting welfare benefits, employment opportunities, or freedom of movement.[3] Interestingly, it is the same concerns that affect national and European civil servants today, as exemplified by the question posed by an EU official I interviewed, aiming to know which level of benefits the Union should provide migrants and asylum seekers with, so that their rights were protected, but others would not exploit the situation.

In this context, the use of the criminal law to address migratory flows has spread in Western countries since the mid-1980s, to the extent that, in 2014, all but two EU member states sanctioned irregular entry and/or stay with a fine or imprisonment.[4] In particular, among EU member states making use of the criminal law to sanction migration offences, France emerges as a country of long-lasting and severe criminalisation, where the punishment mandatorily includes both a fine and imprisonment, and dates back to the 1930s. Italy on the other hand, only approached criminalisation more recently, in a context of intense politicisation, and, while not foreseeing deprivation of liberty, commands one of the highest fines in the whole EU.

Building upon the theoretical discussion of Chapters 2 and 3, and the empirical evaluation of the criminalisation of migration in Italy and France of Chapters 4 and 5, I now compare the two case studies, to draw connections between them and highlight broader patterns. In the first part, I follow the structure of the two previous chapters, starting by contrasting the relevance of discursive, implementation and efficacy gaps, and then focusing on the five evaluation criteria, to highlight and reflect on key similarities and differences. In the second part, I analyse the criminalisation of migration in light of the four pitfalls of deterrence identified in Chapter 3, to test their significance and consequences on the overall effectiveness of the measure.

---

[3] Commission Communication SEC(91) 1857, Note de reflexion sur le droit d'asile, p. 8.

[4] FRA (2014).

Primary and Secondary Migration Flows

Before addressing the policy gaps affecting the criminalisation of migration in France and Italy, it is helpful to highlight the historical patterns characterising the phenomenon in the two countries, as well as the latter's inherent connection, through secondary flows.

Despite geographical proximity, the history of migration to Italy and France is substantially different. France belongs to the group of northwestern European countries which experienced immigration throughout the twentieth century, and then sought to stop it following the 1970s oil shocks, by imposing bans on temporary labour permits.[5] Such restrictions contributed to the transformation of Italy (and other southern European states) into a receiving country, after decades of negative net-migration flows.[6] Today, the two countries, both parties to the Schengen Convention, Dublin Regulation, and Return Directive, under the EU framework, still feel the repercussions of such different historical backgrounds, as exemplified by the timeframe of the adoption of the criminalisation of migration (passed in 1938 in France, 2009 in Italy).

In the last decade, attempted irregular entries have been significantly more visible in Italy than France. Sea arrivals to the former country emerge as substantially higher than the number of people refused entry at French borders, for all years between 2011 and 2017.[7] This is in large part due to Italy's geographical proximity to the southern shore of the Mediterranean. However, it is also dependent on the lack of accurate data on foreigners entering France irregularly (due to the lack of border controls among Schengen states), as refusals of entry can only provide a rough indicator of the actual size of the phenomenon.

Still, when focusing on migrants found irregularly present inside the two EU member states, an opposite situation is revealed. Here, France emerges as having substantially higher figures in the whole period for which comparable data is available (2008–2018), with numbers even overcoming the sum of both irregular entries *and* stays in Italy in several years (Fig. 6.1).[8] According to Eurostat, France was among the five EU

---

[5] Ambrosini (2018, p. 62).

[6] Ibid.

[7] Based on ISMU (2014), MoJ (2017, 2018), and Eurostat migr_eirfs (2018).

[8] Specifically, in 2008–2010, 2012, and 2018. Based on Eurostat migr_eipre (2018), ISMU (2014), and MoJ (2017, 2018).

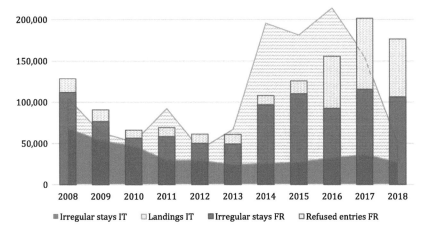

Fig. 6.1 Irregular migration in Italy and France, by apprehension pathway (irregular entry or stay) (stacked graph), 2008–2018 (*Source* Author's elaboration based on Eurostat [migr_eipre, migr_eirfs], ISMU [2014], MoI [2017, 2018])

member states with the highest numbers of cumulative irregular presences in 2008–2018 (together with Greece, Germany, Spain, and Hungary).[9]

As the above hints, it is not uncommon for migrants to enter Italy by sea, and then travel to other countries, including France. For many, Italy has assumed the role of a transit country, on the way to the desired final destination.[10] Although data often shields their visibility, so-called 'secondary flows', namely the movements of migrants among EU member states, are most apparent when focusing on borderlines. A striking example is Ventimiglia, which saw almost 23,000 people passing through the town in 2017, with up to 4000 per month in summertime.[11] Similarly, a little further north, migrants attempt to cross from the surroundings of Bardonecchia into Briançon.[12] According to the mayor

---

[9] Eurostat migr_eipre (2018). See also: https://ec.europa.eu/eurostat/statistics-explained/index.php?title=Enforcement_of_immigration_legislation_statistics#Non-EU_citizens_found_to_be_illegally_present.

[10] See, for example, Camilli (2018), Bartolo (2018), and Peytermann (2018).

[11] Oxfam (2018, p. 8).

[12] Camilli (2018) and Tomasetta (2018).

of Bardonecchia, in March 2018, over 120 migrants passed through the town in just three days, despite the hardships of the mountainous territory.[13]

The visibility of secondary flows from Italy to France has increased in recent years, as exemplified by figures on entry refusals and Dublin requests. Specifically, numbers of third country nationals refused entry at French land borders rose since 2015: Within two years (2015–2017), France registered an increase of 1127%,[14] with most cases occurring at the frontier with Italy.[15] Likewise, French Dublin requests (i.e. requests sent to other EU member states, to take back asylum seekers) saw an upward trend since 2015–2016. Starting with an average of 665 demands sent by France to Italy in 2008–2010, the number spiralled to 16,354 in 2017.[16] In proportional terms, requests to Italy represented 40% of overall French requests in 2017, testifying the significance of secondary flows between the two countries.

Although the data suggest an increase in secondary flows since 2015, it is likely that these were already present before then. Indeed, two events may have contributed to the 2015–2016 rise: The introduction of border controls in France, and of hotspots in Italy. With the end of the possibility of crossing a Schengen border without having one's documents checked, and the simultaneous initiation of the systematic registration of virtually all landed migrants,[17] the secondary movements linking Italy and France have come to light.

Overall, the significance of secondary flows reveals the strict interconnectedness between Italy and France, suggesting an added value of analysing migratory dynamics to the two countries in conjunction, rather than independently. Importantly, the above also points to a key difference between the two EU member states: While deterrence in Italy mainly targets potential migrants in third countries, in France, it also targets those who are already on European soil.

---

[13] Tomasetta (2018).

[14] Based on Eurostat migr_eirfs (2018).

[15] See lanouvellerepublique.fr/a-la-une/de-plus-en-plus-de-migrants-refoules-aux-frontieres-en-france.

[16] Based on Eurostat migr_dubro (2018).

[17] Parliamentary commission on the reception system (2017, p. 58) and interviewees 15, 16, and 17.

Discursive Gaps

Having seen the broader context in which the comparison between Italy and France is inserted, I now turn to the specific norm of criminalisation. In line with the structure used in previous chapters, I begin the analysis by comparing the extent of discursive gaps, i.e. of the distance between the rhetoric surrounding the norms and their actual form once written down. Due to multiple reasons, it is possible to see a very different evolution of the criminalisation of migration in the French and Italian cases.

To begin with, a temporal distance of over 70 years separates the introduction of the norm in the two countries: While France criminalised irregular migration in 1938,[18] Italy only did so in 2009.[19] This implies a very different context as, while France was about to embark on a second World War and exploiting the rhetoric of potential foreign infiltrators,[20] Italy was in a peaceful context, although also relying on similar discursive strategies, as will be discussed later.[21] The significant time lag is however coherent with the different migration history characterising the two countries: As said earlier, migratory flows to France were significant throughout the twentieth century,[22] whereas those to Italy did not intensify until the 1970s, with the country only putting in place an extensive set of legislative measures to address the phenomenon in 1990.[23]

Although criminalisation was introduced at very different times, its adoption occurred in a context of heightened politicisation of migration in both countries. If in France, the recession of the 1930s spurred the rise of xenophobic sentiments and of the far right,[24] in early 2000s Italy, not only was migration control being debated in an increasingly polarised way, but the proposal to adopt criminalisation was also often at the centre of such discussions.[25] Interestingly however, the former country does

---

[18] Decree of 2 May 1938.
[19] Law No. 94/2009, art. 1.
[20] Ben Khalifa (2012, p. 12) and Maga (1982).
[21] See section "The Politicisation of Migration".
[22] Hollifield (2014) and Wihtol de Wenden (2010).
[23] Law No. 39/1990.
[24] Perry (2004, p. 341).
[25] See Law Proposal N. 5808 (1999) and Vesci (2008).

not appear to have witnessed intense direct politicisation of the criminalisation of migration itself (although more historical research would be needed, to draw definite conclusions).[26] Overall, criminalisation seems to have emerged as less severe where its direct politicisation was more intense. While France sanctioned irregular entry and stay with both a fine and a prison sentence, Italy too considered the latter,[27] but then ended up adopting the former only.

Following the introduction of the norm, this was contested in both countries, albeit with very different results. On one hand, despite resistance by public officials,[28] criminalisation was partly repealed in France in 2012 and 2018,[29] following ECJ rulings.[30] On the other hand, regardless of parliamentary[31] and public society requests, the norm was never modified in Italy, in fear of a political repercussions,[32] with the result that criminalisation is still in place, at the time of writing.

As emerges from the above, discursive gaps diverged greatly in Italy and France: The seemingly limited politicisation of criminalisation in France, together with the actual adoption of sanctions including fine, imprisonment, and expulsion, suggests that such a gap was indeed rather weak, if not absent.

In Italy, on the contrary, a gap between the high rhetoric initially proposing imprisonment for irregular entry, and the eventual sanction only foreseeing a fine, clearly emerges. Given, on one hand, the intense rhetorical emphasis on the norm, and, on the other hand, the cut in investigations to tackle underground employment and the adoption of an amnesty shortly thereafter,[33] it is possible to interpret the actions of the Italian Government as pursuing a strategy of *deliberate malintegration*[34]: Adopting very visible measures against irregular migration, while

---

[26] Interviewee 49 and D'Ambrosio (2010).

[27] Law Proposal N. 733/2008, art. 9, para. 1.

[28] Henriot (2013).

[29] See Law 2012–1560, art. 2 and 8 and Circular INTV1824378J.

[30] ECJ Case C-329/11 and ECJ, Case C-47/15.

[31] Law 67/2014, art. 2, c. 3, lett. b.

[32] See, for example, Baer (2016), Bei (2016), and Meli (2016).

[33] Pasquinelli (2009).

[34] The concept of 'deliberate malintegration' was elaborated by Boswell (2007).

Table 6.1 The criminalisation of migration and its discursive gaps

	Italy	*France*
Introduction	2009	1938
Sanctions	€5000–10,000 fine, possible expulsion	€3750 fine, 1-year imprisonment, possible expulsion
Partial repeal	–	2012 (irregular stay), 2018 (irregular entry from Schengen state)
Severity	Medium	High
Discursive gap	High	Low

Source Author's elaboration

in fact quietly allowing for wealth accumulation by avoiding sanctioning irregular work.[35]

Taking this point further, the Italian case may be understood through realist lenses as revealing the varied and contrasting interests affecting migratory policies. Although Boswell proposes her framework as a 'third way' to realist and liberal institutionalist approaches,[36] her focus on states as the key actors, as well as on contrasting interests, suggests that it may be possible to trace a link to Freeman's theoretical understanding of migration policy.[37] In this way, it might be argued that, in the Italian case, the incoherence and limited efficacy of the norm reflected the state's interest in adopting a policy that was in fact less stringent, than what politicians' rhetoric suggested. This, in turn, may be related to the underlying preferences of socio-economic groups and, in particular, of employers and families significantly relying on foreign work.

I will return to issues of coherence and efficacy in later sections. First, however, having seen that the discursive gap was apparent in Italy (where imprisonment was suggested at first but then not pursued), and likely limited in France (where the law took a rather severe form), I investigate the extent of the 'implementation gaps' affecting the norms, from a comparative perspective.

---

[35] For more on this, see sections '(Political) Utility' and 'The Politicisation of Migration'.

[36] Boswell (2007).

[37] See Freeman (1998) and Freeman and Kessler (2008).

IMPLEMENTATION GAPS

In this section, I discuss the implementation of the criminalisation of irregular migration, based on data from national and supranational statistical institutes (Eurostat and Istat), Ministries of Justice and of the Interior, police departments, and Frontex. Generally speaking, both Italy and France made a rather low use of the foreseen sanctions as a way to address irregular entry and/or stay, but previous studies fail to depict an accurate image of the extent to which this is the case.[38] In the next paragraphs, I examine the application of the norm, focusing on the number of registered infractions and condemnations, the demographics of the people sanctioned, and the role of key actors.

Two main differences mark a stark contrast between the procedural implementation of the crime of irregular entry and/or stay in Italy and France. First, while public prosecutors are compelled to criminally pursue offenders in Italy, they have more discretion on whether to do so in France.[39] Second, migrants apprehended irregularly entering or staying are held in custody in France,[40] but left free to move in Italy.[41] Custody in France can either be of criminal nature (*garde à vue*, GAV, applicable to offences of irregular stay until 2012) or of administrative one (*retenue*, introduced in late 2012). In other words, the criminal process may (and often does) start with custody there. The above have key implications, since any disapplication of the norm is in contrast with national provisions in Italy, whereas this is not the case in France. Moreover, lack of custody in Italy implies further efforts to locate the persons, notify them of the trial, and bring them to compliance. Even in France, however, there may be plenty of scope for foreigners to remain irregularly in the country, as exemplified by the high number of apprehensions of unauthorised stay.

In this context, registered cases of irregular entry or stay were very different in the two countries, until 2012. Before that year, France used the *garde à vue* extensively, averaging 112,687 registered migrants per year in 2006–2012 (Fig. 6.2).[42] Following the repeal of the crime of irregular stay in 2012, however, the practice was no longer possible, and

---

[38] With the partial exception of di Martino et al. (2013) and Scherr (2011).

[39] Italian Constitution, art. 112; French Code of Criminal Procedure, art. 79.

[40] See Code of Criminal Procedure, art. 62 and 63 and Circular INTK1300159C.

[41] Interviewee 14.

[42] Departmental figures registered by police and gendarmerie (2018).

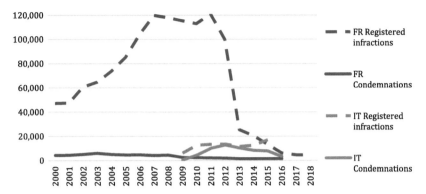

Fig. 6.2 Registered crimes and condemnations for irregular entry and/or stay in Italy and France, 2000–2018 (Author's elaboration on Departmental figures registered by police and gendarmerie [2018], Condemnations by infraction [2016], and ISTAT [2018])

the number of recorded offences became closer to the Italian one (which averaged 13,251 per year, in 2010–2015).[43]

Despite the generally higher number of registered infractions in France, condemnations were lower there than in Italy, between 2010 and 2016. Even at the peak of registered crimes in 2007–2012, France only had, on average, approximately 2700 yearly condemnations for irregular entry and/or stay.[44] The relatively low number of cases that ended up in full criminal processes was due to French provisions that allowed the criminal procedure to be exploited for administrative purposes (to be 'instrumentalised'): Until the law allowed it, the GAV was used to hold apprehended migrants in custody, until they could be transferred to detention centres.[45] At that point, the criminal process would be

---

[43] ISTAT (2018). It should be noted, however, that Italian registered crimes sometimes cluster several migrants together, and the overall number of involved people may thus be higher.

[44] Condemnations by infraction (2016).

[45] See Lochak (2016) and interviewees 49, 52.

suspended, and the administrative one leading to expulsion begin.[46] The extent of the trend was such that twice as many migrants were registered for the crime of irregular entry and stay, than found irregular, in 2010, 2011, and 2012.[47] Due to different legal provisions, instrumentalization did not apply to Italy, which had closer numbers of registered crimes and actual condemnations. These too, however, started diverging very soon and, following 2012, dropping numbers were registered for both initiated trials and convictions.[48]

Compared to the actual number of migrants who should have been subjected to the norms, had they been applied literally, the above figures appear marginally relevant in both countries. In Italy, an average of 15% of the overall number of migrants who should have been subjected to the norm[49] were actually condemned for it, in 2009–2016.[50] Since 2014, the proportion has seen single-digit figures only. Even more strikingly, in France, the percentage of irregular migrants who were condemned, over the total number of migrants found irregularly present, never overcame 4% since 2008, and stabilised at around 1% since 2014.[51] Thus, in both countries, the criminalisation of migration has been applied less than systematically.

Not only did condemnations rarely materialise, but sanctions were also often below the law's provisions (especially in Italy) (Fig. 6.3). Between 2009 and 2018, in Italy, monetary sanctions were on average €4597, consistently below the €5000–10,000 range stated in the Consolidated Text on Migration.[52] In France, penalties were closer to those foreseen by the law, though at times lower: In 2012–2016, the average fine for irregular entry and/or stay was €3663 (while the law foresaw €3750)

---

[46] Ibid.

[47] Based on Departmental figures registered by police and gendarmerie (2018) and Eurostat, migr_eipre.

[48] ISTAT (2018).

[49] Calculated as: Irregularly present migrants + migrants landed by sea - asylum applicants.

[50] Based on: ISMU (2014), MoI (2017, 2018), EUROSTAT migr_eipre and migr_asyapp (2018), and ISTAT (2018).

[51] Based on EUROSTAT migr_eipre (2018) and Condemnations by infraction (2016).

[52] MoJ (2018).

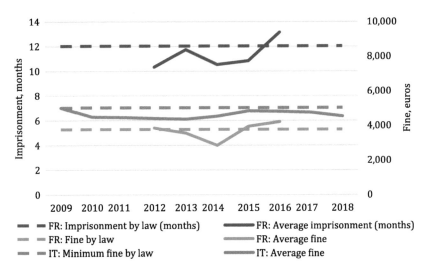

Fig. 6.3 Punishment for irregular entry/stay in Italy and France: Foreseen vs sanctioned, 2009–2018 (*Source* Author's elaboration on Tables of condemnations [2012–2016] and MoJ [2018])

and the average prison term was 11.3 months (while the norm foresaw 12).[53]

Overall, the above brings to light a significant implementation gap affecting both Italy and France, characterised by both low condemnation rates and (at least in the Italian case) sanctions. In this context, it is relevant to note a few points concerning the demographics of the people condemned.

First, women tend to be over-represented in condemnations. Although women are a minority of condemned foreigners in both Italy and France, such ratio is higher than the one representing female migrants apprehended as irregularly staying (over total apprehended third country

---

[53] Both the average prison term and fine were however lower, when considering condemnations involving irregular migration infractions only (and no other crimes). Annuaires statistiques (2006, 2007, 2011–2012) and Tables of condemnations (2012–2016).

nationals, TCNs) (see Fig. 6.4). Such pattern is true for all years for which data is available between 2008 and 2016 (with the only exception of 2009 in Italy). It is starkest in Italy in 2014, and in France in 2015, when the proportion of condemned women was almost double that of apprehended female migrants (respectively, 11.5 and 5.9% in Italy, and 14.8 and 6.9% in France).[54] Part of the explanation may be that foreign women with an irregular status who are found involved in prostitution are then tried for the crime of irregular migration,[55] although more data would be needed, to explore the hypothesis.

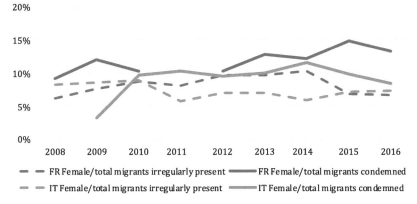

Fig. 6.4 Proportion of women in irregular migratory flows and in condemnations for irregular entry/stay, 2008–2016 (*Source* Author's elaboration based on Annuaires statistiques [2012], Tables of condemnations [2012–2016], ISTAT [2018], Eurostat migr_eipre [2018])[56]

---

[54] Based on Annuaires statistiques (2006, 2007, 2012), Tables of condemnations (2012–2016), ISTAT (2018), and Eurostat migr_eipre (2018).

[55] See, for example: https://questure.poliziadistato.it/it/Imperia/articolo/8705b7edb246e6b4583073356.

[56] Please note: 2008–2010 figures on the proportion of women condemned in France refer to: irregular entry, stay, re-entry after ban, facilitation of entry, and violation of house arrest for migrants subject to expulsion.

Second, few minors are condemned for irregular entry and/or stay. If, in France, on average 12 minors were condemned annually in 2012–2016,[57] in Italy, only 5 were condemned, overall,[58] suggesting a general disapplication of the norm to children.

Third, most of the people condemned in both Italy and France were of African origins. In 2000–2010, African migrants represented an average of 57% of foreigners condemned for migration-related infractions in France; in 2009–2016, they averaged 52% in Italy.[59] This looks consistent with the significant presence of people with African origins among migrants in both states more generally. On the contrary, migrants of Asian origins figure more in Italian statistics than in French ones, representing an average of 30% of condemnations in the former, 17% in the latter, in the above-mentioned timeframes.

Overall, in the context of the criminalisation of migration, most condemned foreigners were male, adults, and of African origins. Although women represented a minority of condemnations, this was often higher than their proportional presence in overall migratory flows.

In sum, it is possible to argue that an implementation gap characterised the criminalisation of migration in Italy and France, although in a slightly different form (Table 6.2). While in France, criminal custody was applied systematically in the years 2006–2012, this did not translate into actual condemnations (which never affected more than 4% of overall migrants found irregularly staying, since 2008). On the contrary, Italy had a somewhat higher absolute and relative number of condemnations, although still contained (affecting about 15% of subjectable migrants), and inserted in a context of general lenience towards the registration of landings (at least until the introduction of hotspots in 2015–2016).[60]

The reasons for this second type of policy gaps appear strongly related to the influence of judicial actors, offering support to the liberal institutionalist thesis. As seen in Chapter 2, according to Hollifield, states are

---

[57] Tables of condemnations (2012–2016).

[58] ISTAT (2018).

[59] Based on Annuaires statistiques (2006, 2007, 2012) and ISTAT (2018). While the figures for Italy refer to irregular entry and/or stay, those for France refer to: irregular entry, stay, re-entry after ban, facilitation of entry, and violation of house arrest for migrants subject to expulsion.

[60] Fargues and Bonfanti (2014, p. 13), interviewees 13, 15, 16, 17, and Parliamentary commission on the reception system (2017, p. 58).

Table 6.2 The criminalisation of migration and its implementation gaps

	Italy	*France*
Registered infractions	~12,000 per year (clustered data) in 2009–2015	~113,000 per year in 2006–2012, then decreasing to ~12,000 per year in 2013–2018
Condemnations	Peaking at 12,646 (2012) ~15% of subjectable migrants in 2009–2016	Peaking at 5884 (2003) Never more than 4% of irregular stays in 2008–2016
Demographics of condemnations	Over-condemnation of women Mainly African nationals	Over-condemnation of women Mainly African nationals
Role of courts	Local justices of the peace and prosecutors often disapplied the norm	ECJ and Supreme Court reduced the scope of the norm

Source Author's elaboration

constrained in their control of migration by their legal systems, to the extent that their degree of openness is determined by the extent to which rights are constitutionally protected.[61]

Indeed, the ECJ's Achughbabian ruling played a key role in leading to the repeal of the crime of irregular stay in France in 2012, despite resistance by the law-enforcement apparatus.[62] At the same time, in the Italian case, although the supranational Court endorsed the applicability of the norm, it also made the use of expulsion dependent on a case-by-case assessment of migrants' risk of absconding, hence significantly limiting judges' ability to return foreigners through a criminal process (and negating the Government's hopes to reach such goal).[63] Furthermore, local public prosecutors and justices of the peace in Italy also crucially contributed to the increasing disapplication of the measure.[64]

Thus, among the different formulations of the liberal institutionalist hypothesis, that stressing the role of judicial review[65] finds the greatest

---

[61] Hollifield (2004).
[62] ECJ, Case C-329/11.
[63] See ECJ, Case C-430/11, Sagor, para. 40–42.
[64] See, for example, Interviewees 14, 47.
[65] Hollifield (2004) and Joppke (1998).

support from the present cases. On the contrary, Guiraudon's emphasis on the greater constraining powers of national judicial actors, rather than supra- or sub-national ones,[66] is contradicted by the key role that the latter played in both France and Italy[67] (Table 6.2).

Evaluation and Efficacy Gaps

(In)Efficacy: Stalling Returns, Undiminishing Migration

Having discussed discursive and implementation gaps, I now turn to the last kind of gaps: Those accounting for the ability of the criminalisation of migration to achieve its goals, given the impact of external factors. To provide an all-round evaluation of the norm, and in line with previous chapters' structure, I do not merely focus on its efficacy, but also consider its efficiency, coherence, sustainability, and utility, comparing how Italy and France fared in each of the parameters.

The aim of the criminalisation of migration was very similar, for Italy and France, and consisted of a twofold goal.

First, both countries aimed to use criminalisation to increase the number and timeliness of returns, as shown by the fact that they both included expulsion as a possible (alternative or additional) sanction. The way in which returns were supposed to be made easier, however, presents differences: As seen above, France took advantage of the criminal GAV to keep migrants in custody while checks were being conducted, so that they could be swiftly transferred to detention centres, to be expelled. Italy, on the contrary, aimed to accelerate returns by exploiting a clause in the EU Return Directive which makes it possible to immediately deport foreigners following a criminal verdict of guilt, without having to grant them time for voluntary return (which is mandatory in most other cases).[68] No evidence of such intention was found in France.

---

[66] Guiraudon (2000).

[67] This is in line with more recent arguments stressing that, since the Treaty of Lisbon, the judicial constraint takes place at the EU level too, through the increasing competencies of the ECJ on migration-related matters (see, for instance, Hadj-Abdou 2016).

[68] See Corriere del Mezzogiorno (2014) and Chamber of Deputies, 15/01/2014, p. 14.

Second, both countries pursued a deterrent outcome. In France, the national penal code expressly mentions that the prevention of new infractions is among the key purposes of the criminal law,[69] and the objective of decreasing the country's attractiveness for migrants was predominant in Daladier's rhetoric in the 1930s.[70] In Italy, on the other hand, the Government made its deterrent goal explicit when presenting the norm in 2009, expressly targeting a 10% decrease in irregular migration, compared to 2007 levels.[71]

Having seen that both France and Italy aimed to (a) increase returns and (b) deter further migration, it remains to be seen to what extent such goals materialised, in practice.

Starting with repatriations, these were not significantly facilitated by criminalisation, as France still figures among the EU member states with the lowest return rates,[72] and Italy proved unable to exploit the criminal law to increase expulsions.

In the former country, looking specifically at the years of intense use of criminal custody, it is possible to see that these were paralleled by a tendential increase in the rate of effective returns to third countries[73] (which rose from 15.5% in 2008 to 19.5% in 2012, to then drop following 2012).[74] If the trend mirrors the changing use of GAV for irregular stay (repealed in 2012), it however also reflects the increased emphasis placed on returns by the Sarkozy administration (e.g. through the establishment of repatriation targets),[75] leading to multiple potential explanations. In any case, even with increased rates, these did not manage to significantly catch up with EU standards, and continued remaining less than, or roughly equal to, half other member states' average (see Fig. 6.5).[76]

Importantly, the effective return rate drops even further when excluding returns to Western Balkan countries, which generally have very

---

[69] Penal Code, art. 130–131.

[70] See Maga (1982, p. 434).

[71] Law Proposal N. 733 (2008, p. 9).

[72] Based on 2016 data (Mananashvili 2017, p. 2).

[73] I.e. the ratio of people actually returned compared to those ordered to leave the country.

[74] Eurostat migr_eirtn (2018).

[75] On the latter, see Marthaler (2008).

[76] Based on data from Eurostat migr_eirtn (2018).

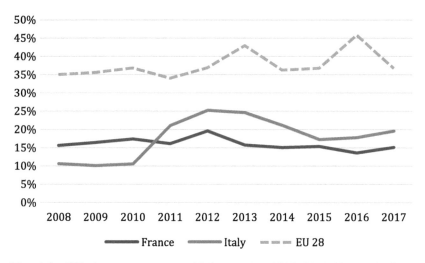

Fig. 6.5 Effective return rates to third countries, 2008–2017 (*Source* Author's elaboration based on Eurostat migr_eirtn [2018])

high compliance ratios. For instance, in France in 2017, consular documents (necessary to return migrants) were obtained within the necessary timeframe in 91% of cases from Albania, but only in 11% of cases from Mali.[77] The trend finds some similarities at the EU level, and in Italy too.[78]

Finally, looking at the expulsions directly stemming from criminal trials, these have been low. In France, expulsions represented on average 3% of condemnations for irregular entry or stay, in 2012–2016.[79] Similarly, in Italy, while it is possible to observe an increase in effective return rates to third countries following 2009 (from 9.9% in 2009 to 25.1% in 2012),[80] these cannot be attributed to the newly introduced crime of irregular migration: On average, only 0.28% of condemnations resulted in

---

[77] Etude d'Impact (2018), section 5.4.2.

[78] See COM(2019) 481 final, p. 15, interviewees 2, 8, and https://www.corriere.it/dataroom-milena-gabanelli/migranti-irregolari-quando-ne-ha-rimpatriati-salvini-8-mesi-governo/accb467e-3c37-11e9-8da9-1361971309b1-va.shtml?refresh_ce-cp.

[79] Based on Tables of condemnations (2012–2016).

[80] Eurostat migr_eirtn (2018).

expulsions in 2010–2016.[81] Indeed, even if the law allowed for the expulsion of migrants, practical constraints, including the difficulty to identify foreigners, the usual absence of the accused at trials, the lack of available means of transport, and difficulties in obtaining return agreements with third countries, made repatriations hard to implement even through criminal sanctions.[82]

To conclude, the possibility to employ returns on a large scale was also limited by the intervention of courts, both in Italy and France, as seen above. In the former, the ECJ judged in 2012 that the repeal of a time for voluntary return should be considered on a case-by-case basis, and could not be applied indiscriminately.[83] In the latter, the Supreme Court ruled in the same year that the GAV could not be applied to irregular stay (since this was no longer sanctionable by imprisonment, following the ECJ's Achughbabian ruling),[84] hence limiting the role of criminal custody as a facilitator of returns.

Looking now at the desired decrease in arrivals, it is possible to argue that this did not materialise either. In Italy, fluctuations in irregular landings and stays do not emerge as related to variations in the norm. To start with, as discussed earlier, actual condemnations were relatively low throughout the period analysed, concerning about 15% of the overall number of migrants to whom they should have been applied. Further, the 2009 fall in landings seems largely related to the agreement signed by the country with Libya,[85] which enabled push-backs to the latter's shores, and is therefore an example of containment, rather than deterrence, insofar as migrants were physically prevented from reaching Italy. Finally, in spite of the possibility of incurring criminal penalties, the number of irregular migrants reaching the country by sea increased in following years, peaking at over 180,000 in 2016.[86]

Similarly, in France, although the number of registered cases of irregular stay decreased between 2008 and 2013 (at the height of the GAV),

---

[81] Based on ISTAT (2018).

[82] Interviewees 17, 30, 32, 39, 41, 46, 54, 55.

[83] Case C-430/11, Sagor, para. 40–42.

[84] Cour de Cassation (2012) C100965.

[85] Chamber of Deputies, 14/04/2011, EMN (2011, p. 2), and Colombo (2011, p. 278).

[86] ISMU (2014) and MoI (2017, 2018).

this is hardly related to the criminalisation of migration, for a number of reasons. First, as seen above, the proportion of unauthorised foreigners who were condemned was extremely low throughout the period (never higher than 4% since 2008, and around 1% since 2014). Second, the decrease in apprehensions following 2008 was mainly due to a drop in Eritrean, Afghani, and Iraqi citizens,[87] most of whom were refugees,[88] and hence exempted from criminalisation and other sanctions for irregular migration, due to the Geneva Convention.[89] Third, a fall in apprehensions does not necessarily equal a fall in presences, and Frontex itself recognised that the closure of Calais and other camps in 2009, together with the removal of border controls with Switzerland, might have simply rendered the flows invisible.[90]

Overall, most interviewees in both Italy and France shared the view that the criminalisation of migration was not successful, in its attempt to deter further irregular migration. Two of them referred to the hope that the norm would act as a deterrent as an 'illusion'.[91] Even senior law-enforcement actors including the French Departmental Director of the Alpes-Maritimes Border Police, and the Italian First Director in the Immigration and Border Control Directorate of the Italian National Police, among others, were of the same viewpoint.[92]

To conclude, efficacy gaps emerged in both the Italian and French cases (Table 6.3). Specifically, both countries aimed to decrease migration and increase returns, but none of such objectives turned out as being successfully accomplished through the criminalisation of irregular migration. Some of the reasons for such gaps have already been hinted to, but will be discussed more in-depth in later sections, in relation to the pitfalls of deterrence. In the meantime, beyond the inefficacy of the norm, it is relevant to evaluate its broader characteristics, from an all-encompassing perspective that builds upon a set of criteria. Indeed, I now turn to the second parameter: The efficiency of the criminalisation of irregular migration.

[87] Eurostat migr_eipre (2018).
[88] MPI (2018).
[89] Refugee Convention (1951), art. 31.
[90] Frontex ARA (2010, p. 30).
[91] Interviewees 30 and 52.
[92] Interviewees 18 and 43.

Table 6.3 The criminalisation of migration and its efficacy gaps

	Italy	France
Goal 1: Increasing returns	• Returns only 0.28% of condemnations in 2010–2016 • Practical constraints • ECJ intervention	• Rate of effective returns to 3rd countries roughly half EU average in 2008–2017 • Returns averaged 3% of condemnations (2012–2016) • Supreme Court intervention
	→ **Returns not dependent on criminalisation**	→ **Slightly higher effective return rates in 2008–2012, but still low compared to EU levels**
Goal 2: Deterring migration	• Landings peaked at over 180,000 in 2016 • Push-backs to Libya made arrivals drop in 2009 • 15% of subjectable foreigners were condemned (2009–2016)	• France in top 5 EUMS for irregular stay in 2008–2018 • Data issues and a drop in refugees made post-2008 apprehensions fall • Less than 4% of apprehended foreigners were condemned (2008–2016)
	→ **Changes in migration unlikely to be dependent on criminalisation**	

Source Author's elaboration

(In)efficiency: The Duplication of Processes and Costs

The 'efficiency' of a measure refers to the extent to which it manages to achieve the greatest effectiveness, at the lowest cost. In the context of the criminalisation of migration, the norm significantly lacks in such parameter, predominantly as a result of a duplication of processes and costs.

Recalling the main expenses and revenues related to the measure in the two countries, in France, expenses relate predominantly to three factors: (a) the cost of legal support to migrants in custody (between €61 and

€450, depending on the length and nature of custody)[93]; (b) the cost of prison (on average €99.49 per person per day, in 2013)[94]; and (c) the cost of expulsions (between €10,000 and €27,000 per person in 2008–2010, depending on estimates).[95] In Italy, on the other hand, the absence of a prison sentence erases related costs, making expenses mainly due to: (a) legal and translation costs (estimated at €650 per migrant)[96]; (b) JPs' honorarium (€35 per hearing, €55 per judgement)[97]; and (c) expulsions (estimated between €3000–5000 each).[98] Since custody is not foreseen by the Italian system, transportation costs for both migrants and police officers may be included in the equation too.

Looking at the sources of revenues, these originate in both Italy and France from the fines paid by migrants, but are not guaranteed. First, as seen above, average sanctions have been below the foreseen levels, particularly in the former country, averaging €4597 in Italy in 2009–2018, and €3663 in France in 2012–2016. Second, although no data was found on effective payments, several of the interviewees were of the opinion that these are seldom made.[99] Indeed, given the situation of destitution that many irregular migrants face, and their general lack of sizeable assets,[100] it is highly doubtful that such fines, even if of criminal nature, are paid.

Adding to the difficulty in collecting fees, the inefficiency of the system is given in both Italy and France by a further problem, which can be conceptualised as the first of a number of paradoxes characterising the criminalisation of irregular migration: The duplication of processes and costs (see Table 6.4). In Italy, this takes the form of the parallel administrative and criminal procedures that are supposed to start at the apprehension of an irregular migrant, both aiming to expel the latter, but in fact simply contributing to the duplication of costs and overwhelming public offices. In France, on the contrary, the first paradox emerges out of the provision foreseeing that expulsion should take place at the end

---

[93] Étude d'impact (2012).

[94] French Senate, avis n. 114 (2014, p. 42).

[95] See Gourevitch (2012, pp. 54–63) and Harzoune (2012).

[96] Law Proposal n. 733/2008, p. 9.

[97] Interviewees 32 and 54.

[98] Gabanelli and Ravizza (2019) and Polchi (2017).

[99] Interviewees 15, 32, 33, 39, 46.

[100] Savio (2016). Similarly, Ambrosini and Guariso (2016).

of, rather than instead of, the sentenced prison term. Here, not only do imprisonment costs fail to be justified on the basis of convicted individuals' future reintegration into the host community (which is one of the goals of criminal punishment),[101] but are also to be added to those of expulsion, duplicating overall expenses.

Overall, in both Italy and France, criminalisation was affected by an inefficient duplication of efforts: While in France, the combination of expulsion and imprisonment made costs for the latter redundant, in Italy, the mandatorily double processing of irregular migrants made the judicial system overloaded and struggling to cope.[102] Moreover, in both cases, fines were hard to collect, due to the general lack of possessions of the people condemned. Both issues are not only efficiency concerns, but also involve matters of coherence, or lack thereof, taking us to the next criterion.

(In)Coherence: The Contradictory Nature of Sanctions, and Other Problems

Strictly linked to the issue of the inefficiency of the criminalisation of migration, is that of its incoherence, both internal (within itself) and external (with other laws and provisions).

Starting with the former, this stems from the duplication of efforts (as just seen), but also from a second paradox (see Table 6.4): The contradictory nature of sanctions themselves. In France, *internal* incoherence emerges from the inconsistency of aiming to expel migrants, while actually forcing them to remain in French territory for up to a year, through imprisonment.[103] In Italy, on the contrary, it stems from the foreseeable inability of foreigners in irregular situations to pay fines as high as €10,000.

Adding to the above, the introduction of the crime of irregular migration led to a third paradox, different however for the two countries. In Italy, it contributed to the worsening of parallel investigations, especially

---

[101] Maugendre (2012).

[102] See Interviewees 34, 38, 39. Ferrero (2010), Public Prosecutor Siracusa, 2017, and Senate, 08/07/2015, p. 6.

[103] Interviewee 43.

against human smugglers.[104] When migrants are investigated in a related case, they gain a number of legal protections (including the right not to answer prosecutors' questions, and to have all documents translated into their own language), which make trials against traffickers longer and more complicated.[105] In France, the above does not take place, due to different geographical and legal factors,[106] but it is possible to see a paradox in the increased marginalisation of migrants caused by their criminalisation, as time in prison has the potential to lead to greater involvement with criminal organisations.[107]

In sum, despite the different nature of penalties, the criminalisation of migration was in neither EU member state an internally coherent measure to address irregular migration.

Focusing now on *external* incoherence, this stemmed in France from the norm's incompatibility with the EU Return Directive, as a consequence of which irregular stay was decriminalised in 2012. The legal change did not completely solve the issue, however, as the Parliament's decision to maintain irregular entry as a crime, leveraging the ECJ's mere focus on the criminalisation of stay, may leave space to regard the norm as still affected by external coherence problems. On the contrary, the above does not appear to be problematic in Italy, where no prison term is foreseen and the monetary sanction is not meant to take place as an alternative to expulsion, but rather in parallel to it.[108] Indeed, both the ECJ and the Constitutional Court endorsed the compatibility of the Italian version of the norm with supranational law.

To conclude, the criminalisation of irregular migration has been affected by high incoherence in both the Italian and French versions. While both suffer from internal inconsistency, however, only the more northern variant has been found by high courts to be externally inconsistent too.

---

[104] Di Natale, interviewed in Ziniti (2016), Roberti, cited in Baer (2016), Savio (2016), and Interviewee 34.

[105] Ibid.

[106] Migrants do not generally enter France through clustered sea landings, and the pursuit of crimes is not always mandatory there.

[107] Saas (2012, p. 41). Similarly, interviewee 22.

[108] See ECJ, case C-430/11, Sagor, as opposed to case C-38/14, Zaizoune.

(Un)Sustainability: The Trade-Off Between Certainty and Efficiency

Moving now to the sustainability of criminalisation, this parameter assesses the extent to which the norm may be continued in the long term.

The sustainability of the norm is affected by a trade-off between cost-efficiency and sanctions' certainty (see Fig. 6.6 and Table 6.4). On one hand, full criminal pursuit of irregular entry and stay demonstrates to potential migrants the certainty of incurring punishment, but also implies high and redundant costs, both in terms of resources and time. In the long run, this places an unnecessary burden on public finances (e.g. through both imprisonment and expulsion costs in France) and on public offices (e.g. through parallel criminal and administrative procedures in Italy). On the other hand, limited implementation keeps state expenditure more contained, but also undermines deterrence by making the threat inconsistently enforced. In the long run, this negatively affects the principle of certainty of the rule of law, potentially in other areas too.

Adding to the above, the criminalisation of irregular migration lacks sustainability for a further reason: The potential of creating vicious cycles, by fostering domestic insecurity through rhetoric or irregular activities.[109] First, by increasing the rhetoric surrounding migration control, criminalisation brings attention to the borders. When it fails, however, the porousness of the latter is exposed, which may, in turn, lead to more public demand for visible—but often chiefly symbolic—measures. Second, by denying migrants' regularisation and affecting their criminal records, criminalisation may push foreigners further underground, making them more reliant on smuggling networks to enter the concerned country and/or support themselves once inside. As a result, not only are the revenues and power of underground organisations boosted, but migrants' vulnerability to violence and trafficking is also increased. Such dynamics will be explored in-depth in sections 'The Politicisation of Migration' and 'Does Criminalisation Deter, or Simply Divert?', in relation to the politicisation of the norm and the substitution effects it may generate.

---

[109] For similar arguments on vicious cycles, either through securitisation or irregularity, see Chebel d'Appollonia (2012) and Andersson (2016), Carling, cited in Ambrosini (2018, p. 108) and Talani (2018).

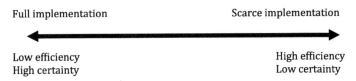

Fig. 6.6 Criminalisation and trade-offs (*Source* Author's elaboration)

(Political) Utility

The last of the five parameters, utility, evaluates who the key actors benefitting from a norm or intervention are. In both the Italian and French cases, significant part of the utility of the criminalisation of migration was of political nature, coupled however with economic and effectiveness concerns, respectively (see Table 6.4).

On one hand, in France, the GAV benefitted law-enforcement officers concerned with effectiveness, e.g. in the form of return targets, by granting them an instrument to exert greater control on migrants and prevent the latter from absconding. Indeed, as convincingly argued by Henriot, once the ECJ ruled that irregular stay should no longer be sanctioned by imprisonment (and hence that the GAV could not be applied to such infraction), French political, administrative, and judicial authorities repeatedly challenged the decision.[110]

Such findings seem to support Guiraudon's emphasis on the role of law and order officials in shaping migration policies[111]: Although, in this case, they did not 'venue-shop' at the supranational level, they proved crucial in delaying the effects of the ECJ's decision. At the same time, however, as seen above, the fact that a supranational court played a key role in leading to the repeal of the norm does not appear to confirm the academic's assumption that national judicial review tends to be stronger.

Adding to the above, beyond the bureaucratic level, the criminalisation of migration in France and its continuation can also be interpreted as beneficial to those political actors, first of whom Sarkozy, who expressly made of returns a key aspect of their electoral campaign.[112]

---

[110] Henriot (2013).
[111] Guiraudon (2000).
[112] See Henley (2003).

In Italy, on the other hand, the criminalisation of migration, introduced following the start of the financial crisis, pursued a strategy of *deliberate malintegration*, as suggested earlier. The norm aimed at pursuing a very visible measure to obtain political consensus, while however also avoiding damaging employers of irregular work, as demonstrated by the regularisation that followed the norm in August 2009. It should be clarified, however, that it was not necessarily that interest groups, such as employers' federations, actively called for the introduction of such a measure. More likely, they took advantage from such a measure capturing the public debate and obfuscating the lax action against irregular employment. Overall, the above well-exemplifies Ambrosini's view of migration policies as providing 'a combination of *symbolic politics*, designed to please anti-immigrant sentiments, and *instrumental politics*, meant to supply cheap migrant labour'.[113]

Overall, the utility of the criminalisation of migration mainly emerged in the form of political and economic benefits in Italy, and of political and effectiveness gains in France. In this context, understanding the politicisation of both migration processes in general, and of criminalisation in particular, is of vital importance to make sense of the contradictions affecting the norm in the two countries, and will be explored in depth in section "The Politicisation of Migration".

To sum up the analysis conducted up to this point, the criminalisation of irregular entry and/or stay proved unable to meet its objectives, with the partial exception of France managing to slightly increase returns, although only by making an instrumental use of criminal custody. More broadly, criminalisation emerged as a norm with severe inconsistencies, at the external level (for France specifically), and at the internal one (for both countries), including contradictory sanctions, negative effects on parallel investigations, and increased marginalisation. This has led to significant inefficiencies too, predominantly related to the duplication of processes and costs, and the un-collectability of fines. As a consequence, if the full implementation of the norm is too costly, a limited one is too uncertain, thus making either decision lack sustainability. Finally, in all such dynamics, benefits have predominantly been of political, economic, and effectiveness-related nature.

[113] Ambrosini (2018, p. 14, emphasis in original).

What were the reasons for the above low scores in most evaluation parameters? Have the norms and processes criminalising migration in Italy and France suffered from issues that are specific to them, or is it possible to highlight broader patterns? In the next section, I address such and related questions, by investigating the role that the four pitfalls of deterrence played in the criminalisation of migration in the two countries.

Table 6.4 Overview of the criminalisation of migration in Italy and France, by evaluation parameter

	Italy	*France*
Efficacy	• Landings peaked at over 180,000 in 2016, despite criminalisation • Returns only 0.28% of condemnations in 2010–2016 → **Stated goals were not met**	• France among top 5 EU MS by irregular migration in 2008–2018, despite criminalisation • Higher effective return rates in 2008–2012, but still about half EU average
Efficiency	• Parallel administrative and criminal process • Overloading of judicial system → **Paradox 1**: Duplication of processes and costs	• Cost of prison plus expulsion • Prison not targeted at reintegration
Internal coherence	• Uncollectable fines → **Paradox 2**: Contradictory nature of sanctions	• Imprisonment in contrast with goal of expulsion • Uncollectable fines
	• Longer investigations vs smugglers → **Paradox 3**: Worsening of parallel investigations	• Possible greater involvement with criminal activities following prison → **Paradox 3**: More marginalisation
External coherence	• Upheld by both ECJ and Italian constitutional court → **Acceptable**	• Imprisonment in contrast with Return Directive (ECJ) → **Weak**
Sustainability	• Full implementation too costly, but scarce implementation too uncertain → **Counter-effect**: Vicious cycle of insecurity through rhetoric and more irregularity	
Utility	• Deliberate malintegration: political and economic gains → **Key role of political factors**	• Political and effectiveness-related gains

Source Author's elaboration

Deterrence Pitfalls and Criminalisation

In Chapter 3, I highlighted four pitfalls of deterrence, hypothesising they are responsible for limiting the success of such strategy. To complete the analysis, in the next pages, I examine how such factors played out in the case of the criminalisation of migration in Italy and France, and whether they had an impact on the evaluation parameters addressed above. Specifically, I focus on the drivers of migration, substitution effects, secondary audiences, and information.

Deterrence and the Drivers of Migration

As seen thus far, the criminalisation of migration did not lead to less migration or more returns to third countries, either in Italy or France. What I have not yet discussed, however, are the reasons for such failure, which are to be found in a number of factors.

To begin with, recalling the criminological studies analysed in Chapter 3, these predominantly highlight the role of the perceived certainty of sanctions (rather than their perceived severity), to enhance deterrence.[114]

In the context of criminalisation, certainty has been scarce in both Italy and France, with mostly moderate numbers of people being condemned (see section "Implementation Gaps"). Interestingly, although several interviewees report Italy as being perceived by migrants as generally laxer on law enforcement than France[115] (which would be consistent with the previously mentioned example of Italian police officers letting migrants go unregistered), the opposite is true with regards to the crime of irregular entry and stay specifically. Although, in 2009–2016, the proportion of migrants condemned for irregular entry or stay has been generally low in both Italy and France, in the former country this stood at an average of 15%, while only averaging 2% in the latter. Recalling the importance of perceptions, rather than actual figures,[116] in determining the success of deterrence, it may be argued that the overall migration enforcement

---

[114] See Beccaria (1764) and Nagin (2013).
[115] Interviewees 22 and 29 and IOM (2011, p. 20).
[116] von Hirsch et al. (1999, p. 33).

system affects third country nationals' understanding of the criminalisation of migration too, and plays a significant role in shaping their decisions.

On the contrary, severity is higher in France, where both a fine and prison term are foreseen as sanctions (potentially, with expulsion too). Still, in line with previous studies, severity does not seem to deeply affect migrants' perspectives: Not only are fines likely to often go unpaid, even when very expensive, but there is also evidence suggesting that prison does not detain a strong dissuasive or stigmatising effect on migrants, who often perceive it as a normal step in their journey to Europe.[117] Indeed, this would support the social-control and labelling theories, according to which, once a sanction becomes normalised in a group, it loses its stigma.[118]

Overall, the certainty of being apprehended and convicted for irregular migration in France or Italy has not been very high in either country. As for the severity of the sanctions, this was greater in the former country, but may not necessarily have played a determinant role to increase deterrence, if indeed detention and migration-related sanctions are increasingly normalised.

Beyond criminological aspects, the fundamental issue at the basis of the ineffectiveness of criminalisation lies in deterrence theory's assumption that migrants' decision-making is rational and that, to reduce migration, it is enough to increase the costs related to it. This includes (at least) a twofold problematic, since not only are economic costs the only ones looked at, but the very fact of only considering costs, and omitting benefits, makes criminalisation unable to address the deeper causes of migration itself.

Indeed, there is evidence that the expected benefits to be derived from mobility may be related to economic goals and fluctuations, such as employment rates, wage levels, or economic crises.[119] Still, such benefits may also not only be of economic nature. As an example, third country nationals offered money as part of voluntary returns packages

---

[117] Interviewee 22. Similarly, interviewees 9, 52.

[118] See Becker (1963).

[119] See, for example, Borjas (1989), Castles and Miller (2011, p. 6), and Fix et al. (2009, p. 19).

often prefer to remain in Europe than accepting it[120]: In 2017, only one-third of people returned by France were so through assisted programs, a proportion which drops to less than one-tenth in Italy.[121] In this context, benefits may also have political character, including fleeing violence or persecution, as in the case of people escaping the Syrian and Libyan crises. Further, migration may offer rewards including the amelioration of one's social status (as demonstrated by Carling and Hernández-Carretero in the case of Senegalese migration to Europe),[122] or the possibility to escape an abusive situation. Finally, benefits may be of legal nature, as hypothesised by the liberal institutionalist thesis, according to which migration to the West often implies significant legal protection.[123] Indeed, among survey respondents in Italy, 18% indicated access to social security services as being among the reasons for which they chose Italy as a country to migrate to, suggesting a role for legal entitlements. Overall, if migration is to be understood as a calculus, following deterrence theory's assumption of rationality, it should be acknowledged that this may not only include economic motivations, but non-economic ones too, including of political, social, or legal nature.

Crucially, however, the criminalisation of migration fails to address the structural factors at the origins of migration. While the goal of the present analysis is not to propose a theory on the causes of international migration,[124] the research hitherto conducted has supported the argument that migratory flows were significantly related to macro-level international political and economic factors, rather than to the criminalisation of migration, in the last decade. Politically, a number of uprisings and conflicts in neighbouring countries encouraged outflows from such territories.[125] This emerged as particularly evident in 2011 and 2014, when increased numbers of people from Tunisia and Syria, respectively,

---

[120] Interviewee 21.

[121] In Italy, 7045 persons were returned, of which 465 through assisted returns. In France, 15,665 persons were returned, of which 4800 through assisted returns. Source: Eurostat migr_eirt_ass (2018).

[122] Carling and Hernández-Carretero (2008).

[123] Hollifield (1998).

[124] For an overview of the principal ones, see de Haas (2011).

[125] See Scherr (2011).

reached both Italy and France,[126] following the outbreak of the Arab Spring in the former,[127] and in parallel with the deadliest year of the civil conflict in the latter.[128] Dreadful conditions in further countries, first of which Libya, Eritrea and Nigeria also contributed to increasing migratory flows. Importantly, the above does not only relate to conflicts, but to dire economic circumstances too. Indeed, the deep inequality between countries in the Southern and Northern hemispheres has significantly stimulated outflows,[129] to the extent that, according to structuralist views, countries' position in the global economy is a determinant factor of migration.[130]

Overall, in both the French and Italian cases, key factors leading to fluctuations in irregular migration were not so much linked to the introduction or amendment of the criminalisation of irregular entry and/or stay, but rather to broader IPE dynamics. As a matter of fact, the surveys revealed that, for 70% of respondents in Italy, knowing the consequences of irregular migration did not make them change their mind on how to reach the country, and the same is true for all respondents in France.

Does Criminalisation Deter, or Simply Divert? on Alternatives and Skinny Balloons

As a consequence of the above structural and deeper dynamics, it is suggested here that, if deterrence is to have an effect, this is predominantly identifiable in the shift of migratory flows towards other routes or means, rather than less migration. Among the four substitution effects proposed by de Haas, namely spatial, categorical, temporal and reverse-flow,[131] the first two seem the most relevant in the Italian and French cases.

---

[126] Eurostat migr_eipre (2018) and Frontex (2019).

[127] See Fargues and Fandrich (2012).

[128] For an estimate of yearly casualties in the Syrian civil war, see: https://ewn.co.za/2019/01/02/nearly-370-000-killed-in-syria-s-civil-war-since-2011.

[129] Castles (2004).

[130] Sassen (1996) and Talani (2015).

[131] For a discussion of substitution effects, see de Haas (2011).

The president of the American Border Patrol Union T. J. Bonner well-described the spatial substitution effect, by comparing it to a balloon. In his words, we can:

> Imagine the border as a big, long, skinny balloon. When you squeeze in one part, it comes out in another. It doesn't disappear.[132]

Indeed, in the context of a nation state, or of the Schengen area, relocating migrants' entry point is not going to prove a successful solution, simply shifting the problem to another region or member state, while failing to reduce the overall number of unauthorised entries.

An example of such 'balloon effect' is provided by France: Once the country decided to temporarily reintroduce border controls in 2015, it started conducting checks on trains and vehicles at the frontier with Italy, including in the Alpes-Maritimes area bordering Ventimiglia. Being met with resistance by French officers, who would turn back people without the necessary documents,[133] third country nationals aiming to leave Italy adopted several strategies to avoid apprehension. In particular, these included people going further underground (e.g. in the hills behind Ventimiglia),[134] attempting riskier routes (e.g. walking on the autoroute),[135] or moving to other segments of the border (e.g. to Bardonecchia).[136] The continued pressure on French borders, exemplified by increasing entry refusals since 2015,[137] shows the persistence of attempts, despite the difficulties and harsh conditions characterising them,[138] and points to the substitution effect caused by the closure of borders.

More broadly, it might also be possible to see a shift in the flows crossing the Mediterranean Sea across time and space: Starting with Spain being the main entry point for irregular migrants in Europe in 2006, the

---

[132] McCombs (2006, p. 10), cited in Fan (2008, p. 707).

[133] Oxfam (2018).

[134] See, for example, https://www.riviera24.it/2017/06/ventimiglia-migranti-nei-boschi-del-grammondo-in-attesa-di-raggiungere-la-francia-258380/.

[135] See, for example, https://www.riviera24.it/2017/07/ventimiglia-migranti-a-piedi-in-autostrada-di-nuovo-chiuso-il-tratto-di-a10-al-confine-di-stato-2-259397/.

[136] See, for example, Camilli (2018).

[137] Eurostat Migr_eirfs (2018).

[138] Oxfam (2018, p. 8).

trend changed following the adoption of a number of restrictive measures in the country, following which, flows to Greece increased.[139] Once both the Central and Eastern Mediterranean routes were closed, through the 2016 EU-Turkey statement and the Italy-Libya 2017 Memorandum of Understanding, arrivals to the Western Mediterranean surged again, increasing by 131% from 2017 to 2018.[140] No straightforward conclusion can be drawn on the causal relationships between such agreements and the direction of the flows, as not all those who would have embarked on the Eastern Mediterranean route necessarily took the Western one instead, or vice versa. Yet, acknowledging (as shown earlier in the chapter) that many migrants consider Italy as an entry point, aiming to travel to other European destinations, it is possible to argue that the relocation of entries does not contribute to a better-managed EU migration system.

In fact, looking back at the French case mentioned above, this highlights the redirection of flows towards not only other entry points, but also other modalities of border crossing, and in particular towards more irregularity. Indeed, a categorical substitution effect can be seen in the country.

In Italy too, such outcome can be appreciated. While, until at least 2006, most migrants became irregular by overstaying their visa,[141] the significant increase in sea landings in the last years, and more contained apprehensions inside the country, suggest a change of trend. Moreover, although, for decades, the most common way to be regularised was through one of the frequently granted amnesties,[142] the last of such schemes (in the timeframe covered by the analysis) was in 2012,[143] and indeed we can see migrants seeking international protection as a way to legally remain in the country.[144] Thus, the lack of regularisation opportunities may have contributed to a redirection of the flows towards other entry/stay categories, to achieve the goal of remaining in Europe.

---

[139] Fargues and Bonfanti (2014, p. 5) and interviewee 4.

[140] EMN (2019, p. 58).

[141] MoI (2007, p. 383).

[142] Ambrosini (2018, pp. 81–82) and interviewee 13.

[143] A new regularisation was adopted in 2020, though this falls outside the scope of the analysis.

[144] For a similar argument, see interviewee 13. See also Interviewee 24.

Indeed, categorical substitution effects are not recent phenomena, and have been acknowledged by the European Commission itself. For instance, it is possible to see evidence of them already in the second half of the twentieth century, when the closure of legal immigration pathways to Europe following the 1970s oil shocks led many to pursue the asylum route, to enter or stay in Europe.[145] In the Commission's words, the '[asylum] procedure [was] increasingly sought by people motivated by economic reasons', becoming a 'parallel means of migration'.[146]

The last two substitution effects also appear easy to imagine in a European context. Although no data is available on circular migration, it is likely that, once a person has entered the Union by crossing the Sahara Desert, Mediterranean Sea, and the Alps, they will not want to risk their lives again, choosing instead to remain in the receiving country until they are either regularised or willing to go back. Similarly, deterrence may lead to a shift in time, for example by incentivising migrants (or smugglers) to wait for the most appropriate moment to embark and attempt the journey, as they do when considering sea and weather conditions in the Mediterranean.[147]

In conclusion, substitution effects may significantly affect the efficacy of the criminalisation of migration and related deterrence measures, contributing to redirecting the flows towards other entry points or categories, rather than stopping them. By doing so, they also play a part in fostering a vicious cycle of insecurity that, in turn, negatively affects the sustainability of the norm, as will be discussed in the following paragraphs.

Among the counterproductive consequences of substitution effects and, in particular, of categorical ones, is the generation of more irregularity.[148] On one hand, the absence of legal routes and the toughening of border controls make migrants increasingly reliant on smuggling networks to reach the EU. On the other, by preventing unauthorised foreigners from accessing regular labour market, healthcare, or education services, they make them more reliant on underground organisations inside the Union too. In particular, by affecting criminal records, the criminalisation

---

[145] Commission of the European Communities (1991, p. 3).

[146] Ibid., p. 9.

[147] As evident from data from MoI (2016, 2017, 2018), landings to Italy drop in winter months, when sea conditions are rougher.

[148] For similar arguments on vicious cycles through irregularity, see Carling, cited in Ambrosini (2018, p. 108) and Talani (2018).

of migration directly contributes to making it harder for the concerned third country nationals to obtain a job or residence permit.[149]

The more criminalisation and deterrence push migrants underground, the more they reinforce insecurity, by fostering irregularity and the profits of shadow economies. Indeed, the widespread underground employment of irregular migrants in Italy is renowned, especially (but certainly not only) in the agricultural sector in southern regions.[150] The above leads to a thriving environment for mafia and criminal groups, which also prosper from 'emergency' situations as occasions in which high amounts of funds are allocated and need distributing quickly, often ending up lacking accountability and control.[151] As shown by Orsini and Sergi, this is indeed what happened in the three Italian regions of Calabria, Sicily, and Lazio, with mafia organisations greatly benefitting from the 'migrant business', and several scandals being brought to light, including *Mafia Capitale* in Rome.[152]

Together with the increase in underground organisations' profits, the lack of regularisation also incentivises migration itself, through the attractiveness of employability in the black market. This is especially the case in southern European countries, and particularly in Italy, where the prosperous shadow economy (estimated by the IMF as between 22.43 and 27.31% of the country's GDP, in 1991–2015)[153] has been found to represent a significant 'pull factor' for migrants (thus challenging Weiner's assumption that ease of access is more relevant than market forces).[154,155] On the contrary, the smaller size of the underground economy in France (thought to range between 11.61 and 16.60% of the GDP in the same period),[156] was found to be less of an attractive factor.[157] It should be mentioned, however, that both countries significantly rely on irregular labour, to the extent that, according to the Financial Times, already

---

[149] See Saas (2012, p. 41).
[150] Talani (2018, p. 13).
[151] Orsini and Sergi (2018).
[152] Ibid. For more on Mafia Capitale, see Abbate and Lillo (2015).
[153] Medina and Schneider (2018, p. 53).
[154] Weiner (1995, pp. 205–207).
[155] Reyneri (2001, p. 22), Talani (2018), and interviewee 9.
[156] Medina and Schneider (2018, p. 52).
[157] Reyneri (2001, p. 49).

in 1990s France irregular migrants 'greased the wheels': 'The sector of construction, including for the Channel Tunnel, rests on them; the sector of textiles would collapse without them; domestic services would vanish'.[158]

As a consequence of the above, especially in Italy, a stronger fight against irregular employment is necessary, to address the phenomenon of unauthorised migration.[159] Following Reyneri, however, this should be considered in its complexity, as countries' *social norms* regarding the acceptability of being employed irregularly largely shape sanctions' effectiveness too.[160] In his words: 'The strength of the underground economy, which allows it to escape any control, comes from its deep roots in the society'.[161] Indeed, whereas in France, irregular employment is largely stigmatised, this is not the case in Italy,[162] a fact that, together with the widespread use of irregular foreign labour not only by firms but by families too,[163] makes the regulation of the phenomenon a much deeper issue.

Importantly, the increase in irregularity also fosters migrants' vulnerability to being exploited, physically assaulted, or trafficked. Be it enough to think about the Nigerian women and girls who arrived in Italy as victims of trafficking in human beings (likely over 8000 in 2016),[164] and to the fact that for all those who attempt to reach Europe through Libya, the line between smuggling and trafficking is an increasingly thin one, with repeated abuses and ill-treatments.[165] Not only did the CNN film an auction where migrants were sold by smugglers,[166] but Lampedusa doctor Pietro Bartolo (among others) also reported some of the horrors

---

[158] Financial Times, 2 February 1990, as cited in Overbeek (1996, p. 70). Author's translation.

[159] Similarly, see interviewee 6.

[160] Reyneri (2001, pp. 29, 62).

[161] Ibid., p. 62.

[162] Ibid., pp. 29, 60, 61.

[163] Ibid., p. 44 and Ambrosini (2018, pp. 65, 80).

[164] In 2016, 11,009 Nigerian women arrived to Italy by sea (UNHCR 2016) and IOM estimates 80% of them to be victims of trafficking (IOM Italy 2017, p. 9).

[165] Amnesty International (2017).

[166] See: https://edition.cnn.com/videos/world/2017/11/13/libya-migrant-slave-auction-lon-orig-md-ejk.cnn.

taking place in such 'camps', referring for example to the story of a young man who had the lower part of his legs skinned, as a 'game' for guards to 'turn a black man white'.[167] Several sources also reveal a new way to extract moneys from migrants in the North African country, by using physical violence and electric cables to make them scream while on the phone with their families, to persuade the latter to pay.[168]

In this context, the criminalisation of migration has been suspected of decreasing migrants' trust in the police and other institutional authorities, with the result that the former feel discouraged from reporting injustices against them or from demanding protection, fearing persecution.[169] A key example is that of victims of trafficking in human beings, who are often too afraid of their captors, and suspicious of the police and other official institutions, to reach out to the latter. Indeed, vulnerability increases not only along the journey, but in Europe too: According to Oxfam, for instance, Ventimiglia may have become a nerve centre for trafficking in human beings.[170] While this is not necessarily the result of criminalisation alone, it is without doubts that the increasing difficulty of crossing the Italian–French border, especially since 2015, has made migrants more reliant on smuggling networks and, with it, also more vulnerable to them.

Overall, the above dynamics constitute a vicious cycle that, starting by criminalising and deterring migration, increases irregular foreigners' dependence on criminal networks, both to enter Europe and its member states, and to support themselves once there. As a consequence, underground organisations thrive, becoming characterised by an increasingly professional network,[171] and by revenues believed to be second only to those generated by drug smuggling,[172] while the availability of jobs in the underground economy attracts further migrants. At the same time, however, foreigners' vulnerability to violence and exploitation also

---

[167] Bartolo (2018, pp. 20–21).

[168] Melissari (2019). Similarly, Bartolo (2018, p. 21).

[169] Delvino (2017, p. 41) and Delvino and Spencer (2014, p. 8). On the consequences of insecure immigration status for women, see McIlwaine et al. (2019, p. 16).

[170] Oxfam (2018, p. 10).

[171] UNHCR (2017).

[172] Di Nicola and Musumeci (2014, p. 40).

increases, and, with it, overall insecurity, thus completing the cycle and making the lack of legal pathways unsustainable, in the long term.

The Politicisation of Migration

In addition to the issues highlighted in the two sections above, several of the problematic aspects related to the criminalisation of migration can be traced back to the third pitfall of deterrence: The politicisation of migration and of the latter's criminalisation.

The 'politicisation' of immigration refers to the process that makes immigrants' integration and control become relevant in electoral terms for party competition, or the object of a political conflict more in general.[173] Until the 1970s, a client system of immigration policy-making tended to prevail in Europe, with decisions taken technically, according to the interests of the élites, and far away from the public opinion.[174] With the 1970s oil shocks, however, economic downturn strongly hit European markets, in particular the heavy industry and manufactory sectors, which were the main employers of immigrant labour,[175] and brought the issue of immigration to the political debate. Foreigners were thus pictured as the cause of most of society's ills,[176] first of which unemployment and criminality,[177] and as a threat to the cultural identity of receiving societies.[178] The quiet client politics system of managing immigration was therefore replaced by heated debates, which not only raised the electoral value of the topic and attracted the interest of political parties and the media, but also led to the oversimplification of the discourse around it.[179]

Broadly speaking, migration is today politicised in both countries under study, as exemplified by data on the proportion of people perceiving

---

[173] Schain (2012, p. 23).
[174] Schain (2012), Joppke (1998, p. 292), and Boswell (2003, p. 9).
[175] Hollifield (2004, p. 895).
[176] Boswell (2003, p. 4).
[177] Weiner (1995, p. 142).
[178] Weiner (1995, p. 140).
[179] Boswell (2003, p. 9; 2011, p. 12).

the phenomenon as among the two most important issues in their countries. According to the Eurobarometer, in 2018, this percentage stood at 35% in Italy, 17% in France.[180]

With specific regards to the criminalisation of migration, however, the degree of public attention to the norm, has been quite different. Today, in France, there seems to be limited politicisation of the measure[181]: The topic is mostly absent from the public debate, and its contestation mainly restricted to specialised lawyers and activists.[182] On the contrary, the norm spurred extremely polarised discussions in Italy, both at the time of its adoption and today,[183] to the extent of being considered by an interviewee as the Northern League's commercial.[184]

More broadly, however, despite contrasting levels of politicisation of the norm, and different historical contexts, the rhetoric surrounding migration at the time of the introduction of the crime was surprisingly similar, hinting towards both the politicisation and securitisation of the phenomenon. Indeed, both countries had politicians and news outlets presenting foreigners as in competition for employment opportunities with citizens, and fostering xenophobic arguments.[185]

Securitisation was also high in both states. Both in 1930s France and 2000s Italy, it is possible to see increasing rhetoric presenting migratory flows as a *'Trojan Horse'*, likely to hide (predominantly German) infiltrates and spies in the former, Islamic terrorists in the latter. Reference to such assumptions was explicitly made in both cases: Daladier straightforwardly referred to the openness to refugees as welcoming a 'Trojan Horse of spies and subversives' in 1933,[186] and several politicians in Italy cautioned against the risk of having terrorist attacks carried out by

---

[180] Eurobarometer (2019).

[181] See, for example, interviewee 49 and D'Ambrosio (2010, pp. 5–6).

[182] See, for example, Gisti (2012), Henriot (2012), and Lochak (2016).

[183] See Chapter 4, section "From Rhetoric to Paper: Discursive Gaps". See also: Interviewees 55, 17, 30.

[184] Interviewee 17. Similarly, Chamber of Deputies, 08/07/2015, p. 14.

[185] See, for example, Maga (1982) and Talani and Rosina (2019).

[186] Maga (1982, pp. 429–430).

migrants.[187] Although no evidence was found to support similar claims in either case,[188] these are likely to have contributed to heightening the sense of weariness against foreigners, as well as the stigma attached to them.

Indeed, the link between criminalisation and securitisation emerges as clearest in a statement made by Alfano in 2016 when, asked by a journalist why the crime of irregular migration should be maintained, despite its lack of implementation, he answered:

> This is not the point. In the field of security, we are playing two related but different matches: *One regards reality, the other the perception of reality.*[189]

In all of the above, the key problem lies in the fact that deterrence targets not one but multiple audiences and, whenever a secondary one is given more relevance than the primary one, emerging measures are likely to result incoherent and counterproductive. This is, indeed, the case for the criminalisation of migration, where the interests of the domestic public were prioritised, compared to those of migrants, as I will show in the next paragraphs.

To begin with, both Italy and France have multiple audiences being targeted by deterrence generally, and criminalisation specifically. While the notional one is represented by *potential and actual irregular migrants*, who are supposed to be discouraged by the norm, the actual addressees are more numerous, chiefly including the *domestic public*. To the latter, the criminalisation of migration aims to show politicians' involvement in the 'fight against migration', to appease public anxieties (real or perceived), and demonstrate action. A third audience also emerges, slightly different in the two countries. In Italy, the strategy of deliberate malintegration characterising the adoption of the norm, exemplified by the tough rhetoric and subsequent regularisation, can be interpreted as meaning to communicate to employers that they would still be able

---

[187] Chamber of Deputies, 08/07/2015, pp. 19, 22–23. Former Italian Minister Calderoli (LN) also explicitly referred to migrants as a 'Trojan Horse for terrorism', https://www.secoloditalia.it/2017/07/amburgo-lo-conferma-limmigrazione-clandestina-cavallo-di-troia-del-terrorismo/.

[188] Maga (1982, pp. 429–430) and Chamber of Deputies, 08/07/2015, pp. 19, 22–23.

[189] Bei (2016). Author's translation, emphasis added.

to draw from the much-needed foreign labour (especially relied upon in the agricultural, construction and housekeeping sectors),[190] with no substantial change. On the contrary, in France, the resistance to the repeal of the GAV can be read as intending to convey the message to law-enforcement officials and bureaucrats that efficacy (hence, their ability to meet set targets of expulsions) would not be affected. Finally, although in Chapter 3, I hypothesised international partners to also be a further target audience, in the case of the criminalisation of migration they do not seem to have played a significant role, as the norm was predominantly domestically oriented.

Among such audiences, the domestic public was the privileged one, in the context of the criminalisation of migration. On one hand, in Italy, neither the repeated criticism against the norm by judges, prosecutors, and other actors,[191] nor its shown ineffectiveness,[192] managed to match the political desire not to appear 'soft' on migration. Importantly, such concerns did not only affect right-wing parties, such as the LN, but also centre-left and other ones, including parts of the *Partito Democratico* (PD) and *Movimento 5 Stelle* (M5S).[193] Matteo Renzi, for instance, while acknowledging that 'according to the magistrates, the crime as such is not necessary, does not make sense and clogs the courts', also stressed that 'there is a perception of insecurity for which we will pursue [decriminalisation] slowly, all together, without haste'.[194] Similarly, Minister Boschi pointed to the need to 'prepare' the people, before pursuing such change.[195]

In the words of former European Commission President and Italian Prime Minister Romano Prodi, the goal of the criminalisation of migration was to predominantly reassure the domestic public, whose concerns assumed precedence over those of efficacy.[196] Politically, criminalising

---

[190] Reyneri (2001).
[191] See Pansa (2016), Baer (2016), and Canzio (2016).
[192] See Commissione Fiorella (2012).
[193] For the latter, see Grillo and Casaleggio (2013).
[194] Renzi, cited in Corriere della Sera (2016). Author's translation.
[195] Boschi, cited in Meli (2016).
[196] Interviewee 20.

migration was an advantageous strategy, and the lack of positive outcomes unimportant.[197]

Even in France, where the criminalisation of entry and stay has not necessarily been a publicly contested issue, the norm's indirect politicisation through the rhetoric surrounding migration in general, and returns specifically, led the establishment to strongly oppose its repeal.[198]

Overall, the predominant emphasis on the interests of an audience that was not the notional one ended up making the norms incoherent, either internally or externally (or both). In Italy, the contradiction inherent in adopting a visible measure to control entry flows, while leaving undamaged the irregular market that employs them, resulted in the norm foreseeing a highly symbolic sanction (an expensive fine of criminal nature), but hence being unable to meet migrants' longer-term concerns (the deeper drivers of mobility). In France, concerns of efficacy (meant to ease repatriations through the GAV) and political gain (related to the electoral value of returns), made the country willing to uphold norms that were in contrast with EU law.

The above ended up being counterproductive too. Indeed, the criminalisation of migration contributed to the aforementioned vicious cycles not only by increasing irregularity, but also through the securitisation and polarisation of migration debates[199]: By placing further attention to border controls, enhancing a securitarian rhetoric, and associating migrants with criminals, the norm increased the visibility of migratory flows. Once it failed at stemming them, however, it also resulted in a heightened sense of uncertainty and insecurity, which led to demands for more visible measures of control. In other words, for more of the same, as demonstrated below.

On one hand, not only was the decriminalisation of irregular stay strongly fought against in France, and that of irregular entry left untouched until 2018, but every step towards the repeal of parts of the norms was paralleled by the introduction of further related crimes. In this way, the law that abrogated irregular stay also created the *délit de mantien*,[200] and the law that repealed irregular entry from a Schengen

---

[197] Ibid.
[198] See Henriot (2013).
[199] Similarly, see Chebel d'Appollonia (2012) and Andersson (2016).
[200] Law 2012-1560, art. 9.

country made the crime of avoiding fingerprinting subject to the interdiction from French territory too.[201] Both subjected the new crimes to similar sanctions foreseen for irregular entry and stay (one-year imprisonment and/or €3750 fine), in what emerges as a spiralling securitisation path.

On the other hand, in Italy, once the criminalisation of entry and stay was adopted in 2009, it became incredibly difficult to repeal, to the extent that Renzi's centre-left government felt unable to revoke it even after it was explicitly found ineffective and counterproductive.[202]

Both the French and Italian cases well-exemplify the political ease with which new securitarian and symbolic measures are adopted, and the inherent difficulties that affect their potential repeal. Indeed, the clash between the short-term measures fostered by political incentives, and the long-term solutions needed to address migration, is increasingly evident, and contributes to enhancing both policies' incoherence and vicious cycles of insecurity. Interestingly, according to a French lawyer interviewed, the tough rhetoric surrounding criminalisation, and subsequent insecurity spread across the population, should be accounted as among the political costs of the norm.[203]

Finally, in the Italian case, the above does not seem justified by the actual desires of the electorate. In this context, it is interesting to examine the attempted repeal of the norm by focusing more in-depth on the M5S's perspective. At the very beginning, it was indeed two M5S MPs, Cioffi and Buccarella, who put forward an amendment suggesting the repeal of the crime of irregular entry and stay, in 2013.[204] The proposal was however met with significant scepticism within the movement, especially by its leaders Grillo and Casaleggio, who viewed it as easily disputable by voters and responded with a blog post criticising the two MPs' initiative.[205] In this context, a survey was conducted among M5S supporters on 13 January 2014, to decide whether or not the

---

[201] Law 2018-778, art. 35.

[202] Commissione Fiorella (2012).

[203] Interviewee 48.

[204] Amendment 1.0.1/3, available at: http://leg17.senato.it/japp/bgt/showdoc/17/Emendc/734317/714470/index.html.

[205] Grillo and Casaleggio (2013).

M5S Senate Group should support the decriminalisation of migration.[206] A third of certified voters expressed their preference (24,932 out of 80,383), with the majority (15,839, i.e. 64%) supporting the repeal of the crime, and only 9093 (36%) its continuation.[207] Although the survey is not representative of the whole Italian population, its results provide key evidence to refute the PD's later argument that 'the people would not understand'.[208] Indeed, the voters of the M5S, which, in Grillo's own words, is a 'populist' party speaking to 'people's bellies',[209] had actually proven ready to accept decriminalisation already in 2014.

To sum up, the politicisation of migration has led to a number of counter-effects in the two countries, despite not being directly concerned with the criminalisation of entry and stay in one of them. Specifically, by privileging the domestic audience, it made political goals assume predominance over efficacy ones, and did not allow for the repeal of the norm, even when it became clear that the latter was ineffective and ripe with contradictions. It also distorted the coherence of the norm, by preventing it from focusing on the target audience's needs, and, in so doing, undermined its sustainability, by fostering vicious cycles of insecurity, as exemplified by spiralling criminalisation in France. Finally, although political utility was considered as among the key reasons for maintaining the norms, it is questionable whether the people would really not have understood a different choice, as hinted to by the 2014 survey among M5S supporters.

A final remark should be made, regarding the role of political actors. According to Schain, mainstream parties have two options when dealing with populist far-right parties: Either isolate their agenda, or co-opt it.[210] In the case of criminalisation, the latter has been the case in both countries, where neither the centre-right nor the centre-left were able to challenge the right's rhetoric, but played instead into the securitisation of migration, by incorporating it.

---

[206] Castigliani (2014).
[207] Ibid.
[208] Alfano as cited in Baer (2016).
[209] Castigliani (2013).
[210] Schain (2006).

In Italy, although then PM Berlusconi withdrew his support for the norm in June 2008,[211] his centre-right Forza Italia was in coalition with the Lega Nord, when the norm was adopted, and did not hinder its passing. The norm was not significantly challenged by Renzi's later centre-left Government either, which, as seen above, declined to repeal it in 2016.

In France, the far right was not involved in the introduction of the norm, which however enjoyed cross-party support over time: If it was Daladier's centre-right cabinet that adopted the norm in 1938, the Socialist Party also supported its reinforcement in 1981, as an alternative to expulsions.[212] A similar attitude characterised the parties once the populist radical right Front National (FN) entered the political scene: Following the success of the FN in ensuring 47% of the vote of the younger cohorts of the working class at the 1997 legislative elections, for example, the Socialists started to promote and introduce tougher measures on immigration control, trying to achieve a central consensus on the issue, but hence also contributing to the shift to the right of French immigration policies.[213] Similarly, given Sarkozy's explicit intention not to lose voters to the FN,[214] the resistance to the repeal of the norm in 2012 might be understood with reference to the fear that either the change itself, or its impact on returns, would be exploited by populist right-wing actors.

Overall, the strong presence of a populist radical right party in both countries, be it the FN or LN, is likely to have contributed to making it harder for mainstream parties to depoliticise the issue, especially in periods of high salience of migration such as the last decade. This would be in line with Schain's finding that the former's indirect impact on the political sphere, once they gain electoral breakthrough, is significant, taking place in a variety of ways, including by shaping policy agendas and influencing the public's priorities.[215] Yet, in the context of criminalisation, the inability of the mainstream to contain populist rhetoric

---

[211] http://www.antigone.it/component/content/article/76-archivio/1940-immigrazione-berlusconi-frena-qclandestinit-sia-solo-aggravanteq-la-repubblica-030608.

[212] Law 81-973, art. 4.

[213] Schain (2006, pp. 276, 283).

[214] Marthaler (2008, pp. 388–389).

[215] Schain (2006).

was conducive to significant drawbacks, including maintaining an internally and/or externally incoherent norm, and fostering vicious cycles of insecurity.

The Role of Information

As a consequence of the prioritisation of the domestic audience, the communication strategy related to the criminalisation of migration appears as mainly targeted at the domestic audience, with the result that potential migrants were scarcely aware of the potentiality of being subject to criminalisation.

From the surveys conducted with third country nationals in Italy and France, it emerged that most respondents were oblivious to the possibility of incurring criminal sanctions. This was especially the case in Italy, where only 2% identified 'fines' as sanctions foreseen by the country for irregular migration (among a given list of measures), but even in France the trend is similar. There, none of the respondents were aware of the possibility of being imprisoned for entering or staying irregularly.

Even beyond the criminalisation of migration, the interviews and surveys suggest a significant overestimation of migrants' degree of knowledge, on both Europe and its migration policies.[216] In Italy, among the surveyed migrants, information emerged as being predominantly acquired through personal networks (exploited in 58% of cases), with public authorities and, perhaps surprisingly, the internet, lagging behind (used by 13 and 12% of respondents, respectively). Knowledge of the possibility of being returned was found to be somewhat more widespread in both countries, compared to other measures (22% of respondents in Italy, 60% in France) but, even then, few migrants reported having learnt information on sanctions *before* leaving for Europe (35% in Italy, 20% in France).

The above seem to question the relevance of the several information campaigns that have been carried out, at both the national and EU level. According to the EMN, 'over €23 million have been devoted to information and awareness raising campaigns since 2015' under the framework of the EU Action Plan against Migrant Smuggling.[217] Yet, as argued in

---

[216] See interviewees 12, 21, 50.
[217] EMN (2019, p. 59).

Chapter 3, these are not necessarily effective and, according to Tjaden and others, only about half of the campaigns initiated generally lead to a change in attitudes.[218] Indeed, various interviewees working with foreigners in irregular situations reported a general lack of interest or belief, among both actual and potential migrants, in information aiming to discourage departures.[219]

In this context, information-awareness efforts also seldom include reference to criminalisation. As an example, both Italy and France participated in the IOM's campaign 'Aware Migrants', launched in 2016 and targeting six African countries. The part of the campaign addressing migrants directly is articulated through a number of videos, social media posts, and webpages.[220] Interestingly, however, no mention of the sanctions implied by criminalisation is made, on either the French[221] or Italian pages.[222] If the above confirms the lack of prioritisation of the migrant audience, it also highlights the difficulty of the norm to achieve its deterrent effect, and inherently contradictory nature.

Conclusion

According to a survey conducted by the IOM in 2011, the first two European countries that young Egyptians aspired to migrate to (following Gulf and Arab ones) were Italy and France, respectively at the 5th and 6th place.[223] What went wrong? Should the criminalisation of migration not have contributed to dissuade such aspirations? From the above analysis, the criminalisation of irregular entry and stay emerged as an ineffective norm, ripe with contradictions and counterproductive outcomes, but nonetheless of key symbolical significance.

To begin with, the criminalisation of migration appears to have taken a milder form where its politicisation was higher. In Italy, the intense discursive emphasis on the criminalisation of migration diverged greatly from its

[218] Tjaden et al. (2018).
[219] Interviewees 15, 12.
[220] https://awaremigrants.org/fr.
[221] https://awaremigrants.org/fr/les-voies-regulieres-pour-entrer-en-france.
[222] See: https://awaremigrants.org/regular-channels-enter-italy, updated August 4, 2016, accessed 31 May 2019.
[223] IOM (2011, p. 13).

actual form on paper, which ended up not adopting imprisonment, but only a fine. The gap can be interpreted as a strategy of *deliberate malintegration*, aiming to propose a visible and symbolic norm, while however leaving unaffected those employers who significantly rely on foreign work. In France, on the other hand, not only does the degree of direct politicisation of the measure seem to have been more limited, but the norm also assumed a more severe nature (foreseeing imprisonment for up to a year), thus suggesting that discursive gaps were not particularly significant.

Beyond different rhetorical dynamics, the enforcement of the norm was of limited nature in both countries. As an example, average condemnation rates for irregular entry and stay represented only a fraction of the overall national ones for the whole period under study, and were significantly lower in France than in Italy. The only exception to limited implementation concerned the extensive use of criminal custody in 2007–2012 in France, which however did not lead to full criminal trials in most cases, due to the instrumentalisation of the measure to promote returns. In this context, judges and prosecutors emerged as key in limiting the application of the norm, either through rulings sanctioning the norm's incompatibility with supranational law (in France), or through direct disapplication (in Italy), in what may be seen as offering support to the liberal institutionalist thesis.

Partly, but not only, due to the above, the norm resulted ineffective and fared poorly on most of the evaluation parameters assessed. To begin with, it did not succeed at reducing irregular migration, whose fluctuations were unrelated to changes in the use of the criminal law in both countries. Concerning the goal of increasing repatriations, this was not met in Italy, and only partly in France, although even in the latter case the instrumental use of criminal custody did not manage to significantly bring effective return rates above half of the EU average. Part of the reasons for the lack of efficacy relate to the incoherence of the norms, and specifically of their sanctions, as exemplified by the difficulty in making migrants comply with monetary fines. The issue of (in)coherence went however beyond that, as the criminalisation of migration also implied inefficient duplications of procedures and costs, taking the form of parallel administrative and criminal processes in Italy (also leading to the overwhelming of public offices), and of redundant expenses to keep migrants in prison while aiming to expel them, in France (also leading to potentially more marginalisation). If the high costs implied in the full implementation of the norm were superfluous and too high, however, a more limited application made the reception system highly uncertain, undermining certitude and any potential deterrent effect, and resulting in a trade-off that would backfire in the long term. In all of the above, the chief utility stemming from the criminalisation of migration seems to have been of political nature, as demonstrated by the fact that electoral concerns repeatedly gained prominence over effectiveness ones.

Altogether, the reasons for the lack of effectiveness of the norm can be related to the inherent weaknesses of deterrence as a strategy to address migration. Indeed, criminalisation was found to have a marginal role, when compared to the deeper drivers of migration. Variations in flows to Italy and France were not explained by changes in the norm, but rather by international political and economic factors, first of which dire conditions in countries of origin and transit (including Libya, Syria, and several sub-Saharan African countries). Both conflicts and economic conditions appeared to have a deeper effect on migrants' decision-making, leaving them unaffected by the potentiality of incurring a fine or prison term. In fact, rather than discouraging migration, criminalisation may have incentivised irregularity, by affecting concerned foreigners' criminal records and making it harder for them to obtain regular employment or residence status. Together with other deterrence measures and the lack of regularisation opportunities, the above also fostered vicious cycles between restrictiveness and irregularity.

As I have argued throughout the chapter, the criminalisation of migration contributed to vicious cycles not only by fostering irregularity, but also by increasing the rhetoric and insecurity surrounding migration, and thus making of political gains a key concern of involved actors. Indeed, as a result of the politicisation of migration, criminalisation privileged the domestic audience, rather than the notional one (made up of potential and actual migrants), resulting not only intrinsically incoherent, insofar as it was unable to address the latter's concerns, but also, by extension, of scarce efficacy and efficiency. The de-prioritisation of the foreign audience also resulted in communication strategies being predominantly targeted at the domestic electorate, and hence unable to inform and persuade migrants of the sanctions foreseen, as shown by the very low level of information held by surveyed third country nationals.

In this context, the policy gaps emerging from the Italian and French cases may be interpreted in different ways. On one hand, the ineffectiveness of the norm in the former country might be understood in light of Freeman's realist understanding of migration policies as being shaped by multiple and diverging interests: Contrasting preferences on the optimal level of restrictiveness (particularly by employers of foreign workers) led to a policy that was intentionally less strict than what the rhetoric surrounding it would have suggested, thus reducing its potential for success.

On the other hand, the French case may be explained with reference to the liberal institutionalist focus on the role of legal and supranational institutions: Both the European Court of Justice and the Supreme Court were pivotal in bringing about the partial disapplication of the norm, just like security agencies were key in delaying such disapplication. As a matter of fact, the liberal institutionalist perspective might be applied to the case of Italy too, where both the ECJ and local courts proved crucial in leading to a looser enforcement of the norm. However, if the above supports Hollifield's emphasis on courts' intervention (in both Italy and France), and Guiraudon's focus on security agents (in France), it also challenges the latter's view that local and supranational courts have more limited influence in constraining states' migration-control efforts, since these proved, in fact, key actors in both case studies.

Finally, if one accepts the transnationalist view of migration being the result of changes in globalisation dynamics, the ineffectiveness of criminalisation in reducing the phenomenon may then be explained in both the French and Italian cases by policies' inherent inability to control the flows. From this perspective, because migration is structural, the policy gap resulting from the norm's failure is not so much due to the influence of interest groups in asking for more openness, nor to the mitigating role of courts, but rather to the intrinsic impossibility for state policies to regulate the phenomenon. I will discuss each of these three interpretations in more detail in the concluding chapter, proposing to link them to the threefold understanding of policy gaps used throughout the book.

Overall, while the deterrent effect of criminalisation emerged as largely irrelevant, the norm was rife with contradictions, including contributing to creating more irregularity, placing an excessive burden on courts, and fostering the insecurities associated to migration. Yet, its symbolic nature prevented governments from supporting its repeal, even in the case of declared paradoxical effects (in Italy), or limited direct politicisation (in France), thus making of criminalisation an ineffective and counterproductive, albeit hard to repeal, measure.

References

Abbate, Liro, and Marco Lillo. 2015. *I re di Roma: Destra e sinistra agli ordini di mafia capitale*. Milano: Chiarelettere (ebook version).

Ambrosini, Maurizio. 2018. *Irregular Immigration in Southern Europe: Actors, Dynamics and Governance*. Cham: Palgrave Macmillan.

Ambrosini, Maurizio, and Alberto Guariso. 2016. Immigrazione illegale: perché è un reato inutile. *LaVoce*, February 2. http://www.lavoce.info/archives/39544/immigrazione-illegale-perche-e-un-reato-inutile/. Accessed 10 Jan 2017.

Amnesty International. 2017. *Libya's Dark Web of Collusion: Abuses Against Europe-Bound Refugees and Migrants*. London: Amnesty International Ltd.

Andersson, Ruben. 2016. Europe's Failed 'Fight' Against Irregular Migration: Ethnographic Notes on a Counterproductive Industry. *Journal of Ethnic and Migration Studies* 42 (7): 1055–1075.

Annuaire Statistique de la Justice. 2006. Secrétariat Général, Direction de l'Administration générale et de l'Équipement, Sous-Direction De La Statistique, Des Études Et De La Documentation, Paris. http://www.justice.gouv.fr/art_pix/1_annuairestat2006.pdf. Accessed 13 Nov 2018.

Annuaire Statistique de la Justice. 2007. Secrétariat Général, Direction de l'Administration générale et de l'Équipement, Sous-Direction De La Statistique, Des Études Et De La Documentation, Paris. http://www.justice.gouv.fr/art_pix/1_annuaire2007.pdf. Accessed 13 Nov 2018.

Annuaire Statistique de la Justice. 2011–2012. Secrétariat Général, Direction de l'Administration générale et de l'Équipement, Sous-Direction De La Statistique, Des Études Et De La Documentation, Paris. http://www.justice.gouv.fr/art_pix/stat_annuaire_2011-2012.pdf. Accessed 13 Nov 2018.

Baer, Giovanna. 2016. Clandestinità: Le Sette Vite di un Reato Inutile. *PaginaUno*, N. 46, February–March. http://www.rivistapaginauno.it/reato-clandestinita.php. Accessed 1 May 2018.

Bartolo, Pietro. 2018. *Le stelle di Lampedusa: La storia di Anila e di altri bambini che cercano il loro futuro fra noi*. Milano: Mondadori.

Beccaria, Cesare. 1973 [1764]. *Dei Delitti e delle Pene*. Milan: Letteratura Italiana Einaudi.

Becker, Howard S. 1963. *Outsiders: Studies in the Sociology of Deviance*. New York: The Free Press.

Bei, Francesco. 2016. Alfano: clandestinità, una norma sbagliata, ma ora deve restare. *Repubblica*, January 10. https://www.repubblica.it/politica/2016/01/10/news/angelino_alfano_riconosce_che_la_norma_anti-irregolari_del_2009_fu_un_errore_renzi_e_io_non_ideologici_usiamo_buon_sens-130935370/. Accessed 19 Sept 2019.

Ben Khalifa, Riadh. 2012. La Fabrique des Clandestins en France, 1938–1940. *Centre d'information et d'études sur les migrations internationales*, 2012/1

N° 139, pp. 11–26. https://www.cairn.info/revue-migrations-societe-2012-1-page-11.htm. Accessed 19 Sept 2019.

Borjas, George J. 1989. Economic Theory and International Migration. *International Migration Review* 23 (3): 457–485.

Boswell, Christina. 2003. *European Migration Policies in Flux: Changing Patterns of Inclusion and Exclusion.* Oxford: Blackwell.

Boswell, Christina. 2007. Theorizing Migration Policy: Is There a Third Way? *International Migration Review* 41 (1): 75–100.

Boswell, Christina. 2011. Migration Control and Narratives of Steering. *British Journal of Politics and International Relations* 13: 12–25.

Camilli, Annalisa. 2018. Da Bardonecchia a Briançon, in viaggio con i migranti sulle Alpi. *Internazionale*, January 9. https://www.internazionale.it/reportage/annalisa-camilli/2018/01/09/bardonecchia-briancon-alpi-migranti. Accessed 7 Jan 2020.

Canzio, Giovanni. 2016. Intervento del Primo Presidente Dott. Giovanni Canzio per la cerimonia di inaugurazione dell'anno giudiziario. *Corte Suprema di Cassazione*, January 28. http://www.cortedicassazione.it/corte-di-cassazione/it/inaugurazioni_anno_giudiziario.page. Accessed 1 May 2018.

Carling, Jørgen, and Maria Hernández-Carretero. 2008. Kamikaze Migrants? Understanding and Tackling High-Risk Migration from Africa. Paper presented at the Narratives of Migration Management and Cooperation with Countries of Origin and Transit, Sussex Centre for Migration Research, University of Sussex.

Castigliani, Martina. 2013. Grillo, confessione a eletti M5S: 'Finzione politica l'impeachment di Napolitano'. *Il Fatto Quotidiano*, October 30. https://www.ilfattoquotidiano.it/2013/10/30/grillo-confessione-a-porte-chiuse-napolitano-sotto-accusa-e-finzione-politica/760888/. Accessed 20 Sept 2019.

Castigliani, Martina. 2014. Abolizione reato immigrazione clandestina, gli iscritti M5S votano sì al referendum. *Il Fatto Quotidiano*, January 13. https://www.ilfattoquotidiano.it/2014/01/13/abolizione-reato-immigrazione-clandestina-gli-iscritti-m5s-votano-si-al-referendum/841806. Accessed 20 Sept 2019.

Castles, Stephen. 2004. Why Migration Policies Fail. *Ethnic and Racial Studies* 27 (2): 205–227.

Castles, Stephen, and Mark J. Miller. 2011. Migration and the Global Economic Crisis: One Year on. http://www.age-of-migration.com/uk/financialcrisis/updates/migration_crisis_april2010.pdf. Accessed 20 Sept 2019.

Chebel d'Appollonia, Ariane. 2012. *Frontiers of Fear: Immigration and Insecurity in the United Sates and Europe.* London: Cornwell University Press.

Colombo, Asher. 2011. Gli Stranieri e la Sicurezza. In *Rapporto sulla criminalità e sicurezza in Italia 2010*, ed. Marzio Barbagli and Asher Colombo, 269–340. Milano: Il Sole 24 ORE and Fondazione ICSA.

Commission of the European Communities. 1991. Communication de la Commission des Communautés européennes au Conseil et au Parlement européen sur le droit d'asile' SEC(91) 1857 final, October 11. Brussels, available at the archives of the *Fondation Jean Monnet pour l'Europe*, document PDB 4/7/11.

Commissione Fiorella per la revisione del sistema penale. 2012. *Relazione*, released on the website of the Italian Ministry of Justice on 23 April 2013. https://www.giustizia.it/giustizia/it/mg_1_12_1.page;jsessionid=Cnr qkdgEhbOXrgVen91FlTUP?facetNode_1=3_1&facetNode_2=3_1_3&fac etNode_3=3_1_3_1&facetNode_4=4_57&contentId=SPS914197&previsiou sPage=mg_1_12. Accessed 19 Sept 2019.

Commissione Parlamentare di Inchiesta sul Sistema di Accoglienza, di Identificazione ed Espulsione, nonché sulle Condizioni di Trattenimento dei Migranti e sulle Risorse Pubbliche Impegnate. 2017. Relazione sul Sistema di Protezione e di Accoglienza dei Richiedenti Asilo (Parliamentary commission on the reception system, 2017). *Doc. XXII-bis N. 21*, Referee: on. Paolo BENI, Approved by the Commission on 20th December 2017. http://documenti.camera.it/_dati/leg17/lavori/documentiparlament ari/IndiceETesti/022bis/021/INTERO.pdf. Accessed 19 Sept 2019.

Condemnations by infraction. 2016. French Ministry of Justice. http://www.jus tice.gouv.fr/art_pix/stat_condamnations_infractions_2016.ods. Accessed 13 Nov 2018.

Corriere del Mezzogiorno. 2014. Immigrazione, Alfano: "Problema strutturale e non più emergenziale". *Corriere del Mezzogiorno*, April 16. http://cor rieredelmezzogiorno.corriere.it/catania/notizie/cronaca/2014/16-aprile-2014/immigrazione-alfano-problema-strutturale-non-piu-emergenziale-223 82555678.shtml. Accessed 10 Jan 2017.

Corriere della Sera. 2016. Renzi: "Il reato di clandestinità non sarà nel prossimo Cdm. Aspettiamo". *Corriere della Sera*, January 20. https://www.corriere.it/ politica/16_gennaio_10/renzi-il-reato-clandestinita-non-sara-prossimo-cdm-aspettiamo-2a4226ba-b7cd-11e5-8210-122afbd965bb.shtml?refresh_ce-cp. Accessed 5 Mar 2020.

Council of the European Communities. 1976. Resolution of 9 February 1976 on an [1974] action programme for migrant workers and members of their families (Council Resolution, 1976). *Bulletin of the European Communities*, Supplement 3/76. http://aei.pitt.edu/1278/1/action_migrant_work ers_COM_74_2250.pdf. Accessed 7 Jan 2020.

Cour de Cassation. 2012. Arrêt n° 965 du 5 juillet 2012 (11–30.530), Première Chambre Civile - ECLI : FR : CCASS : 2012 : C100965. https://www. courdecassation.fr/jurisprudence_2/premiere_chambre_civile_568/965_5_2 3804.html. Accessed 27 Feb 2019.

Czaika, Mathias, and Hein de Haas. 2011. The Effectiveness of Immigration Policies: A Conceptual Review of Empirical Evidence. International Migration Institute Working Paper 33. Oxford: IMI.

D'Ambrosio, Luca. 2010. Quand l'immigration est un délit. *La vie des idées*, November 30. https://laviedesidees.fr/Quand-l-immigration-est-un-delit.html. Accessed 12 Apr 2019.

de Haas, Hein. 2011. The Determinants of International Migration: Conceptualising Policy, Origin and Destination Effects. International Migration Institute Working Paper 32, IMI, Oxford.

Delvino, Nicola. 2017. *The Challenge of Responding to Irregular Immigration: European, National and Local Policies Addressing the Arrival and Stay of Irregular Migrants in the European Union*. Global Exchange on Migration and Diversity Report. https://www.compas.ox.ac.uk/wp-content/uploads/AA17-Delvino-Report.pdf. Accessed 29 Mar 2019.

Delvino, Nicola, and Sarah Spencer. 2014. *Irregular Migrants in Italy: Law and Policy on Entitlements to Services*. ESRC Centre on Migration, Policy and Society (COMPAS), University of Oxford.

Departmental figures registered by police and gendarmerie. 2018. https://www.data.gouv.fr/fr/datasets/chiffres-departementaux-mensuels-relatifs-aux-crimes-et-delits-enregistres-par-les-services-de-police-et-de-gendarmerie-depuis-janvier-1996/. Accessed 12 Nov 2018.

di Martino, Alberto, Francesca Biondi Dal Monte, Ilaria Boiano, and Raffaelli Rosa. 2013. *La criminalizzazione dell'immigrazione irregolare: legislazione e prassi in Italia*. Pisa: Pisa University Press.

Di Nicola, Andrea, and Giampaolo Musumeci. 2014. *Confessioni di Un Trafficante di Uomini* (Confessions of a People-Smuggler). Milano: Chiarelettere.

EMN. 2011. Ad-Hoc Query on Illegal Migration in the Mediterranean Sea Basin: Responses from France, Italy, Malta and Spain (4 in Total), Requested by PL EMN NCP on 23 March 2010 (to CY, FR, GR, IT, MT and ES only), Compilation produced on 21 November 2011. https://ec.europa.eu/home-affairs/sites/homeaffairs/files/what-we-do/networks/european_migration_network/reports/docs/ad-hoc-queries/illegal-immigration/210.emn_ad-hoc_query_illegal_migration_in_the_mediterranean_sea_basin_updated_wider_dissemination.pdf. Accessed 5 Jan 2020.

EMN. 2019. *Annual Report on Migration and Asylum 2018*. European Migration Network.

Eurobarometer. 2019. 'What do you think are the two most important issues facing our country at the moment?' https://ec.europa.eu/commfrontoffice/publicopinion/index.cfm/Chart/getChart/chartType/lineChart/themeKy/42/groupKy/208/savFile/54. Accessed Feb 2019.

European Commission, Communication from the Commission to the European Parliament, the European Council and the Council, *Progress Report on the*

Implementation of the European Agenda on Migration, 16 October 2019, COM(2019) 481 final.

European Union: Court of Justice, Case C-329/11, Alexandre Achughbabian v Préfet du Val-de-Marne, Judgment of the Court (Grand Chamber) of 6 December 2011, European Court Reports 2011-00000. http://eur-lex.eur opa.eu/legal-content/EN/TXT/?uri=CELEX%3A62011CJ0329. Accessed 29 Mar 2017.

European Union: Court of Justice, Case C-38/14, Judgment of the Court of Justice of 23 April 2015, Zaizoune, Case, Judgment of the Court (Fourth Chamber) of 23 April 2015. http://curia.europa.eu/juris/document/document.jsf;jsessionid=5B901EAE866FB67775FF28A7247FFBA1?text=&docid=163877&pageIndex=0&doclang=EN&mode=lst&dir=&occ=first&part=1&cid=5292996. Accessed 6 Jan 2020.

European Union: Court of Justice, Case C-430/11, Judgment of the Court of Justice of 6 December 2012, Sagor, Case, Md Sagor, Judgment of the Court (First Chamber) of 6 December 2012, ECLI:EU:C:2012:777. https://eur-lex.europa.eu/legal-content/GA/SUM/?uri=CELEX:62011CJ0430. Accessed 1 May 2018.

European Union: Court of Justice, Case C-47/15, Sélina Affum v Préfet du Pas-de-Calais and Procureur général de la Cour d'appel de Douai, Judgment of the Court (Grand Chamber) of 7 June 2016. https://eur-lex.europa.eu/legal-content/en/TXT/PDF/?uri=uriserv%3AOJ.C_.2016.296.01.0011.01.ENG. Accessed 29 Apr 2019.

Eurostat migr_dubro. 2018. Statistics on enforcement of immigration legislation, *Eurostat*. http://appsso.eurostat.ec.europa.eu/nui/show.do?dataset=migr_eipre&lang=en. Accessed Dec 2018.

Eurostat migr_eipre. 2018. Statistics on enforcement of immigration legislation, *Eurostat*. http://appsso.eurostat.ec.europa.eu/nui/show.do?dataset=migr_eipre&lang=en. Accessed Dec 2018.

EUROSTAT migr_eipre and migr_asyapp. 2018. Statistics on enforcement of immigration legislation, *Eurostat*. http://appsso.eurostat.ec.europa.eu/nui/show.do?dataset=migr_eipre&lang=en. Accessed Dec 2018.

Eurostat migr_eirtn. 2018. Statistics on enforcement of immigration legislation, *Eurostat*. http://appsso.eurostat.ec.europa.eu/nui/show.do?dataset=migr_eipre&lang=en. Accessed Dec 2018.

Eurostat migr_eirfs. 2018. Statistics on enforcement of immigration legislation, *Eurostat*. http://appsso.eurostat.ec.europa.eu/nui/show.do?dataset=migr_eipre&lang=en. Accessed Dec 2018.

Eurostat migr_eirt_ass. 2018. Statistics on enforcement of immigration legislation, *Eurostat*. http://appsso.eurostat.ec.europa.eu/nui/show.do?dataset=migr_eipre&lang=en. Accessed Dec 2018.

Fan, Mary. 2008. When Deterrence and Death Mitigation Fall Short: Fantasy and Fetishes as Gap-Fillers in Border Regulation. *Law & Society Review* 42: 701.

Fargues, Philippe, and Christine Fandrich. 2012. *Migration After the Arab Spring*. MPC Research Report 2012/09.

Fargues, Philippe, and Sara Bonfanti. 2014. *When the Best Option Is a Leaky Boat: Why Migrants Risk Their Lives Crossing the Mediterranean and What Europe Is Doing About It*. Migration Policy Centre, EUI.

Ferrero, Giancarlo. 2010. *Contro il Reato di Immigrazione Clandestina: Un'Inutile, Immorale, Impraticabile Minaccia*, 2nd ed. Roma: Ediesse.

Fix, Michael, Demetrios G. Papademetriou, Jeanne Batalova, Aaron Terrazas, Serena Yi-Ying Lin, and Michelle Mittelstadt. 2009. Migration and the Global Recession. Migration Policy Institute.

FRA. 2014. *Criminalisation of Migrants in an Irregular Situation and of Persons Engaging with Them*. Vienna: FRA.

Freeman, Gary P. 1998. The Decline of Sovereignty? Politics and Immigration Restriction in Liberal States. In *Challenge to the Nation-State: Immigration in Western Europe and the United States*, ed. Christian Joppke, 86–108. Oxford: Oxford University Press.

Freeman, Gary P., and Alan E. Kessler. 2008. Political Economy and Migration Policy. *Journal of Ethnic and Migration Studies* 34 (4): 655–678.

French Ministry of the Interior. 2012. *Etude d'Impact: Projet de loi relatif à la retenue pour vérification du droit au séjour et modifiant le délit d'aide au séjour irrégulier pour en exclure les actions humanitaires et désintéressées*, September 21 (Etude d'Impact 2012). http://www.senat.fr/leg/etudes-impact/pjl11-789-ei/pjl11-789-ei.html. Accessed 10 Jan 2017.

French Ministry of the Interior. 2013. Circular INTK1300159C, January 18.

French Ministry of the Interior. 2018a. Circular INTV1824378J, September 11. http://circulaire.legifrance.gouv.fr/pdf/2018/09/cir_43960.pdf. Accessed 10 Jan 2019.

French Ministry of the Interior. 2018b. *Etude d'Impact: Projet de loi pour une immigration maîtrisée et un droit d'asile effectif*, February 20 (Etude d'Impact 2018). https://www.gisti.org/IMG/pdf/pjl2018_etude-impact_20180221.pdf. Accessed 10 Jan 2019.

French Parliament, Code of Criminal Procedure, as of 17 February 2017. https://www.legifrance.gouv.fr/affichCode.do?cidTexte=LEGITEXT0 00006071154. Accessed 23 Feb 2017.

French Parliament. Décret sur la police des étrangers. *Journal Officiel* of May 1st, 2nd, 3rd 1938 (Decree of 2nd May 1938), pp. 4967–4969. https://gallica.bnf.fr/ark:/12148/bpt6k20313224/f23.image. Accessed 10 Jan 2019.

French Parliament, Law 81-973 of 29 October 1981, Relative Aux Conditions D'entree Et De Sejour Des Etrangers En France. *JORF*, October 30.

French Parliament, Law 778-2018 of 10 September 2018. *Pour une immigration maîtrisée, un droit d'asile effectif et une integration réussie. JORF* n' 0209 of 11 September 2018. https://www.legifrance.gouv.fr/affichTexte.do?cidTexte=JORFTEXT000037381808&categorieLien=id. Accessed 29 Feb 2020.

French Parliament, Law 2012-1560 of 31 December 2012. Loi relative à la retenue pour vérification du droit au séjour et modifiant le délit d'aide au séjour irrégulier pour en exclure les actions humanitaires et désintéressées. https://www.legifrance.gouv.fr/affichTexte.do;jsessionid=1368A8F4515099719D3F22684307AB15.tpdila10v_2?cidTexte=JORFTEXT000026871211&dateTexte=20140525. Accessed 10 Jan 2017.

French Parliament, Penal Code, as of 1 January 2017. https://www.legifrance.gouv.fr/affichCode.do?cidTexte=LEGITEXT000006070719. Accessed 10 Jan 2017.

French Senate. 2014. SESSION ORDINAIRE DE 2014–2015 Enregistré à la Présidence du Sénat le 20 novembre 2014, AVIS PRÉSENTÉ *au nom de la commission des lois constitutionnelles, de législation, du suffrage universel, du Règlement et d'administration générale (1) sur le projet de* loi de finances *pour* 2015, *ADOPTÉ PAR L'ASSEMBLÉE NATIONALE* (Avis n. 114). http://www.senat.fr/rap/a14-114-8/a14-114-8_mono.html#toc144.

Frontex. 2019. *Detections of illegal border-crossings statistics.* https://frontex.europa.eu/along-eu-borders/migratory-map/. Accessed 6 Aug 2019.

Frontex. 2010. *Annual Risk Analysis 2010* (ARA 2010). Warsaw: Frontex. https://frontex.europa.eu/assets/Publications/Risk_Analysis/Annual_Risk_Analysis_2010.pdf. Accessed 30 April 2019.

Gabanelli, Milena, and Simona Ravizza. 2019. Migranti irregolari: quanti ne ha rimpatriati Salvini in 8 mesi di governo? Corriere della Sera, March 3. https://www.corriere.it/dataroom-milena-gabanelli/migranti-irregolari-quando-ne-ha-rimpatriati-salvini-8-mesi-governo/accb467e-3c37-11e9-8da9-1361971309b1-va.shtml. Accessed 7 Jan 2020.

Gisti, ed. 2012. *Immigration: un régime pénal d'exception.* Paris: Gisti. https://www.gisti.org/publication_som.php?id_article=2781#1lutte. Accessed 23 Feb 2017.

Gourevitch, Jean Paul. 2012. L'immigration en France: dépenses, recettes, investissements, rentabilité. *Contribuables Associés*, Monographie n°27, December 2012, 63.

Grillo, Beppe, and Gianroberto Casaleggio. 2013. Reato di clandestinità. *Il Blog delle Stelle*, October 10. www.ilblogdellestelle.it/2013/10/reato_di_clandestinita_4.html. Accessed 26 Apr 2018.

Guiraudon, Virginie. 2000. European Integration and Migration Policy: Vertical Policy-Making as Venue Shopping. *Journal of Common Market Studies* 38 (2): 251–271.

Hadj-Abdou, Leila. 2016. The Europeanisation of Immigration Policies. In *An Anthology of Migration and Social Transformations*, ed. A. Amelina. IMISCOE Research Series.

Harzoune, Mustapha. 2012. Combien coûte une expulsion? histoire-imm igration.fr/questions-contemporaines/politique-et-immigration/combien-coute-une-expulsion. Accessed 10 Jan 2019.

Henley, Jon. 2003. France Sets Targets for Expelling Migrants. *The Guardian*, October 28. https://www.theguardian.com/world/2003/oct/28/france.jon henley. Accessed 13 Apr 2019.

Henriot, Patrick. 2012. Les formes multiples de l'enfermement, une nouvelle forme de 'punitivité'. In *Immigration: un régime pénal d'exception*, ed. Gisti, 60–71. Paris: Gisti. https://www.gisti.org/publication_som.php?id_article=2781#1lutte. Accessed 23 Feb 2017.

Henriot, Patrick. 2013. Dépénalisation du séjour irrégulier des étrangers : l'opiniâtre résistance des autorités françaises. *La Revue des droits de l'homme* [online], 3: 1–14.

Hollifield, James F. 1998. Migration, Trade, and the Nation-State: The Myth of Globalization. *UCLA Journal of International Law and Foreign Affairs* 3: 595–636

Hollifield, James F. 2004. The Emerging Migration State. *International Migration Review* 38 (3): 885–912.

Hollifield, James F. 2014. Immigration and the Republican Tradition in France. In *Controlling Immigration: A Global Perspective*, ed. James F. Hollifield, Philip L. Martin, and Pia M. Orrenius, 3rd ed., 157–187. Stanford: Stanford University Press.

Huysmans, Jef. 2000. The European Union and the Securitization of Migration. *Journal of Cutaneous Medicine and Surgery: Incorporating Medical and Surgical Dermatology* 38 (5): 751–777.

IOM. 2011. *Egypt After January 25. Survey of Youth Migration Intentions*. Cairo. http://www.migration4development.org/sites/default/files/iom_2011_egypt_after_january_25_survey_of_youth_migration_intentions.pdf. Accessed 31 May 2019.

IOM Italy. 2017. *La Tratta di Esseri Umani attraverso la Rotta del Mediterraneo Centrale: Dati, Storie e Informazioni Raccolte dall'Organizzazione Internazionale per le Migrazioni*. http://www.italy.iom.int/sites/default/files/news-documents/RAPPORTO_OIM_Vittime_di_tratta_0.pdf. Accessed 9 Oct 2017.

ISMU. 2014. 'Sbarchi e Richieste di Asilo: Serie Storica Anni 1997–2014' (Ismu elaboration upon MoI data). http://www.ismu.org/wp-content/uploads/2015/02/Sbarchi-e-richieste-asilo-1997-2014.xls. Accessed 1 May 2018.

ISTAT. 2018. Data related to article 10-bis TUI, 2009–2016 (obtained in May 2018).

Italian Constitution. [1947] 2012. https://www.senato.it/documenti/reposi tory/istituzione/costituzione.pdf. Accessed 20 Mar 2020.

Italian Ministry of the Interior (MoI). 2007. *Rapporto sulla criminalità in Italia, analisi, prevenzione e contrasto*. https://www.poliziadistato.it/statics/45/rap porto_criminalita_appendice_2006.pdf. Accessed 10 Jan 2019.

Italian Ministry of the Interior (MoI). 2016, 2017, 2018, 2019. Cruscotto Giornaliero (update on landings) as of 31/12/2016, 31/12/2017, 31/12/2018, 30/6/2019. http://www.libertaciviliimmigrazione.dlci.interno.gov.it/it/doc umentazione/statistica/cruscotto-statistico-giornaliero. Accessed 2017–2019.

Italian Parliament. 2011. Urgent Interpellation n. 2/01053, presented by Gabriella Carlucci, meeting n.464, April 14 (Chamber of Deputies, 14/04/2011). http://banchedati.camera.it/sindacatoispettivo_16/showXh tml.asp?highLight=0&idAtto=37940&stile=6. Accessed 5 Jan 2017.

Italian Parliament. 2014. Resoconto Stenografico dell'Assemblea n. 166, meeting n. 166, Wednesday, January 15 (Chamber of Deputies, 15/01/2014). http://www.senato.it/service/PDF/PDFServer/BGT/00735336.pdf. Accessed 5 Jan 2018.

Italian Parliament, Law 39/1990. (Law Martelli) Conversione in legge, con modificazioni, del decreto-legge 30 dicembre 1989, n. 416, recante norme urgenti in materia di asilo politico, di ingresso e soggiorno dei cittadini extracomunitari e di regolarizzazione dei cittadini extracomunitari ed apolidi gia' presenti nel territorio dello Stato. Disposizioni in materia di asilo. *Gazzetta Ufficiale*, N. 49, February 28.

Italian Parliament, Law 67/2014. Deleghe al Governo in materia di pene detentive non carcerarie e di riforma del sistema sanzionatorio. Disposizioni in materia di sospensione del procedimento con messa alla prova e nei confronti degli irreperibili. *Gazzetta Ufficiale*, N. 100, May 28. http://www.gazzettau fficiale.it/eli/id/2014/05/02/14G00070/sg. Accessed 10 Jan 2017.

Italian Parliament, Law 94/2009. "Disposizioni in Materia di Sicurezza Pubblica" (Dispositions on Public Security). *Gazzetta Ufficiale*, No. 170, Ordinary Supplement No. 128 (24 July 2009) Law 94/2009. http://www. asgi.it/wp-content/uploads/public/legge.15.luglio.2009.n.94.pdf. Accessed 16 June 2016.

Italian Parliament, Law Proposal N. 5808. Modifiche al testo unico delle disposizioni concernenti la disciplina dell'immigrazione e norme sulla condizione dello straniero, emanato con decreto legislativo 25 luglio 1998, n.286. Presented on 15 March 1999. http://briguglio.asgi.it/immigrazione-e-asilo/ 1999/maggio/pdl-fini.html. Accessed 1 May 2018.

Italian Parliament, Law Proposal N. 733/2008. 'Disposizioni in materia di sicurezza pubblica', communicated to the President on 3rd June 2008. http://www.senato.it/service/PDF/PDFServer/BGT/00302495.pdf, accessed 10 January 2017.

Italian Senate. 2015. Resoconto Stenografico del Senato n. 14, Indagine Conoscitiva sui Temi dell'Immigrazione, meeting n. 295, Wednesday, July 8 (Senate, 08/07/2015).

Joppke, Christian. 1998. Why Liberal States Accept Unwanted Immigration. *World Politics* 50 (2): 266–293.

Lochak, Danièle. 2016. 'Pénalisation', in 'L'étranger et le Droit Pénal'. *AJ Pénal*, Janvier, 10–12. https://hal-univ-paris10.archives-ouvertes.fr/hal-01674295/document. Accessed 5 Jan 2019.

Maga, Timothy. 1982. Closing the Door: The French Government and Refugee Policy, 1933–1939. *French Historical Studies* 12 (3): 424–442.

Mananashvili, Sergo. 2017. EU's Return Policy: Mission Accomplished in 2016? Reading Between the Lines of the Latest EUROSTAT Return Statistics. ICMPD Policy Brief.

Marthaler, Sally. 2008. Nicolas Sarkozy and the Politics of French Immigration Policy. *Journal of European Public Policy* 15 (3): 382–397.

Maugendre, Stéphane. 2012. Interdiction du territoire : histoire d'une exception. In *Immigration: un régime pénal d'exception*, ed. Gisti, 43–56. Paris: Gisti. https://www.gisti.org/publication_som.php?id_article=2781#1lutte. Accessed 23 Feb 2017.

McIlwaine, C.J., L. Granada, and I. Valenzuela-Oblitas. 2019. *The Right to Be Believed: Migrant Women Facing Violence Against Women and Girls (VAWG) in the 'Hostile Immigration Environment' in London*. London: Latin American Women's Rights Service.

Medina, Leandro, and Friedrich Schneider. 2018. Shadow Economies Around the World: What Did We Learn Over the Last 20 Years? IMF Working Paper, WP/18/17.

Meli, Maria Teresa. 2016. Boschi e le unioni civili: «Io dico sì alla norma sulle adozioni». *Corriere della Sera*, January 10. https://www.corriere.it/politica/16_gennaio_10/boschi-intervista-unioni-civili-stepchild-adoption-adozioni-senato-no-scambi-6c6fa974-b762-11e5-8210-122afbd965bb.shtml. Accessed 7 Jan 2020.

Melissari, Laura. 2019. Sea Watch, uno dei migranti a bordo: "Carola ci ha salvati, ma Salvini ha in parte ragione". *TPI*, June 30. https://www.tpi.it/cronaca/sea-watch-testimonianza-migrante-20190630358115/. Accessed 7 Jan 2020.

MPI. 2018. Asylum Recognition Rates in the EU/EFTA by Country, 2008–2017, Migration Policy Institute. https://www.migrationpolicy.org/programs/data-hub/charts/asylum-recognition-rates-euefta-country-2008-2017. Accessed 5 Jan 2020.

Nagin, Daniel S. 2013. Deterrence in the Twenty-First Century: A Review of the Evidence. Carnegie Mellon University Research Showcase @ CMU, Working Paper. Heinz College Research.

Orsini, Giacomo, and Anna Sergi. 2018. The Emergency Business. Migrants Reception, Mafia Interests and Glocal Governance: From Lampedusa to Rome. In *Cross Border Crime Colloquium 2017*, ed. P.C. Van Duyne et al. Oisterwijk: Wolf Legal Publishers.

Overbeek, Henk. 1996. L'Europe en quête d'une politique de migration : Les contraintes de la mondialisation et de la restructuration des marchés du travail. *Études Internationales* 27 (1): 53–80.

Oxfam. 2018. *Nowhere but Out: The Failure of France and Italy to Help Refugees and Other Migrants Stranded at the Border in Ventimiglia*. Oxfam Briefing Paper, June.

Pansa, Alessandro. 2016. Interview with Repubblica: "Reato di clandestinità, Pansa: 'Abolirlo? Meglio riformarlo'". *Repubblica*, January 10. http://www.repubblica.it/politica/2016/01/10/news/clandestinita_resta_reato_pansa_c osi_com_e_intasa_procure_-130954064/. Accessed 1 May 2018.

Pasquinelli, Sergio. 2009. Perché la Sanatoria Ha Fatto Flop. *La voce*, October 9. http://www.lavoce.info/archives/25930/perche-la-sanatoria-ha-fatto-flop/. Accessed 1 May 2018.

Perry, Matt. 2004. 'Sans Distinction de Nationalité?' The French Communist Party, Immigrants and Unemployment in the 1930s. *European History Quarterly* 34 (3): 337–369.

Peytermann, Lucie. 2018. "Pour rien au monde, je ne le referai": le récit sans tabou d'un migrant ivoirien. *France Soir*, March 7. http://www.francesoir.fr/actualites-france/pour-rien-au-monde-je-ne-le-referai-le-recit-sans-tabou-dun-migrant-ivoirien. Accessed 7 Jan 2020.

Polchi, Vladimiro. 2017. In 74 per scortare 29 migranti: così funzionano le espulsioni. *Repubblica*, January 18. https://www.repubblica.it/cronaca/2017/01/18/news/in_74_per_scortare_29_migranti_cosi_funzionano_le_espulsioni-156271202/. Accessed 7 Jan 2020.

Public Prosecutor Siracusa. 2017. Relazione Annuale sull'Andamento della Giustizia nel Distretto di Catania (Periodo 1 Luglio 2016–30 Giugno 2017), October 15.

Reyneri, Emilio. 2001. 'Migrants' Involvement in Irregular Employment in the Mediterranean Countries of the European Union. International Migration Papers (IMP) Working Paper Series.

Saas, Claire. 2012. L'immigré, cible d'un droit pénal de l'ennemi. In *Immigration: un régime pénal d'exception*, ed. Gisti, 32–42. Paris: Gisti. https://www.gisti.org/publication_som.php?id_article=2781#Ilutte. Accessed 23 Feb 2017.

Sassen, Saskia. 1996. *Losing Control? Sovereignty in an Age of Globalization*. New York: Columbia University Press.

Savio, Guido. 2016. Reato di Clandestinità. Inutile e Dannoso, Ecco Perché Va Cancellato. *Stranieri in Italia*, January 16. stranieriinitalia.it/attualita/attual ita/attualita-sp-754/reato-di-clandestinita-inutile-e-dannoso-ecco-perche-va-cancellato.html. Accessed 1 May 2018.

Schain, Martin A. 2006. The Extreme-Right and Immigration Policy-Making: Measuring Direct and Indirect Effects. *West European Politics* 29 (2): 270–289.

Schain, Martin A. 2012. *The Politics of Immigration in France, Britain, and the United States*, 2nd ed. New York: Palgrave Macmillan.

Scherr, Mickaël. 2011. L'évolution des phénomènes d'immigration illégale à travers les statistiques sur les personnes mises en cause par la police et la gendarmerie nationale, 144–158. Dossier INHESJ. https://inhesj.fr/sites/default/files/ondrp_files/contributions-exterieures/m_scherr.pdf. Accessed 12 Dec 2018.

Tables of condemnations. 2012–2016. Ministry of Justice. http://www.justice.gouv.fr/statistiques-10054/donnees-statistiques-10302/les-condamnations-27130.html. Accessed 13 Nov 2018.

Talani, Leila Simona. 2015. International Migration: IPE Perspectives and the Impact of Globalisation. In *Handbook of the International Political Economy of Migration*, ed. Leila Simona Talani and Simon McMahon, 17–36. UK: Edward Elgar.

Talani, Leila Simona. 2018. *The Political Economy of Italy in the Euro: Between Credibility and Competitiveness*. London: Palgrave Macmillan.

Talani, Leila Simona, and Matilde Rosina, eds. 2019. *Tidal Waves? The Political Economy of Populism and Migration in Europe*. Bern: Peter Lang.

Tjaden, Jasper, Sandra Morgenstern, and Frank Laczko. 2018. Evaluating the Impact of Information Campaigns in the Field of Migration: A Systematic Review of the Evidence and Practical Guidance. Central Mediterranean Route Thematic Report Series. Geneva: International Organization for Migration.

Tomasetta, Lara. 2018. Il mio viaggio tra le Alpi sulle tracce dei migranti in fuga da Bardonecchia verso la Francia. *TPI*, April 6. https://www.tpi.it/2018/04/06/viaggio-alpi-migranti-bardonecchia-francia/. Accessed 7 Jan 2020.

UN General Assembly, *Convention Relating to the Status of Refugees* (Refugee Convention), 28 July 1951, United Nations, Treaty Series, Vol. 189, p. 137. Available at: http://www.refworld.org/docid/3be01b964.html. Accessed 20 Aug 2015.

UNHCR. 2016. Italy: UNHCR Update #10, December 2016. http://reliefweb.int/sites/reliefweb.int/files/resources/2016_12_UNHCRCountryUpdateItaly-December_2016_V8.pdf. Accessed 24 Jan 2018.

UNHCR. 2017. Mixed Migration Trends in Libya: Changing Dynamics and Protection Challenges. https://www.unhcr.org/595a02b44.pdf. Accessed 5 Mar 2020.

Vesci, Pietro. 2008. L'immigrazione nei programmi elettorali per le elezioni politiche del 2008: UDC, Lega Nord, La Destra e La Sinistra Arcobaleno.

Neodemos.info, April 2. http://www.neodemos.info/pillole/limmigrazione-nei-programmi-elettorali-per-le-elezioni-politiche-del-2008udc-lega-nord-la-destra-e-la-sinistra-arcobaleno/?print=pdf. Accessed 1 May 2018.

von Hirsch, Andrew, Anthony E. Bottoms, Elisabeth Burney, and P-o. Wikström. 1999. *Criminal Deterrence and Sentence Severity: An Analysis of Recent Research*. University of Cambridge Institute of Criminology. Oxford, UK: Hart Publishing Ltd.

Weiner, Myron. 1995. *The Global Migration Crisis: Challenge to States and to Human Rights*. New York: Harper Collins College Publishers.

Wihtol de Wenden, Catherine. 2010. Irregular Migration in France. In *Irregular Migration in Europe: Myths and Realities*, ed. Anna Triandafyllidou, 115–124. Farnham, Surrey: Ashgate Publishing Limited.

Ziniti, Alessandra. 2016. Il procuratore di Agrigento, Renato Di Natale: 'Con il reato di clandestinità abbiamo gli uffici ingolfati'. *Repubblica – Palermo*, May 29. http://palermo.repubblica.it/cronaca/2016/05/29/news/il_pro curatore_di_agrigento_renato_di_natale_con_il_reato_di_clandestinita_abbi amo_gli_uffici_ingolfati_-140854592/. Accessed 1 May 2018.

CHAPTER 7

Conclusion

Introduction

> Picture the illegal flow of whatever – people, drugs – as a long, skinny balloon that stretches from one part of the border, the Pacific Ocean, to the Gulf of Mexico. You squeeze it in one spot, and it's going to migrate to another spot.[1]

With these words, the president of the American Border Patrol Union T. J. Bonner depicted what he saw as being states' difficulties in controlling migratory phenomena, often leading to a mere shift in their direction. If we were to translate the above into the European context, we could visualise the flow of migrants crossing the Mediterranean Sea as the 'long, skinny balloon', stretching from the Aegean Sea to the Strait of Sicily, and from there to the Pillars of Hercules. To what extent can states control such flows, in a context of globalisation? What are the consequences of restrictive migration controls, and who benefits from them?

With such questions at heart, the core aim of this research has been to shed light on the functioning and potential consequences of specific

---

[1] Bonner, President of the National Border Patrol Council, as reported in United States Congress (2006, p. 47).

measures that have seen a surge in popularity in recent years: Those criminalising irregular migration. Throughout Europe, multiple governments have made use of criminal sanctions, in an attempt to discourage would-be migrants from reaching (or remaining in) their territories. Concurrently, through the rhetoric of deterrence, irregular migration has been presented to the domestic public as a phenomenon that can not only be controlled, but also pre-emptively avoided.

In light of such developments, theoretically, this book has aimed to contribute to the international political economy (IPE) debate on states' ability to control migration, by shedding light on the effectiveness and repercussions of understudied criminalisation measures, from an interdisciplinary perspective. Empirically, it has focused on the criminalisation of irregular migration in Italy and France (namely, the use of the criminal law, to sanction irregular entry and stay), seeking to evaluate the success and ramifications of such policies, with particular emphasis on the last two decades.

In short, the study has led to the conclusion that the criminalisation of migration, as an example of deterrence, has not been a successful strategy in either Italy or France. Not only did it score poorly in most, if not all, of the evaluation parameters considered, but it also resulted in significant counterproductive effects, including the generation of vicious cycles, through both heightened rhetoric and increased irregularity.

This concluding chapter first briefly outlines the theoretical context of the study, the methodology used, and the main findings regarding the criminalisation of irregular migration in France and Italy. Subsequently, it discusses how such results may be interpreted by different schools of thought, and suggests a way to understand the relationship between such approaches as related to a threefold conceptualisation of policy gaps. Finally, the chapter concludes by proposing some policy alternatives to criminalisation and deterrence.

THE CRIMINALISATION OF IRREGULAR MIGRATION: BETWEEN POLICY GAPS AND VICIOUS CYCLES

The discussion on the effectiveness of criminalisation and deterrence has its theoretical roots in the IPE controversy over the extent to which states are able to exert authority over migratory flows, in a context of globalisation. Specifically, if realist authors maintain that nation states still retain significant regulatory capacity in the area of human mobility, the idea is significantly challenged by both liberal institutionalist and transnationalist scholars. While institutionalists regard such power as being constrained by legal and supranational institutions, however, transnationalist authors view national sovereignty as being eroded by the process of globalisation itself, to the extent that migration becomes a largely uncontrollable phenomenon.

In this context, deterrence is strongly based on the realist assumption that states can control migration, if they desire to. It is also rooted in the neoclassical expectation that mobility (even when irregular) is the result of a rational decision-making process, based on individuals' calculation of costs and benefits.

Criminological studies offer helpful insight into the way in which such strategy works, leading to the identification of three key factors. First, traditional deterrence theories stress that certainty of apprehension and punishment is more relevant than the latter's severity. Second, labelling and social-control theories underline that, in addition to the above, social costs are of vital importance to restrain potential offenders, through the threat of group marginalisation and stigmatisation. Third, perceptual deterrence theories highlight that the focus should not be on the actual costs implied by sanctions, but on how these are perceived by potential offenders themselves.

The theoretical analysis of deterrence has led the author to hypothesise that the strategy, when applied to migration control, is likely to suffer from a number of weaknesses. To start with, by focusing on the introduction of negative incentives, and disregarding the deeper and more structural drivers of migration, deterrence is unable to comprehensively tackle the phenomenon. Additionally, even if it was to succeed in reducing some of the arrivals, in doing so, it may also lead to unintended substitution effects, resulting in a diversion of flows towards other directions or categories (the 'balloon effect' mentioned in the opening of the chapter). Further, the political dimension of deterrence may backfire: The more policy-makers are concerned with the interests of the domestic audience, rather than those of potential migrants, the more incoherent and ineffective the deterrent threat is likely to result. Finally, migrants may not have accurate information concerning the foreseen sanctions, or they may discount its relevance.

In this context, the criminalisation of irregular migration, that is to say the adoption of the criminal law to address migration offences, emerges as a key example of deterrence, thanks to its double aim to simultaneously increase the legal and social costs of migration. Indeed, this happens through penalties foreseeing fines or imprisonment, but also through the moral condemnation implied by the criminal law itself.

Building upon the above conceptual framework, and aiming to shed light on how it materialised in the case of the criminalisation of migration, the book has adopted a comparative policy-evaluation approach. Drawing on Czaika and de Haas' threefold conceptualisation of policy gaps,[2] it has analysed the distance between the rhetoric surrounding the norm and its form once written down (discursive gaps), that between the latter and its implementation (implementation gaps), and, finally, that between implementation and the overall efficacy of the intervention (efficacy gaps). More specifically, the review of the EU ex-post policy-evaluation model has led to the identification of five parameters (efficacy, efficiency, coherence, sustainability, and utility) as the most relevant to assess the magnitude of efficacy gaps, in the context of the criminalisation of migration.

In carrying out the above, the investigation has relied on a range of primary and secondary sources, including interviews, questionnaires, official datasets, laws and court rulings of national and supranational nature, parliamentary debates, and academic writings. Delving deeper into some of the above, 54 interviews were carried out with key stakeholders (among whom politicians, judicial and law-enforcement actors, humanitarian organisations, and others), to portray their viewpoints and identify relevant patterns. Information gathered through such conversations was then triangulated with both quantitative and qualitative data. Finally, 104 written questionnaires were conducted with third country nationals in Italy and France, to report migrants' own perceptions of the norms under review.

Based on the evidence analysed, the criminalisation of migration in Italy emerged as being characterised by significant discursive, implementation and efficacy gaps. Indeed, the adoption of criminal sanctions against irregular entry and/or stay had long been suggested by far-right parties, before being formalised by the Berlusconi IV Government in 2009, through the adoption of a criminal fine of up to €10,000. Although the fine is among the highest foreseen in the EU for crimes of irregular

---

[2] Czaika and de Haas (2013).

migration, it fell short of the initial proposal to introduce imprisonment, and was eventually coupled with an amnesty one month after its adoption, thus making the distance between the rhetorical emphasis on the instrument, and its final form, significant.

The implementation of the norm was also weaker than what it should have been. Although the number of migrants condemned progressively increased in the first four years since the introduction of the crime (peaking at 12,646), it started dropping soon after 2012 (falling to 2952 in 2016), and there is reason to believe it may have further plummeted to three-digit figures in 2018.

Finally, the efficacy of the norm was meagre. Firstly, the drop in arrivals in 2009–2013 is likely to be explained by factors unrelated to the criminalisation of migration, including push-backs to Libya and the decreased attractiveness of Europe more broadly, following the economic crisis. Furthermore, landings to the country increased in following years, peaking at over 180,000 in 2016, despite the possibility of undergoing a criminal trial. Finally, only a yearly average of 15 criminal cases resulted in expulsions in 2010–2016 (i.e. 0.28% of overall condemnations). Altogether, the above demonstrates the very limited effect that the norm had on both reducing arrivals and increasing returns.

In this context, the Italian criminalisation of migration emerged as lacking in most of the evaluation criteria considered. In addition to being unable to meet its goals, it failed to make an efficient use of resources: Offices were overwhelmed by the burden of pursuing migrants both criminally and administratively, anti-smuggling investigations were lengthened, and there is little evidence confirming that fines were ever paid. While these issues also made the norm internally inconsistent, the only parameter that does not emerge as problematic is that of external coherence. Indeed, both national and supranational courts supported the norm's compatibility with existing legislation.

The above matters led to an unsustainable choice in the long term: Either the full application of the norm, with high certainty but little efficiency, or its more limited implementation, with higher cost-effectiveness but scarce reliability of the legal system. Based on the interviews conducted and ISTAT data, the latter was the option increasingly favoured by police, prosecutors and judges, since 2013.

On balance, it appears that the symbolic but ineffective form that criminalisation took in Italy ended up profiting two main groups of stakeholders. First, political actors in pursuit of electoral support were able to leverage the norm to show concern and responsiveness to the insecurities spread among part of the population. Second, employers of irregular foreign workers (including both firms and families) profited from the fact that the norm actually left their source of labour largely untouched.

In France, on the other hand, the criminalisation of migration was characterised by significant implementation and efficacy gaps, but more moderate discursive ones. Starting with the latter, irregular migration was turned into a criminal offence already in the late 1930s, and available data suggests that this has not been particularly controversial, with most of the public debate focusing today on the criminalisation of helpers instead (the so-called *délit de solidarité*). In spite of the seemingly little politicisation of the norm, the sanctions it foresees are of notable severity, as France is among the few countries in the EU punishing irregular entry or stay with both imprisonment and a fine. Overall, discursive gaps do not seem to have been prevalent in the French case: If, on one hand, the rhetorical emphasis on the norm appears limited, on the other, the sanctions it introduced were rather strict.

On the contrary, the distance between how the norm should have been implemented, and how it was put into practice, was noteworthy. Specifically, the analysis of data released by the French Government confirms previous studies' hypothesis that the criminalisation of migration was largely used in an *instrumental* way: While criminal custody (*garde à vue*, GAV) was employed extensively, especially in 2006–2012, full criminal trial was only rarely pursued (on average, less than 3% of registered foreigners were eventually condemned in those years—in absolute terms, 2983 annually).

Finally, and partly as a result of the above, the efficacy of the norm in reducing migration and increasing returns was limited (like in Italy). On the one hand, France was consistently among the top five EU member states by number of third country nationals found irregularly present in 2008–2018, despite being among the very few imposing both imprisonment and a fine. On the other hand, the use of the criminal law did not significantly increase returns either, which, on average, represented only 3% of annual condemnations in 2012–2016 (in absolute terms, 46 per year). The relevance of the GAV in increasing expulsions also appears to have been limited. Although the rate of effective returns to third countries saw a slight tendential increase until 2012, it remained roughly half the EU average throughout the period analysed (with a national yearly average of 16% of migrants sanctioned with expulsion actually being returned, compared to the EU's 38%, in 2008–2017).

Not only was the criminalisation of migration unsuccessful in accomplishing its objectives, but also affected by a number of related weaknesses, including incoherence, inefficiency, and unsustainability. Virtually all such issues can be related to the contradictory requirement that expulsions should take place at the expiration of the prison term. Indeed, as a consequence of it, the norm resulted incoherent and incompatible with the EU Return Directive, its costs redundant, and its long-term enforcement unsustainable.

In this context, like in Italy, a trade-off emerged between the extensity of the application of the measure (hence its certainty), and its cost-effectiveness. If initially (in 2006–2012), the trade-off was resolved through an instrumental use of the norm (largely employing criminal custody but not trial), since 2012 a less systematic enforcement of the norm has been preferred.

In light of the above, it seems that the criminalisation of migration in France benefitted two main categories of actors. First, its security apparatus, which vastly relied on the possibility of placing migrants in criminal custody to meet expulsion targets, until the partial repeal of the crime in 2012. Second, political actors who discursively emphasised repatriations and the maintenance of a firm approach, seeking electoral gains.

In sum, the deterrent effect of the criminalisation of migration emerged as weak in both France and Italy. Still, while the norm proved unsuccessful and counterproductive, its symbolic value prevented it from being easily repealed.

Indeed, it seems possible to argue that, through both heightened rhetoric and increased irregularity, the criminalisation of migration contributed to fostering vicious cycles of insecurity. On one hand, both Italy and France struggled to decriminalise migration, despite acknowledging the problematics of the norm, to the extent that, when the latter accepted to proceed with the repeal, it only did so while also introducing further criminalising measures (for example, against the refusal of being fingerprinted). On the other hand, by making irregular migration a criminal infraction, and coupling it with further deterrence measures, governments ended up increasing migrants' dependence on (and vulnerability to) irregular networks and routes.

How can the ineffectiveness and counterproductivity of criminalisation be explained? Specifically, what role did the weaknesses of deterrence (mentioned in previous paragraphs) play? From a criminological perspective, it is possible to see that, while in France the certainty of being held in criminal custody was high, especially in 2006–2012, that of being sanctioned was significantly more limited. This was not necessarily the case in Italy (especially until 2012), but interviews suggested that migrants tended to perceive the country's migration management and legal systems as significantly looser than the French ones. As a result, the relatively higher degree of certainty is unlikely to have contributed to making the threat more effective. As for the severity of the norm, this was higher in France, where imprisonment is foreseen, but may not have been as relevant in achieving the expected results, as none of the foreigners interviewed were aware of the potential sanction. Finally, the social costs related to the criminalisation of migration appeared negligible in migrants' home communities, where being detained for some time while in Europe was sometimes seen as normal, and to whom migrants are inclined not to convey negative messages.

Beyond this, however, the criminalisation of migration was affected by a deeper issue, in both countries: The inability of the threatened costs to match the expected benefits of migration. Neither financial costs, nor the deprivation of liberty, nor the stigmatisation derived from having a criminal record, managed to significantly affect potential migrants' calculations of costs. Instead, international conflicts and upheavals (including the Arab Spring, the Syrian and Libyan civil wars), economic fluctuations (among which the financial and Eurozone crises), and inequality in living standards (such as between the Global North and South), all emerged from the analysis as having played a significantly greater role in shaping the composition and size of migratory flows to the two countries.

Not only was criminalisation largely ineffective, but also likely to have contributed to further irregularity. By giving criminal records to concerned migrants, the norm worsened their perspectives of social and economic integration, increasing instead their marginalisation. This effect can be understood by referring to the 'balloon' (substitution) effect, which may be further exemplified, in geographical terms, by the emergence of migrants' hubs in Ventimiglia first, and close to Bardonecchia then, following the reintroduction of border controls in France in 2015.

In this context, in introducing and maintaining the 'crime of irregular migration', both French and Italian politicians showed the substantial prioritisation of their respective domestic audiences, as opposed to the migrant ones. This emerged as most evident from debates surrounding the potential abrogation of the norms. Statements by politicians in both countries revealed their understanding that repealing the crime would invite people to think less of their efforts to address irregular migration, and not be appreciated as the coherence and efficiency ameliorating change that it would have, in fact, represented. Indeed, despite the significant temporal distance, the rhetoric surrounding migration at the time of the adoption of the norm in the two countries was surprisingly similar, with references to foreigners as potential 'Trojan horses' in both cases—hiding enemy spies in 1930s France, Islamic terrorists in 2000s Italy. As a matter of fact, criminalisation emerges as a clear instance of the securitisation of migration.

Partly as a result of the secondary importance attributed to the notional audience (i.e. to potential migrants), the latter may be only scarcely aware of the existence of criminalisation. Conducting a questionnaire among third country nationals, including in Italian and French reception facilities, it emerged that in neither country migrants were familiar with the criminal norms in place to sanction irregular entry and/or stay: Only 2% of respondents in Italy knew about them, and none of those in France did. Confirming previous studies' results, most information was found to be gathered through personal networks.

Overall, all four of the hypothesised weaknesses of deterrence appear to have played a key role in contributing to the lack of effectiveness of the criminalisation of migration.

CRIMINALISATION, MIGRATION, AND GLOBALISATION: THEORETICAL OBSERVATIONS

Looking back at the different perspectives on states' ability to control migration, how would the three main IPE schools of thought interpret the above findings? In this section, I discuss the results from different viewpoints, and then suggest an alternative way of framing the discussion among them.

Starting with the first IPE perspective discussed, given the realist nature of the key assumption on which deterrence rests (that migration can be controlled), authors subscribing to this theoretical viewpoint may make one of two arguments: (1) That migration would have been higher, had criminalisation not been introduced,[3] and (2) that the limited effectiveness of the norm reflects states' actual preference for looser measures, based on pressures from socio-economic interests.[4] While the former statement is difficult to prove, due to the lack of counter-factual evidence, the second may be more testable.

With realist lenses, it may be possible to say that, while Italian politicians sought electoral support through symbolic and visible norms, they meanwhile avoided directly penalising the vast sectors of the economy relying on irregular foreign labour, including both employers (especially in the construction and agricultural businesses) and families (especially for elderly care and domestic support). This emerges as most evident when considering the 2009 cut in work-site inspections, explicitly meant by the Government to reduce firms' hurdles in the aftermath of the financial crisis,[5] as well as the regularisation of 294,000 migrants working as carers or domestic workers,[6] adopted one month after the introduction of the crime of irregular migration.

Whether the above was the result of more, or less, explicit pressures from such socio-economic groups, the simultaneous pursuit of contrasting goals (what Boswell calls 'deliberate malintegration')[7] was responsible for the overall incoherence of the norm and, in realist terms, for its ineffectiveness—or, perhaps more appropriately, for its actual *effectiveness* in meeting its true objectives. These, in turn, may be understood as twofold. First, addressing migration rhetorically, but leaving it largely unaffected in practice, may be seen as supporting the interests of pressure groups (either regular, such as employers, or irregular, such as organised criminality) to have cheap, foreign labour available. Second, the mismatch between rhetoric and practice may also be interpreted as a strategy employed by right-wing governments to gain electoral consensus, by capitalising on public insecurities (thus recalling the idea of vicious cycles mentioned above).

Realist theories would struggle a bit more in explaining the criminalisation of migration in France. Such an approach may suggest two

---

[3] Cf. Espenshade (1994).

[4] Cf. Freeman (1998) and Freeman and Kessler (2008).

[5] Italian Ministry of Work, Health and Social Policies (2008, pp. 4–5).

[6] Pasquinelli (2009).

[7] Boswell (2007).

explanations. First, regarding the introduction of the norm, trade unions' oppositional stance to foreign labour[8] might have induced the Government to pursue stricter measures in 1938. Second, concerning the lack of implementation (and going into the technicalities of how the legal system works), by making and maintaining the offence of irregular migration as a '*délit*' rather than a '*crime*',[9] the legislator enabled public prosecutors to autonomously decide whether and when to criminally pursue infractions, thus shielding employers from a drastic reduction in the foreign labour force.

Both viewpoints are however problematic. Although more data would be needed to buttress or reject the first interpretation of events, existing studies seem to suggest that then Prime Minister Daladier largely saw migration as a security matter.[10] Moreover, the second argument would in fact offer more support to the liberal institutionalist thesis, according to which the legal (rather than the political) system is vital in determining a state's degree of openness to migration.[11]

Altogether, a realist interpretation would likely explain the ineffectiveness of criminalisation, as an example of deterrence, as a result of the influence of contrasting interests on domestic politics, hence highlighting that the norm, once adopted, differed from politicians' stated objectives. As argued here, while such interpretation finds support in the Italian case, it is applied with more difficulty to the French one.

The relevance of the above may be disputed by liberal institutionalist approaches which, together with transnationalist ones, challenge the idea that states are able to control migration. Specifically, adopting the former's perspective, it could be argued that the criminalisation of migration failed due to legal and institutional constraints.[12]

As a matter of fact, despite the political unwillingness to repeal the criminalisation of migration, the norm was softened in practice by the action of courts, in both France and Italy. In the former country, the European Court of Justice (ECJ) declared in 2011 the incompatibility of

---

[8] See Perry (2004, p. 343) and Wihtol de Wenden (2010, p. 121).

[9] In the French system, 'crimes' involve more serious offences (and, hence, sanctions) than 'délits'.

[10] See Maga (1982).

[11] See Hollifield (2004, p. 904).

[12] Cf. Hollifield (2004), Hollifield and Wong (2014, p. 243), and Joppke (1998).

the criminalisation of irregular stay with the EU Return Directive, arguing that its prioritisation of imprisonment constituted an impediment to the requirement set by EU law to give precedence to return procedures.[13] Notwithstanding the resistance with which the ruling was met domestically,[14] the Court was of key importance for the unfolding of the norm, since it initiated the process leading to its partial repeal in 2012.

In Italy, on the other hand, both the ECJ and the Constitutional Court upheld the validity of the criminalisation of migration,[15] which would initially suggest a weaker role of the judiciary. Yet, the former court also prevented Italian judges from sanctioning expulsion (after criminal trial) without a case-by-case assessment of foreigners' risk of absconding,[16] thus substantially reducing the scope of application of the measure. Furthermore, local justices of the peace and public prosecutors often opted for the disapplication of the norm, either because overwhelmed by work, or for genuine belief in the lack of culpability of the accused. In this context, the liberal institutionalist emphasis on the restraining effect of the legal system finds support in both the Italian and French cases.

A further argument may be put forward by authors including Guiraudon, by emphasising the role played by law and order officials in shaping migratory agendas,[17] and hence benefitting from them. Indeed, the introduction of return targets by Sarkozy meant that the French security apparatus stood to greatly profit from the possibility of placing migrants in criminal custody, before transferring them to detention centres and expelling them. By doing so, the possibility that foreigners may abscond and impede the process was significantly reduced, which explains the resolute resistance[18] of political, administrative, and judicial actors against the repeal of the norm in 2011–2012.

This meets however some countering evidence. To begin with, Guiraudon's argument rests on the assumption that judicial review is stronger at the national level (which is the reason why security agents would prefer to take issues to the supranational level, where judicial

---

[13] ECJ, Case C-329/11, Achughbabian.

[14] See Henriot (2013).

[15] Case C-430/11, Sagor; and Constitutional Court, sentence n. 250/2010.

[16] See ECJ, Case C-430/11, Sagor, para. 40–42.

[17] Cf. Guiraudon (2000).

[18] Henriot (2013).

control is milder).[19] Yet, as just mentioned, the supranational ECJ in France, and the local justices of the peace in Italy, played a crucial role in leading to the disapplication of the norm, thus challenging one of the core premises of her argument. Furthermore, in the Italian case, security actors often emerged as being more inclined to turn a blind eye to migrants' irregularity. Thus, if in France the security apparatus played an important role, this seems to have been more limited in the Italian case.

Overall, a liberal institutionalist reading of the ineffectiveness of the criminalisation of migration, as an instance of deterrence, would attribute it to the intervention of the legal apparatus, underlining the distance between the norm as written on paper, and its actual enforcement. As seen above, the Italian and French cases seem to offer support to the idea that legal systems are crucial in limiting states' attempts to control migration, although they question the assumption that supra- and sub-national courts are less able to do so.

Shifting to the last school of thought, the transnationalist perspective would—like the liberal institutionalist one—substantially challenge the realist notion of states as being able to control migration through deterrence and criminalisation (or in any other way). Differently from the viewpoint just discussed, however, the main argument would be that globalisation processes have made migration a structural phenomenon, and therefore an incontrollable one.[20] The inability of criminalisation to bring about a reduction in migratory flows would thus be seen as inevitable.

Specifically, the above can be explained with reference to both the structural causes of the phenomenon, and the measure's intrinsic inability to tackle them, in an argument that recalls two of the pitfalls of deterrence mentioned above. First, structuralist viewpoints would stress that the costs associated to migration, although increased by the threat of criminal fines or prison sentences, were unable to counterweigh the deeper drivers of the phenomenon. From this perspective, being migration the result of

---

[19] Guiraudon (2000, pp. 261–262).
[20] Cf. Sassen (1996), Castles (2004a, b), Léon and Overbeek (2015), and Talani (2015).

processes related to globalisation and the global reallocation of production,[21] the introduction of monetary penalties or of detention measures plays an irrelevant role in shaping the flows.

Further, the idea of substitution effects may also be proposed by this theoretical perspective. While transnationalist authors hold diverging views on the extent to which state sovereignty is reduced in the contemporary world order, those more open to the possibility that nation states still retain some importance[22] would emphasise that, even if criminalisation had managed to affect some potential migrants, the result would have probably been a diversion of the migratory attempts towards other directions or entry-categories, rather than the suppression of such endeavours altogether.

While the existence and intensity of substitution effects (either in terms of a shift in entry points or methods) are difficult to statistically demonstrate, it is plausible that flows are, to a certain extent, diverted rather than stopped. In the European context in particular, the actual entry point may be of subordinate importance, due to the lack of internal frontiers, and as testified by the presence of significant secondary movements[23] among Schengen countries.

To sum up, a transnationalist perspective would see the ineffectiveness of the criminalisation of migration, and of deterrence measures more broadly, as inevitable, due to the phenomenon's structural nature. In doing so, it would stress how factors other than policy interventions are in fact responsible for shaping the flows. If we accept the systemic roots of migration, then both the Italian and French case support the globalisation thesis.

As the above highlights, different IPE perspectives would attribute the ineffectiveness of the criminalisation of migration, as an example of deterrence, in reducing migratory flows, to very different factors: The role of domestic politics and contrasting interests, the constraining effect of legal and institutional factors, and the deeper significance of structural changes related to globalisation processes.

Recalling the threefold interpretation of policy gaps suggested by Czaika and de Haas, it seems that a parallelism may be traced between

---

[21] Cf. Overbeek (2002) and Talani (2010, p. 39).
[22] See, for example, Mittelman (2000, p. 65).
[23] See, for example, COM(2019) 481 final, p. 4.

such model and the three IPE approaches outlined above, with each of the theories focusing, in fact, on a different level of gaps. To start with, in attributing the lack of efficacy to the presence of diverging interests at the policy-making level, realists tend to focus on the *discursive* gaps affecting norms, that is, on the distance between political rhetoric and actual measures. Further, by stressing the role of the judiciary and of liberal norms in restraining the full application of measures, liberal institutionalists often actually emphasise the *implementation* gaps conditioning policies, namely the distance between the norms on paper and their eventual application. Finally, in highlighting the greater role of structural dynamics, transnationalist authors underline in fact *efficacy* gaps, or how external factors ultimately affect the outcomes of migration policies.

By interpreting the debate through such a framework, the realist, liberal institutionalist, and transnationalist approaches appear as not necessarily strictly mutually incompatible, but as providing useful analytical lenses to the understanding of the different kinds of policy gaps affecting states' attempts to control migration. This does not aim to belittle the very different ontological and epistemological assumptions of the three perspectives, nor to say that all authors who subscribe to one of the three viewpoints would feel perfectly represented in such schematisation. There may be—and there are—significant differences within each of the three perspectives, just like there are partial overlaps among them. Yet, the above hopes to suggest an alternative way through which the three schools could interact with each other, in a productive way.

Overall, if the threefold understanding of gaps adds methodological granularity to previously less-structured studies, its deeper contribution lies in being able to effectively bring into communication three different schools of thought, highlighting how each of them tends to focus, in fact, on a different level of policy gaps.

To conclude, despite their different analytical focus, all three IPE perspectives would concede that the criminalisation of migration, as applied in Italy and France, did not manage to meet the expectation that it would reduce unauthorised border crossings. If such strategy did not work, however, one may be left to wonder what may instead succeed, as a potential solution to address unauthorised movements?

What Alternatives to Criminalisation and Deterrence?

Building upon the theoretical and empirical knowledge gathered through these pages, deterrence and criminalisation have emerged as affected by intrinsic weaknesses. To avoid the 'balloon effects' and vicious cycles that may be incentivised by them, this book concludes by proposing some policy alternatives.

To begin with, irregular migration should be decriminalised. This is the case for the states that have been object of the analysis (Italy and France), but likely for further European countries too. Not only do most European countries make use of the criminal law to address migration, but some are also proposing strengthening the sanctions, at the time of writing (e.g. the United Kingdom).[24] Yet, both criminal fines and imprisonment are deeply incoherent sanctions, in the context of irregular migration. If the former are hard to collect, the latter keep migrants on the national territory, while in fact hoping to expel them. Furthermore, based on the study conducted, deterrence is likely to remain limited, while however generating a number of counterproductive consequences, including pushing migrants further underground, and increasing insecurity.

In this context, despite criminalisation being a national norm, the EU may play a role too. The European Court of Justice has already pronounced itself on criminalisation, leading to legislative and/or implementation changes in the countries analysed. At the European Commission, officials are aware that criminalisation is scarcely enforced in most member states,[25] and the institution could promote the decriminalisation of irregular migration, just like it has recently done for humanitarian assistance and rescues at sea.[26]

Decriminalisation needs to be part of a substantial re-thinking of migration management, shifting away from deterrence logics, and towards the enabling of legal migration. Thus, as an alternative to restrictive measures, legal pathways should be opened. This would involve increasing humanitarian corridors for refugees (who should not, in any case, be

---

[24] See FRA (2014) and UK New Plan for Immigration (2021).
[25] See Interviewee 11.
[26] See COM(2020) 6470 final, section 4.

the target of deterrence measures, in light of their right to seek international protection), as well as work permits for both high- and low-skilled, seasonal and longer-term migrants. Several countries already have programs in place for refugees. EU member states committed to 50,039 resettlements for 2018–2019, and to 30,000 for 2020.[27] Such schemes are to be welcomed, but also strengthened. According to the UNHCR, in 2018, only about 4.7% of the world's resettlement needs were met, with 55,692 people being resettled, out of the 1.2million refugees needing so.[28] As for work permits, in some countries, they have seen a dramatic contraction. In Italy, for instance, they dropped from 230,000 in 2008 to 30,850 since 2015 (per year).[29] Post-Brexit Britain set a minimum £25,600-salary for work entries, similarly reducing possibilities for low-skilled migration.[30] Legal mobility opportunities for low-skilled workers need to be increased, for people to have an alternative to the 'leaky boats',[31] and in view of Europe's need for migration.

Overall, strengthening legal pathways would reduce the irregularity and insecurity that are often associated with migration journeys. For migrants, this would decrease vulnerability and exploitation. For citizens, it would reduce the irregular networks operating in the territory, and expand the benefits from migrants' work (e.g. through taxes).[32] For both, it would cut the profits of the irregular side of the 'migration industry', and strengthen overall security. Indeed, it has been the argument of this book that restrictive migration measures contribute to substitution effects, more than stopping the flows. Other studies have shown that growing numbers of visa rejections lead to more irregular entries, and that the regularisation of migrants' status can instead reduce their engagement with irregular activities.[33] Substitution effects are hard to prove,

---

[27] See https://ec.europa.eu/home-affairs/sites/default/files/what-we-do/policies/european-agenda-migration/201912_delivering-on-resettlement.pdf.

[28] See https://www.unhcr.org/news/briefing/2019/2/5c6bc9704/5-cent-global-refugee-resettlement-needs-met-year.html.

[29] See Corrado et al. (2018, p. 19).

[30] Except for seasonal agricultural work. See https://migrationobservatory.ox.ac.uk/resources/primers/policy-primer-the-uks-2021-points-based-immigration-system/.

[31] Cf. Fargues and Bonfanti (2014).

[32] See Dustmann and Frattini (2014).

[33] Czaika and Hobolth (2016) and Pinotti (2017).

but necessary to avoid, to reduce the irregularity currently associated to migratory flows.

To enable the opening of legal pathways, a change in rhetoric is necessary, addressing the misinformation surrounding migration, and acknowledging both the structural drivers of the phenomenon and Europe's need for foreign work. Misinformation on migration is widespread. A New York Times survey showed respondents in several European countries (including the United Kingdom, Italy, France, Spain, Germany, and the Netherlands) substantially overestimating the proportion of population made up of immigrants—sometimes by three times or more.[34] A 2017–2018 large-scale survey similarly found respondents in multiple countries overestimating not only migrants' numbers, but also their reliance on public support, unemployment levels, and cultural distance.[35]

To address misinformation, Europe's need for migration must be acknowledged. Several sectors strongly rely on it, with foreigners accounting for 13% of key workers on average in the EU, up to 30% in some sectors.[36] This has become perhaps most evident during Covid-19 lockdowns, when several European countries resorted to a variety of measures (including regularisations and charter flights) to address labour shortages in agriculture, health, and other sectors.[37]

In this context, the structural origins of migration must also be recognised, including both the conflicts making people flee (such as in Syria and Libya), and the deep inequalities often at the roots of migratory flows. Overall, we must move beyond the emergency framing, to address widespread misperceptions and normalise migration.

Ultimately, this will require not only significant political courage on the part of the actors caught in the vicious cycle of insecurity generated by deterrence logics, but also, especially in Italy and other southern European countries, serious emphasis on addressing the underground economy and the immense profits it generates. As a consequence, it is simultaneously necessary to strengthen the enforcement of labour control

---

[34] See http://www.nytimes.com/2011/12/06/world/europe/perceptions-of-migration-clash-with-reality-report-finds.html.

[35] Alesina et al. (2018). The study referred to France, Germany, Italy, Sweden, the United Kingdom, and the United States.

[36] Fasani and Mazza (2020).

[37] See https://migrants-keyworkers-covid-19.odi.digital and Rosina and Thielemann (2021).

and related measures, in order to tackle the strong reliance on irregular work, the exploitation that often comes with it, and the pull factor that the underground economy represents.

All the above steps are necessary to break the vicious cycle of insecurity and irregularity caused by criminalisation and deterrence. Migration policies are deeply intertwined with broader political and economic developments in countries of origin as well as destination. While this study has focused primarily on the former, a comprehensive policy response must ultimately address the profound national and global inequalities at the roots of migration itself.

References

Alesina, Alberto, Armando Miano, and Stefanie Stantcheva. 2018. Immigration and Redistribution. NBER Working Paper 24733.

Boswell, Christina. 2007. Theorizing Migration Policy: Is There a Third Way? *International Migration Review* 41 (1): 75–100.

Castles, Stephen. 2004a. Why Migration Policies Fail. *Ethnic and Racial Studies* 27 (2): 205–227.

Castles, Stephen. 2004b. The Factors That Make and Unmake Migration Policies. *International Migration Review* 38 (3): 852–884.

Corrado, Alessandra, with contributions from Francesco Saverio Caruso, Martina Lo Cascio, Michele Nori, Letizia Palumbo, and Anna Triandafyllidou. 2018. Is Italian Agriculture a 'Pull Factor' for Irregular Migration—And, If So, Why? Open Society Foundations Report.

Czaika, Mathias, and Hein de Haas. 2013. The Effectiveness of Immigration Policies. *Population and Development Review* 39 (3): 487–508.

Czaika, Mathias, and Mogens Hobolth. 2016. Do Restrictive Asylum and Visa Policies Increase Irregular Migration into Europe? *European Union Politics* 17 (3): 345–365.

Dustmann, Christian, and Tommaso Frattini. 2014. The Fiscal Effects of Immigration to the UK. *The Economic Journal* 124 (580): F593–F643.

Espenshade, Thomas J. 1994. Does the Threat of Border Apprehension Deter Undocumented US Immigration? *Population and Development Review* 20 (4): 871–892.

European Commission. 2020. Commission Guidance on the Implementation of EU Rules on Definition and Prevention of the Facilitation of Unauthorised Entry, Transit and Residence, C(2020) 6470 final, 23.9.2020. https://ec.europa.eu/info/sites/default/files/commission-guidance-implementation-facilitation-unauthorised-entry_en.pdf.

European Commission, Communication from the Commission to the European Parliament, the European Council and the Council. Progress Report on the Implementation of the European Agenda on Migration, 16 October 2019, COM(2019) 481 final.

European Union: Court of Justice, Case C-329/11, Alexandre Achughbabian v Préfet du Val-de-Marne, Judgment of the Court (Grand Chamber) of 6 December 2011, European Court Reports 2011 -00000. http://eur-lex.eur opa.eu/legal-content/EN/TXT/?uri=CELEX%3A62011CJ0329. Accessed 29 Mar 2017.

European Union: Court of Justice, Case C-430/11, Judgment of the Court of Justice of 6 December 2012, Sagor, Case, Md Sagor, Judgment of the Court (First Chamber) of 6 December 2012, ECLI:EU:C:2012:777. https://eur-lex.europa.eu/legal-content/GA/SUM/?uri=CELEX:62011CJ0430. Accessed 1 May 2018.

Fargues, Philippe, and Sara Bonfanti. 2014. When the Best Option Is a Leaky Boat: Why Migrants Risk Their Lives Crossing the Mediterranean and What Europe Is Doing About It. Migration Policy Centre, EUI.

Fasani, Francesco, and Jacopo Mazza. 2020. Immigrant Key Workers: Their Contribution to Europe's COVID-19 Response. IZA Policy Paper No. 155.

FRA. 2014. *Criminalisation of Migrants in an Irregular Situation and of Persons Engaging with Them*. Vienna: FRA.

Freeman, Gary P. 1998. The Decline of Sovereignty? Politics and Immigration Restriction in Liberal States. In *Challenge to the Nation-State: Immigration in Western Europe and the United States*, ed. Christian Joppke, 86–108. Oxford: Oxford University Press.

Freeman, Gary P., and Alan E. Kessler. 2008. Political Economy and Migration Policy. *Journal of Ethnic and Migration Studies* 34 (4): 655–678.

Guiraudon, Virginie. 2000. European Integration and Migration Policy: Vertical Policy-Making as Venue Shopping. *Journal of Common Market Studies* 38 (2): 251–271.

Henriot, Patrick. 2013. Dépénalisation du séjour irrégulier des étrangers: l'opiniâtre résistance des autorités françaises. *La Revue des droits de l'homme* [online] 3: 1–14.

Hollifield, James F. 2004. The Emerging Migration State. *International Migration Review* 38 (3): 885–912.

Hollifield, James F., and Tom K. Wong. 2014. The Politics of International Migration: How Can "We Bring the State Back In"? In *Migration Theory: Talking Across Disciplines*, 3rd ed., ed. Caroline B. Brettell and James F. Hollifield, 227–288. New York: Routledge.

Italian Constitutional Court. 2010. Sentenza 250/2010. *Gazzetta Ufficiale*, No. 28, July 14. https://www.cortecostituzionale.it/actionSchedaPronuncia.do?anno=2010&numero=250. Accessed 24 May 2018.

Italian Minister of Work, Health and Social Policies. 2008. Direttiva del Ministro (Maurizio Sacconi), Servizi Ispettivi e Attività di Vigilanza, 18 September 2008, Annex II to Circular n. 111 of 17 December 2008. https://www.inps.it/circolariZip/Circolare%20numero%20111%20del%2017-12-2008_Allegato%20n%202.pdf. Accessed 5 Mar 2020.

Joppke, Christian. 1998. Why Liberal States Accept Unwanted Immigration. *World Politics* 50 (2): 266–293.

León, Alba I., and Henk Overbeek. 2015. Neoliberal Globalisation, Transnational Migration and Global Governance. In *Handbook of the International Political Economy of Migration*, ed. Leila Simona Talani and Simon McMahon, 37–53. Cheltenham, UK: Edward Elgar.

Maga, Timothy. 1982. Closing the Door: The French Government and Refugee Policy, 1933–1939. *French Historical Studies* 12 (3): 424–442.

Mittelman, James H. 2000. *The Globalization Syndrome: Transformation and Resistance*. Princeton: Princeton University Press.

Overbeek, Henk. 2002. Globalisation and Governance: Contradictions of Neo-Liberal Migration Management. Discussion Paper Series, Hamburg Institute of International Economics.

Pasquinelli, Sergio. 2009. Perché la Sanatoria Ha Fatto Flop. *La voce*, October 9. http://www.lavoce.info/archives/25930/perche-la-sanatoria-ha-fatto-flop/. Accessed 1 May 2018.

Perry, Matt. 2004. 'Sans Distinction de Nationalité?' The French Communist Party, Immigrants and Unemployment in the 1930s. *European History Quarterly* 34 (3): 337–369.

Pinotti, Paolo. 2017. Clicking on Heaven's Door: The Effect of Immigrant Legalization on Crime. *American Economic Review* 107 (1): 138–168.

Rosina, Matilde, and Eiko Thielemann. 2021. Migration Governance in Times of Crisis: Enhancing National Security Through Enhanced Migrant Rights? Paper Presented at 10th Conference of the ECPR Standing Group on the European Union, 10–12 June.

Sassen, Saskia. 1996. *Losing Control? Sovereignty in an Age of Globalization*. New York: Columbia University Press.

Talani, Leila Simona. 2010. *From Egypt to Europe: Globalisation and Migration Across the Mediterranean*. London: I.B. Tauris.

Talani, Leila Simona. 2015. International Migration: IPE Perspectives and the Impact of Globalisation. In *Handbook of the International Political Economy of Migration*, ed. Leila Simona Talani and and Simon McMahon, 17–36. Cheltenham, UK: Edward Elgar.

UK Secretary of State for the Home Department. 2021. New Plan for Immigration: Policy Statement, March. https://assets.publishing.service.gov.uk/government/uploads/system/uploads/attachment_data/file/972472/CCS207_CCS0820091708-001_Sovereign_Borders_FULL_v13__1_.pdf.

United States Congress. 2006. Weak Bilateral Law Enforcement Presence at the U.S.-Mexico Border: Territorial Integrity and Safety Issues for American Citizens. Joint Hearing Before the Subcommittee on Crime, Terrorism, and Homeland Security and the Subcommittee on Immigration, Border Security, and Claims of the Committee on the Judiciary, House of Representatives, One Hundred Ninth Congress, First Session, November 17, 2005. Washington: U.S. G.P.O. https://babel.hathitrust.org/cgi/pt?id=pst.000058167854&view=1up&seq=53. Accessed 14 Feb 2020.

Wihtol de Wenden, Catherine. 2010. Irregular Migration in France. In *Irregular Migration in Europe: Myths and Realities*, ed. Anna Triandafyllidou, 115–124. Farnham, Surrey: Ashgate Publishing.

Appendices

Appendix A: List of Interviewees

Interviewee 1: Coordinator of the Internal Security Fund (ISF)—Police at the European Commission, DG Home, Brussels, 14 June 2017
Interviewee 2: Anonymous, Brussels, 18 July 2017
Interviewee 3: Official at the European Commission, DG Home, Brussels, 19 July 2017
Interviewee 4: Anonymous, Brussels, 20 July 2017
Interviewee 5: Anonymous, Brussels, 25 July 2017
Interviewee 6: Official at the Italian Representation in Brussels, Brussels, 27 July 2017
Interviewee 7: Official at the European Commission, Brussels, 28 July 2017 (a)
Interviewee 8: Official at the European Commission, Brussels, 28 July 2017 (b)
Interviewee 9: Official at the European Commission, Brussels, 31 July 2017
Interviewee 10: Senior official at the European Commission, 1 August 2017
Interviewee 11: Official at the European Commission, 2 August 2017
Interviewee 12: Anonymous, Genoa, 13 November 2017

Interviewee 13: Journalist active in Ventimiglia, Phone interview (Ventimiglia), 15 November 2017
Interviewee 14: Senior official of the Italian National Police, phone, 11 December 2017
Interviewee 15: Official at the European Commission, Rome, 18 December 2017
Interviewee 16: Daniela Pompei, Responsible for immigrant services at the Community of Sant'Egidio, Rome, 18 dicembre 2017
Interviewee 17: Marco Pacciotti, Coordinator of the Democratic Party's Immigration Forum, Rome, 19 December 2017
Interviewee 18: Tommaso Palumbo, First Director in the Central Directorate for Immigration and Border Police of the Italian National Police (Primo Dirigente della Polizia di Stato, Direttore 1^ Divisione, Dipartimento della Pubblica Sicurezza, Direzione Centrale dell'Immigrazione e della Polizia delle Frontiere, Servizio Immigrazione), Phone interview (Rome), 19 December 2017
Interviewee 20: Romano Prodi, Former Italian Prime Minister and European Commission President, Bologna, 20 December 2017
Interviewee 21: Coordinator of Antoniano Onlus Counselling Centre, Bologna, 4 January 2018
Interviewee 22: Ahmed Osman, Cultural Mediator, Alessandria, 9 January 2018
Interviewee 23: Vice Admiral Giovanni Pettorino, Commander-in-chief (elect) of the Italian Corps of the Coast Guards, Genoa, 11 January 2018
Interviewee 24: Staff member at a centre for asylum seekers, Genoa, 13 January 2018
Interviewee 26: Livia Turco, Former Minister of Social Affairs and of Health, Rome, 15 January 2018
Interviewee 27: Judge in Rome Criminal Court, Rome, 16 January 2018 (a)
Interviewee 28: Judge in Rome Criminal Court, Rome, 16 January 2018 (b)
Interviewee 29: Judge in Rome Criminal Court, Rome, 16 January 2018 (c)
Interviewee 30: Domenico Manzione, Undersecretary of the Interior on Migration Matters, Rome, 17 January 2018
Interviewee 31: President of a Territorial Commission for International Protection, Italy, 19 February 2018

Interviewee 32: Justice of the Peace (JP), Agrigento, 20 February 2018 (a)
Interviewee 33: Justice of the Peace (JP), Agrigento, 20 February 2018 (b)
Interviewee 34: Deputy Public Prosecutor, Ragusa, 21 February 2018 (a)
Interviewee 35: Informal talk with Deputy Public Prosecutor, Ragusa, 21 February 2018 (b)
Interviewee 37: Staff member at Interservice Group for Contrasting Irregular Migration (CIGIC) at the Public Prosecution of Siracusa, Siracusa, 23 February 2018
Interviewee 38: Informal talk with Head Prosecutor in Siracusa, Siracusa, 23 February 2018
Interviewee 39: Justice of the Peace (JP), Siracusa, 23 February 2018
Interviewee 40: Judge at Minors' Court in Palermo, Palermo, 24 February 2018
Interviewee 41: Magistrate in the Public Prosecution of Imperia, Imperia, 5 April 2018
Interviewee 43: Departmental Director of the Alpes-Maritimes Border Police (Directeur Départemental de la Police aux Frontières des Alpes-Maritimes), Nice, 6 April 2018
Interviewee 44: Departmental Chief of Staff of the Alpes-Maritimes Border Police (Chef d'état-Major Départemental, DDPAF of Alpes-Maritimes), 6 April 2018
Interviewee 45: Lawyer and President of Avocats poir la Défense de Droits des Etrangers (ADDE), Phone Interview (Toulouse), 20 April 2018
Interviewee 46: Deputy Honorary Prosecutor in Imperia, Imperia, 11 July 2018
Interviewee 47: Deputy Prosecutor in Genoa, Genoa, 27 July 2018
Interviewee 48: Lawyer specialised in Immigration and Criminal Law, Phone interview (Toulouse), 4 September 2018.
Interviewee 49: Patrick Henriot, Former Judge and National Secretary of the Syndicat de la Magistrature, Paris, 6 September 2018.
Interviewee 50: Staff member of Association d'Aide aux Jeunes Travailleurs (AAJT), Telephone (Marseille), 9 September 2018
Interviewee 51: Rapporteur for the National Court for Asylum Right (CNDA), Paris, 10 September 2018
Interviewee 52: Stéphane Maugendre, Lawyer specialised in Immigration and Criminal Law, Former President of GISTI, Paris, 10 September 2018

Interviewee 53: Lawyer formerly operating in an Asylum-Seeking Centre, Phone Interview (Paris), 13 September 2018
Interviewee 54: Justice of the Peace (JP), Catania, 20 September 2018 (a)
Interviewee 55: Justice of the peace (JP), Catania, 20 September 2018 (b)
Interviewee 56: Responsible for Prison-Related Matters at La Cimade, Phone Interview (Paris), 21 September 2018
Interviewee 57: Anonymous, President of Refugiés Bienvenues, London, 24 October 2018

Appendix B: Databases

Annuaire Statistique de la Justice: édition 2006, 2007 and 2011–2, Secrétariat Général, Direction de l'Administration générale et de l'Équipement, Sous-Direction De La Statistique, Des Études Et De La Documentation, Paris, http://www.justice.gouv.fr/art_pix/1_annuairestat2006.pdf, http://www.justice.gouv.fr/art_pix/1_annuaire2007.pdf, http://www.justice.gouv.fr/art_pix/stat_annuaire_2011-2012.pdf (downloaded on 13 November 2018).

Chiffres départementaux mensuels relatifs aux crimes et délits enregistrés par les services de police et de gendarmerie depuis janvier 1996 ('Departmental figures registered by police and gendarmerie, 2018'), https://www.data.gouv.fr/fr/datasets/chiffres-departementaux-mensuels-relatifs-aux-crimes-et-delits-enregistres-par-les-services-de-police-et-de-gendarmerie-depuis-janvier-1996/ (downloaded on 12 November 2018).

Condamnations selon la nature de l'infraction de 2009 à 2016 (Condemnations by infraction, 2016), French Ministry of Justice, http://www.justice.gouv.fr/art_pix/stat_condamnations_infractions_2016.ods (downloaded on 13 November 2018).

Database on Genoa public prosecution activity concerning article 10-bis, for years 2009–2017, obtained from the public prosecutor offices in Genoa (data extracted on 31 July 2018).

Database on Imperia public prosecution activity in front of the justice of the peace, concerning article 10-bis of decree 286/98, for years 2015–2017, obtained from the public prosecutor offices in Imperia (data extracted on 17 May 2018).

Database on legal proceedings in the office of the justice of the peace, concerning the crime of irregular migration, for 2014–2017, obtained from the office of the justice of the peace in Agrigento (data extracted on 23 February 2018).

Délits à la police des étrangers - Évolution de l'action des services - index 69 2008–11, and 2010–13 ('Crimes to foreigners' police, 2011 and 2013'), https://www.data.gouv.fr/fr/datasets/15064-delits-a-la-police-des-etrangers-evolution-de-laction-des-services-index-69/, https://www.data.gouv.fr/fr/datasets/2013-40-delits-a-la-police-des-etrangers-evolution-de-laction-des-services-index-69/ (downloaded on 12 November 2018).

Eurobarometer (2019), 'What do you think are the two most important issues facing our country at the moment?', https://ec.europa.eu/commfrontoffice/publicopinion/index.cfm/Chart/getChart/chartType/lineChart/themeKy/42/groupKy/208/savFile/54 (downloaded February 2019).

Eurostat (2018), Statistics on enforcement of immigration legislation (including migr_eipre, migr_eirfs, migr_eirtn, migr_eiord, migr_dubro, migr_asyapp, migr_asyapp, migr_eirt_ass), http://appsso.eurostat.ec.europa.eu/nui/show.do?dataset=migr_eipre&lang=en (downloaded December 2018).

Eurostat (2019), lfsa_urgan, http://appsso.eurostat.ec.europa.eu/nui/show.do?dataset=lfsa_urgan&lang=en (downloaded February 2019).

Frontex (2017, 2019), *Detections of illegal border-crossings statistics*, https://frontex.europa.eu/along-eu-borders/migratory-map/ (downloaded on 7 October 2017 and 6 August 2019).

Ismu (2014) 'Sbarchi e Richieste di Asilo: Serie Storica Anni 1997-2014' (Ismu elaboration upon MoI data), http://www.ismu.org/wp-content/uploads/2015/02/Sbarchi-e-richieste-asilo-1997-2014.xls (accessed 1 May 2018).

ISTAT (2018), Data related to article 10-bis TUI, 2009-2016 (obtained in May 2018).

Italian Ministry of Justice (MoJ) (2018), Data related to article 10-bis TUI, 2009-2018 (obtained in May 2018).

Tableaux des Condamnations en 2012, 2013, 2014, 2015, 2016 ('Tables of condemnations, 2012-6'), Ministry of Justice, http://www.justice.gouv.fr/statistiques-10054/donnees-statistiques-10302/les-condamnations-27130.html (downloaded on 13 November 2018).

Index

A
Agriculture, 213, 320
Alfano, Angelino, 102, 123, 279
Alpes-Maritimes, 188, 271
Arab Spring, 146, 214, 270
Audiences, 75, 157, 279

B
Bardonecchia, 242, 271
Berlusconi, Silvio, 96, 141, 284

C
Calais, 188, 190, 258
Church, 96
Counterproductive effects, 153, 206, 208, 260, 273, 279, 281
Courts, 32, 178, 253
Covid-19, 320
Criminal organisations, 274

D
Daladier, Eduard, 174, 175, 177, 255
Decriminalisation, 100, 123, 161, 179, 180, 245, 281, 318
Deliberate malintegration, 33, 104, 157, 159, 160, 245, 265, 279, 287, 312
Délit de solidarité, 211
Detention, 44, 59, 64, 70, 106, 133, 138, 187
Deterrence, 44
   certainty, 61, 131, 206, 263, 267
   definition, 58
   of criminalisation, 81, 203, 255, 257
   severity, 61, 131, 206, 268
Duplication of costs and processes, 153, 259, 263

E
Economic crisis, 142, 159, 213, 244, 312

© The Editor(s) (if applicable) and The Author(s), under exclusive license to Springer Nature Switzerland AG 2022
M. Rosina, *The Criminalisation of Irregular Migration in Europe*, Politics of Citizenship and Migration, https://doi.org/10.1007/978-3-030-90347-3

Effectiveness
  coherence, 10, 155, 220, 261
  efficacy, 9, 121, 123, 202, 203, 254
  efficiency, 10, 149, 216, 259
  sustainability, 10, 156, 221, 263
  utility, 10, 157, 222, 264
Employers, 222, 246, 265, 287, 312
Enlargement, 142, 212
European Court of Human Rights, 141
European Court of Justice (ECJ), 99, 122, 123, 155, 160, 178, 220, 245, 262, 313, 318
EU-Turkey Statement, 77, 272
Exploitation, 139, 275
Expulsion, 96–98, 106, 113, 151, 175, 187, 195, 202, 220, 254, 260, 261, 280, 314

F

Fine, 96, 103, 114, 152, 155, 175, 194, 249, 260
Foreign work, 26, 158
Forza Italia (FI), 101
Frontex, 12, 105
Front National (FN), 284

G

Garde à vue (GAV), 179, 181, 183, 185, 189, 202, 247, 254, 264, 280
Globalisation, 22, 34, 37, 316

H

Hotspots, 105, 128, 243, 252

I

Inequality, 39, 270

Information, 45, 78, 133, 135, 207, 285
Information campaigns, 136, 285
Insecurity, 263, 274, 281, 312, 319
Instrumentalisation/Instrumentalization, 182, 223, 248, 287
International Political Economy (IPE), 4, 140, 212, 270, 311
Internet, 135, 136
Irregularity, 263, 273
Irregular work, 246, 274, 312

L

Lampedusa, 101, 120, 151
Legal pathways, 318
Lega Nord (LN), 95, 101, 103, 124, 278, 284
Libya, 77, 135, 141, 142, 144–146, 161, 214, 257, 270, 272
Local actors, 119, 156
  courts, 207
  justices of the peace, 120
  police, 105, 119, 207
  prosecutors, 119

M

Marginalisation, 68, 138, 209, 262
Minors, 107, 115, 156, 252
Movimento Cinque Stelle, 100, 103, 124, 125, 280, 282

N

Neoliberal institutionalism, 29, 252, 287, 289, 313

P

Paris, 188
Partito Democratico (PD), 102, 280
Perceptions, 65, 267, 279

Policy evaluation, 9
Policy gaps, 3, 41, 103, 316
    discursive gaps, 42, 95, 182, 244
    efficacy gaps, 43, 121, 198, 254
    implementation gaps, 43, 104, 184, 247
Politicisation, 245, 277
Populism, 284
Prison, 95, 96, 103, 122, 138, 149, 175, 186, 192, 250, 260, 261
Push-backs, 59, 141, 257

R
Realism, 22, 59, 246, 289, 311
Regularisation, 96, 103, 142, 158, 160, 222, 272, 312, 319
Renzi, Matteo, 102, 280, 282, 284
Return Directive, 97, 98, 155, 178, 220, 254, 262, 314
Returns, 59, 99, 113, 123, 133, 151, 161, 178, 223, 254, 255, 264

S
Salvini, Matteo, 102
Sarkozy, Nicolas, 183, 203, 223, 228, 264
Search and Rescue (SAR), 99, 104, 111, 154
Secondary flows, 241
Securitisation, 57, 84, 139, 279, 281
Security, 58, 84, 151, 157, 161, 319
Sicily, 111, 115

Smugglers, 79, 130, 262
Smuggling, 26, 105, 153, 154, 263, 273, 275
Social control, 68, 268
Social networks, 129, 135
Sovereignty, 25, 31, 32, 34, 38, 59
Stigmatisation, 67, 138, 208
Substitution effects, 73, 270, 316, 319
Syria, 127, 144, 161, 214, 269

T
Trafficking, 105, 154, 263, 275
Transnationalist thesis, 34, 289, 315
Trojan Horse, 176, 278
Tunisia, 144, 214, 269

U
Underground economy, 158, 274, 320
Unemployment, 143, 210

V
Ventimiglia, 271, 276
Vicious cycles, 68, 263, 281, 288, 312
Visa overstaying, 126, 272
Vulnerability, 139, 263, 275, 319

W
Women, 115, 250